Nelson Second
Maths

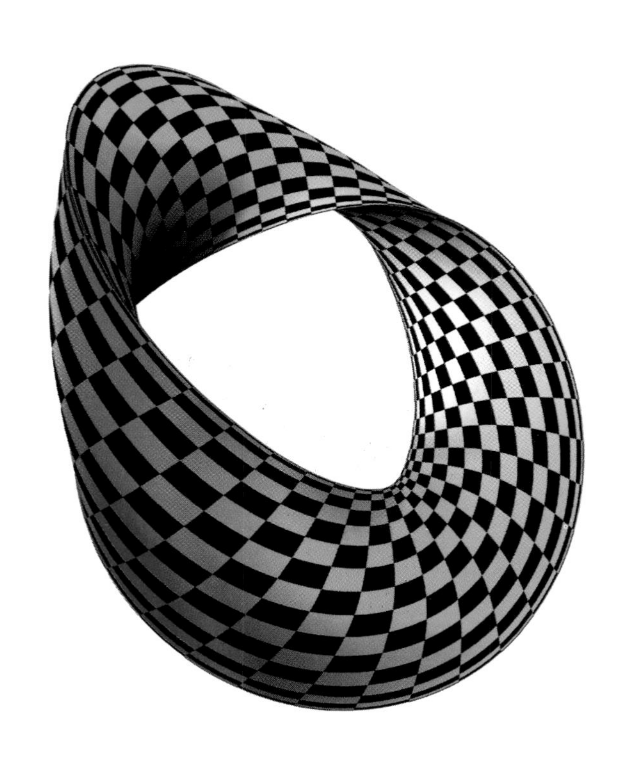

Nelson

Thomas Nelson and Sons Ltd
Nelson House Mayfield Road
Walton-on-Thames Surrey
KT12 5PL UK

Thomas Nelson Australia
102 Dodds Street
South Melbourne
Victoria 3205 Australia

Nelson Canada
1120 Birchmount Road
Scarborough Ontario
MIK 5G4 Canada

First published by Thomas Nelson and Sons Ltd 1996
I(T)P Thomas Nelson is an International Thomson Publishing Company
I(T)P is used under licence

ISBN 0-17-431453-1
NPN 9 8 7 6 5 4 3 2 1

Printed in Spain

Nelson Secondary Maths Student's Book 1 is written by:

Jim Noonan (Series Coordinator)
Terry Bevis
Gay Cain
Brian Martin
Christine Mitchell
Robert Powell
Gwen Wood

The authors are grateful to the schools in Avon, Birmingham, Co. Antrim, Cornwall, Cumbria, Essex, Gloucestershire, Gwent, Hampshire, Hereford & Worcester, Nottingham, Salford, South Glamorgan, Tyneside and Wiltshire involved in trials of material for this book

Contents

Preface

This book forms part of the Nelson Secondary Maths series, designed to cover the Programme of Study for Key Stages 3 and 4 of the National Curriculum in England and Wales. The Scottish 5–14 Guidelines and Northern Ireland Curriculum are also referenced.

Starting Points at the beginning of each chapter give an indication of the background knowledge and skills needed to tackle the chapter content. Assignments covering all aspects of 'Using and Applying Mathematics' occur as separate tasks and as parts of many exercises. These exercises give plenty of opportunity for consolidation and practice of work in Number; Algebra; Shape, Space and Measures and Handling Data. Each chapter ends with a review of the work covered.

The material is designed to support you in acquiring skills and knowledge on the pathway to gaining a better understanding of mathematics.

Handling data

In this chapter you will learn:

- ➜ **how to read information from graphs**
- ➜ **how to collect and display information**
- ➜ **how to calculate different averages**

Starting points

Before you start this chapter you will need to know how to:

- take information from simple tables and lists
- record information using a tally chart
- draw bar charts and pictograms.

Exercise 1

This is a timetable for class 7M. Single lessons are 30 minutes long. Most lessons are doubles lasting an hour.

<table>
<tr><th></th><th>8.50–9.00</th><th>9.00–10.00</th><th>10.00–10.30</th><th>10.30–11.00</th><th>11.00–11.20</th><th>11.20–11.50</th><th>11.50–12.20</th><th>12.20–1.20</th><th>1.20–1.40</th><th>1.40–2.40</th><th>2.40–3.10</th><th>3.10–3.40</th></tr>
<tr><td>Monday</td><td rowspan="5">REGISTRATION</td><td>Maths</td><td colspan="2">English</td><td rowspan="5">BREAK</td><td>Music</td><td>R.E.</td><td rowspan="5">LUNCH</td><td rowspan="5">REGISTRATION</td><td>Science</td><td colspan="2">PE</td></tr>
<tr><td>Tuesday</td><td>Geography</td><td colspan="2">Science</td><td colspan="2">French</td><td>Design</td><td>Geog.</td><td>History</td></tr>
<tr><td>Wednesday</td><td>PE</td><td colspan="2">Maths</td><td>PSD</td><td>Geog.</td><td>English</td><td colspan="2">Science</td></tr>
<tr><td>Thursday</td><td>History</td><td colspan="2">Design</td><td colspan="2">Science</td><td>French</td><td colspan="2">Maths</td></tr>
<tr><td>Friday</td><td>Info. Tech.</td><td>English</td><td>Maths</td><td colspan="2">Art</td><td>English</td><td>RE</td><td>PSD</td></tr>
</table>

1. On which days do 7M have PE?
2. How many lessons of maths do they have?
3. How many lessons of PSD (personal social development) do they have?

A double lesson is the same length as two singles.

4. How long is the lunch break?
5. How many single lessons do they have in a week?
6. What time does the school day start?
7. What time does the school day finish?
8. Which subject has the most lessons?
9. How long is the school day?
10. What time does morning break finish?
11. Look carefully at your own timetable.
    (a) On which days do you have PE?
    (b) How many lessons of science do you have?
    (c) How long is the lunch break?
    (d) What time does the school day start?
    (e) What time does the school day finish?
    (f) Which of your subjects have most lessons?
    (g) How long is your school day?
    (h) What time does morning break finish?

Exercise 2

1. Look at the calendar for September, on the next page.
    (a) What day is the first of September?
    (b) School starts on the 8th of September.
    What day of the week is that?

(c) How many Saturdays are there in September?

(d) How many full weeks are there in September?

(e) What will the date be on the first Saturday in October?

(f) What date was the last Saturday in August?

September						
M	T	W	Th	F	S	Su
		1	2	3	4	5
6	7	8	9	10	11	12
13	14	15	16	17	18	19
20	21	22	23	24	25	26
27	28	29	30			

2 Repeat the same questions using a calendar for September this year.

Exercise 3

This is part of the timetable for the buses the pupils in 7M use to get to school.

Mondays to Saturdays						
	AM	AM	AM	Then every hour until	PM	PM
Station Road	7:20	8:20	9:20	-------	4:30	5:35
Old Post Office	7:24	8:24	9:24	-------	4:34	5:39
Junior School	7:30	8:30	9:30	-------	4:40	5:45
High Street	7:35	8:35	9:35	-------	4:45	5:50
Church Street	7:37	8:37	9:37	-------	4:50	5:55
Main Road	7:40	8:40	9:40	-------	4:55	6:00
School Road	7:45	8:45	9:45	-------	5:02	6:07

Key fact

Bus and train times are described using their starting times, e.g. 8:20, 9:20 etc.

1 When does the first bus start?

2 What day of the week does this bus not run?

3 What time does the 7:20 a.m. bus from Station Road get to the High Street?

4 If you miss the 9:20 bus in the morning what time is the next bus from Station Road?

5 You have to be in school at 8:50. What time is the last bus you can catch at the High Street to get you to school on time?

Exercise 4

This table displays the distances between some cities. To read this diagram read across from one city and down from the other.

	London	Belfast	Birmingham	Cardiff	Edinburgh	Glasgow	Liverpool	Manchester	Newcastle	Sheffield
Belfast	380									
Birmingham	105	270								
Cardiff	152	338	105							
Edinburgh	380	193	292	394						
Glasgow	397	148	292	395	45					
Liverpool	202	219	93	198	221	220				
Manchester	185	254	80	188	215	215	35			
Newcastle	274	221	207	318	110	148	155	132		
Sheffield	157	293	78	205	243	250	74	37	135	
Southampton	76	397	128	117	426	426	236	207	315	200

The distance from Newcastle to Birmingham is 207 miles.

1 How far is it from London to Newcastle?

2 Which two cities are 117 miles apart?

3 Which city is 80 miles from Manchester?

4 Which two cities are 293 miles apart?

5 Which two cities are both the same distance from Southampton?

6 Which two cities are closest together?

7 You have to travel from Newcastle to Belfast, Glasgow, Southampton and Cardiff. The order that you visit the cities is not important.

- Use the mileage chart to investigate the shortest total distance you could travel.
- Starting from the city nearest to your home, what is the shortest journey you can make to visit Birmingham, London, Edinburgh and Cardiff?

Tally charts

Key fact
In a tally chart you record information in groups of five like this.
卌 卌 II = 12

Key fact
The total number for each type of entry is called its frequency.

This tally chart shows how many sisters the pupils in class 7M have.

Number of sisters	Tally	Frequency
0	卌 III	8
1	卌 卌 II	12
2	卌	5
3	III	3
4	I	1
5		0
	Total	29

Exercise 5

1 Use the tally chart above to answer these questions.

(a) How many pupils in 7M have only one sister?

(b) How many pupils have three or more sisters?

(c) Calculate the total number of sisters the pupils in 7M have altogether.

2 These questions are based on this tally chart. It shows the shoe sizes of the pupils in 7M.

(a) What is the smallest shoe size in 7M?

(b) What is the largest shoe size in 7M?

(c) What is the total number of pupils in 7M?

(d) What is the most common shoe size?

Shoe size	Tally	Frequency
2	I	1
3	卌	5
4	卌 I	6
5	卌 III	8
6	卌I	5
7	II	2
8	I	1
9	I	1
	Total	

Drawing a tally chart

When you draw a tally chart from a list of data, or when you are gathering information, you should note the following.

Frequency means 'how many'.

- The values in your table should range from the smallest result to the largest result.
- When you have completed your tally, you should write the total for each type of result. This is called the **frequency**.
- Add all the frequencies at the bottom of the tally chart. This total will help you to check that you have recorded all the data correctly.

Exercise 6

1 This list shows how many brothers the 29 pupils in 7M have.

0 1 2 0 0 1 0 2 1 1

2 1 4 0 3 1 0 3 0 5

1 0 0 1 2 1 0 2 0

(a) Copy this tally chart and complete it to show how many brothers the pupils in 7M have.

Brothers	Tally	Frequency
0		
1		
2		
3		
4		
5		
	Total	

Add up the frequencies to check you have recorded all the values.

(b) How many of these pupils have just two brothers?

(c) What is the highest number of brothers?

(d) How many brothers do these pupils have altogether?

2 (a) This list shows the favourite lessons of the pupils in 7M.
Use the information in this list to draw a tally chart of the pupils' favourite lessons.

Maths	**English**	**Design**
Information Technology	**French**	**History**
Design	**Information Technology**	**English**
Design	**Maths**	**English**
History	**Information Technology**	**Maths**
Maths	**Information Technology**	**Design**
History	**English**	**Maths**
PE	**Information Technology**	**English**
Maths	**PE**	**Maths**
French	**Geography**	

(b) Which is the most popular subject?

(c) Which subject is the least popular?

(d) Which results are the same?

Assignment 1 A class like ours

Before you begin this assignment, you could try making up a data sheet for the pupils in your own class.

Class	**Data sheet**							
Gender	Birth month	Number of brothers	Number of sisters	Number of pets	Junior school	Favourite subject	How you get to school	Pocket money

Try setting up your data on a computer spreadsheet.

- Use the data you have gathered about your own class (or else use the data sheet about 7M, on page 24) to draw a tally chart of the junior schools the pupils in your class (or 7M) came from.
- Which junior school did most pupils come from?
- Are there any junior schools that more than three pupils from the class went to? How many?

Bar charts

Key fact
On a bar chart the frequency is placed on the lines, not in the spaces.

Bar charts are used to show the frequency of each type of data. The frequency is shown only by the height of the column or the length of the bar.

The columns or bars in a bar chart are all the same width.

Exercise 7

Key fact
The vertical axis is up the page.

1 This bar chart show the hours of sunshine recorded by pupils in 7M.

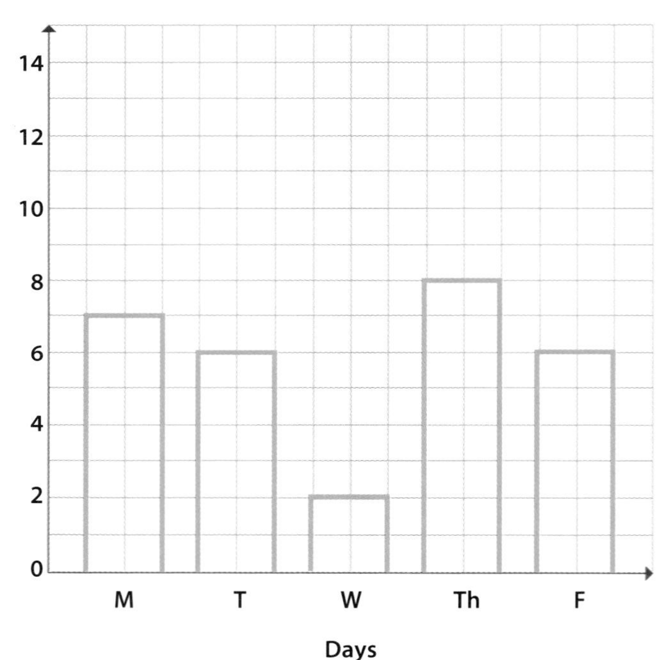

Numbers are placed on the vertical axis.

Days are labelled at the centre of each column.

(a) What do the numbers on the vertical axis represent?

(b) What does the height of each bar represent?

(c) What do you think happened on Wednesday?

(d) Which day had most sunshine?

(e) What was the total number of hours of sunshine shown by the bar chart?

2 In this bar chart, lines are used instead of bars. They represent the number of hours pupils in 7M spent on homework.

(a) What was the most common amount of time spent on homework?

(b) How many pupils have been surveyed?

(c) How many pupils only spent one hour on their homework?

(d) What was the most time spent on homework?

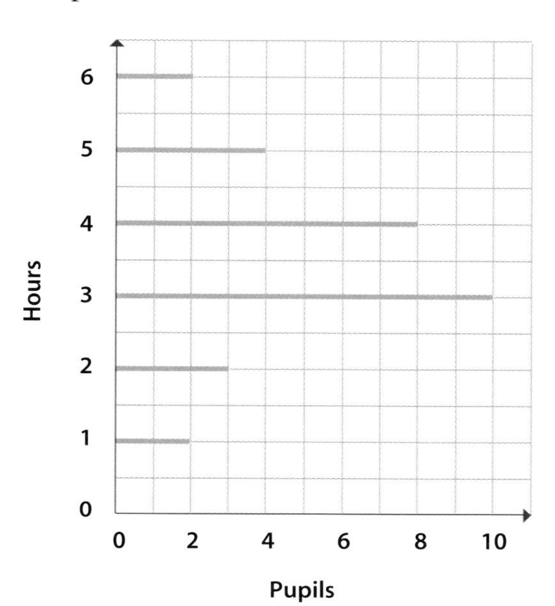

3 The pupils in 7M counted all the coins they had and recorded their results on a bar line chart.

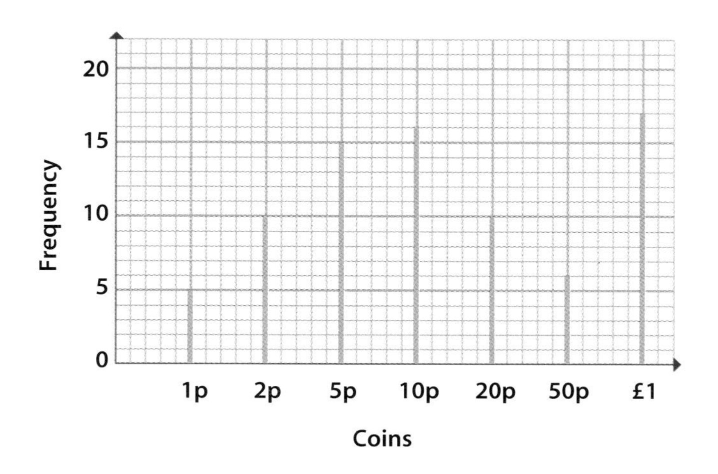

(a) Which was the most common coin?

(b) Which was the least common coin?

(c) How many different types of coin were there?

(d) How many 20p coins were there?

(e) What was the total number of coins?

Key fact

If the data has been gathered by counting, labels will go in the middle of columns.

4 This bar chart shows the days of the week that pupils in 7M were born on.

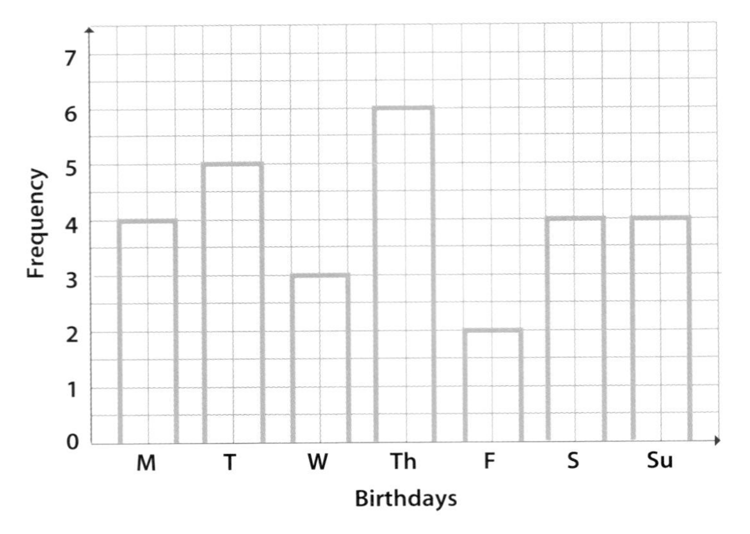

(a) On which day of the week were the least pupils born?

(b) On which day were most pupils born?

(c) On which days were the same numbers of pupils born?

(d) Using the information on your data sheet (or the data sheet for 7M), draw a bar line chart to show which months the pupils were born in.

5 This bar chart displays the junior schools that the pupils of 7M came from.

(a) How many different junior schools did the pupils in class 7M come from?

(b) Which junior school sent most pupils to 7M?

(c) Which school sent least pupils?

(d) Which two schools sent the same numbers of pupils?

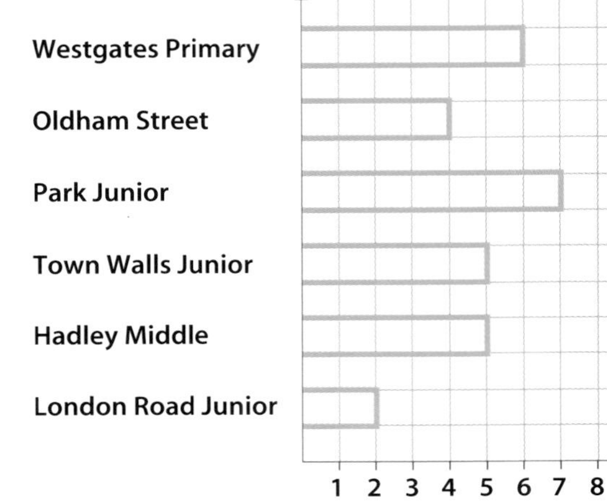

6 This bar chart records the number of films the pupils in 7M watched in one week.

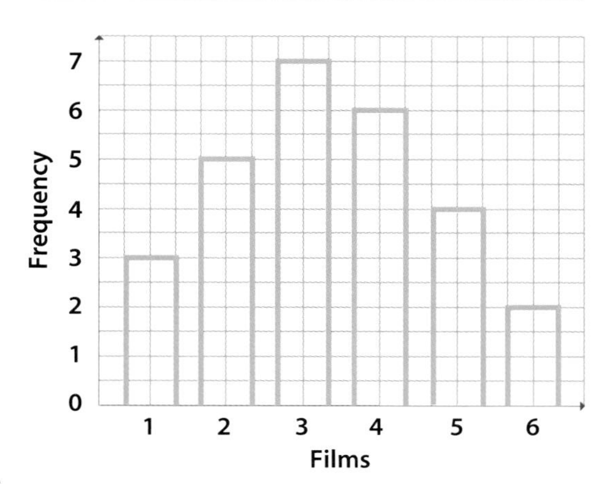

(a) What was the highest number of films that pupils watched?

(b) How many pupils watched more than four films in this week?

(c) What was the number of films watched by the majority of pupils?

(d) How many films were watched in total?

(e) Explain how you worked out the answer to (d).

More difficult bar charts

Exercise 8

1 This bar chart shows single lessons and the number of homeworks for some 7M lessons.

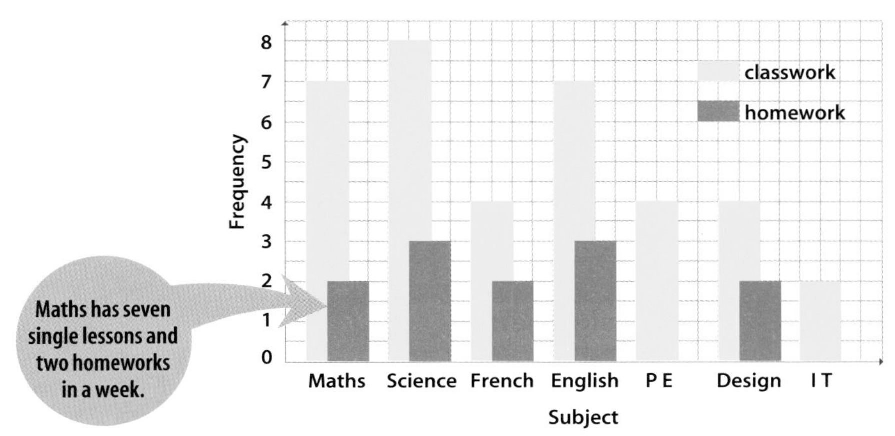

(a) How many lessons does science have in a week?

(b) How many homeworks does science have?

(c) How many homeworks are shown in the seven subjects in this bar chart?

(d) Which lessons set no homework?

2 This bar chart displays the ages of books used in five different subjects.

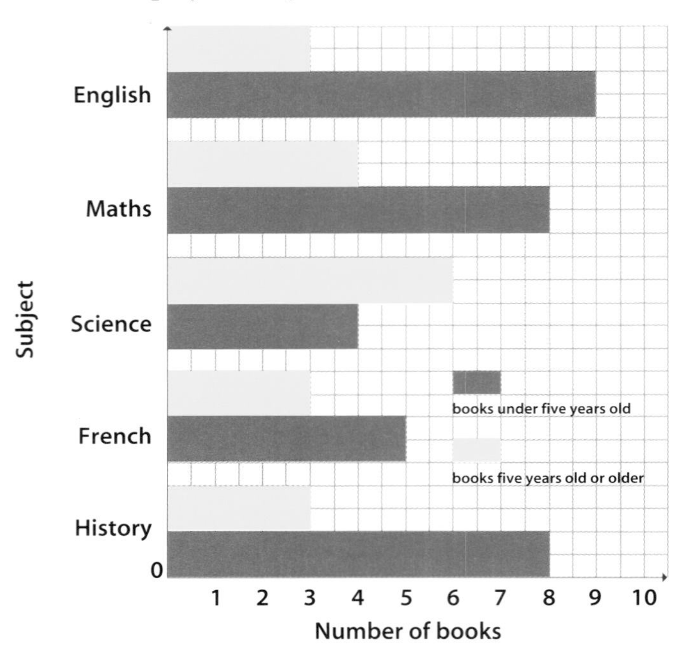

(a) Which subject has the most books less than five years old?

(b) Which subject uses least books five or more years old?

(c) How many science books less than five years old are used?

(d) How many history books five or more years old are used?

(e) How many books less than five years old do the five subjects use?

(f) What is the difference between the number of books less than five years old and the number five or more years old used in maths?

Assignment 2 How many brothers and sisters?

Try using a computer spreadsheet.

- Using either the data you have gathered about your own class, or the data about 7M, draw a bar chart showing how many brothers and sisters pupils in the class have.
- How many pupils have no brothers or sisters?
- How many pupils have more than three brothers or sisters?
- How many children are there in the largest family?

Pictograms

Key fact

In a pictogram each picture can represent a number of things.

A key shows how many.

= 2 people

Pictograms use pictures to represent data. A key must always be given to show how many pieces of data each picture represents.

One picture can represent one piece of data but it is more usual for each picture to represent more than one piece of data.

This pictogram shows the ways that pupils in 7M get to school.

Walk	
Bus	
Car	
Bike	

Key

= 2 pupils

Exercise 9

1 These questions are based on the pictogram above.

(a) What does represent?

(b) How many pupils walk to school?

(c) How many pupils cycle to school?

(d) How many pupils come to school by bus?

(e) What does represent?

(f) How many pupils come to school by car?

(g) How many pupils are there in class 7M?

(h) Draw a pictogram showing the days of the week that the pupils of 7M were born on.

The information you need is given in the bar chart on page 10.

2 This pictogram shows the total number of televisions in Year 7 pupils' homes.

(a) How many televisions do 7P have?

(b) How many televisions do 7Q have?

(c) Which class has the most televisions?

(d) Which class has the fewest televisions?

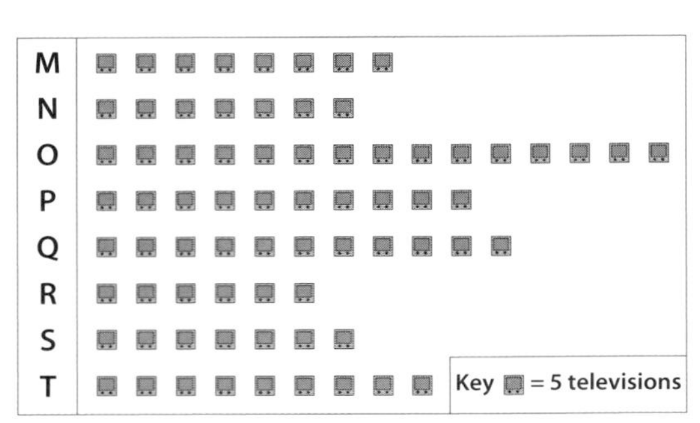

Data

Key fact
Discrete data is data gathered by counting.

Information or data which has individual values is called **discrete data**. Here are some examples of discrete data.

- How many English lessons you have
- How many pupils were born in April
- How many sisters you have

Exercise 10

If a question asks for 'the number of ...' it will involve counting.

1 Which of the following data are discrete?

(a) The numbers of brothers and sisters you have

(b) The speed you walk home from school

(c) The numbers of maths lessons you have in a week

(d) The heights of the pupils in your class

(e) The temperature

(f) The stopping distance of a car

(g) The number of pens in your bag

(h) The weight of the books in your bag

(i) The amount of money in your pocket

(j) The numbers of cars in the car park

Grouping data – class intervals

You should avoid having too many or too few groups.

When you have a wide variety of data it is often useful to collect it together into groups. You should have from four to eight groups of data.

Example

This bar chart shows how much pocket money the pupils in 7M receive. This data has not been grouped.

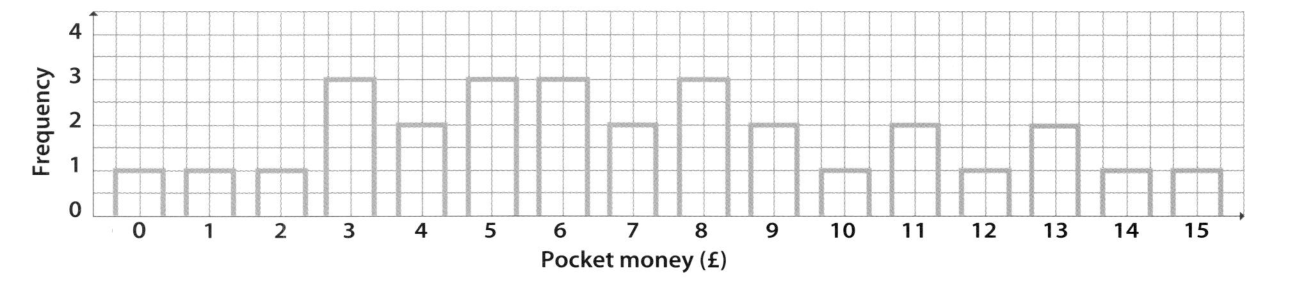

Grouping the data in different ways displays the information in different ways.

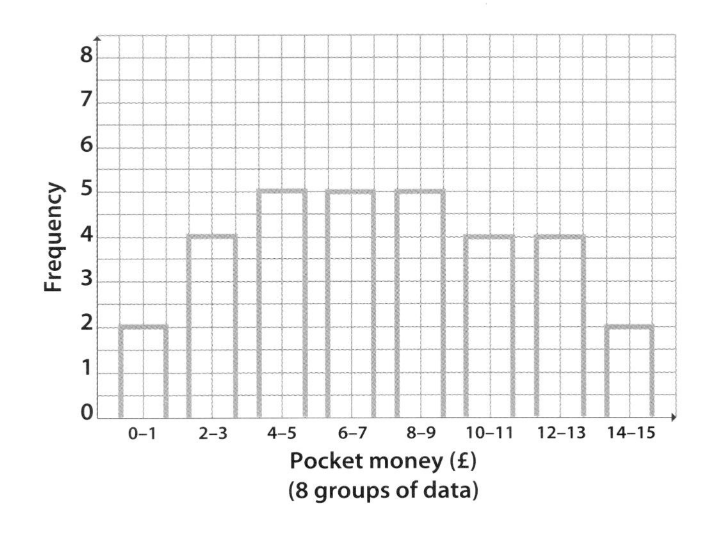

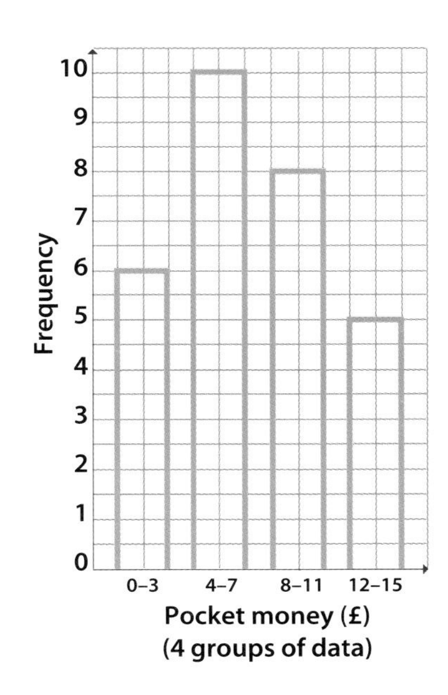

By grouping the data into four groups, you can see clearly that the most common amount of pocket money for this class is from **£4** to **£7**.

Exercise 11

The data in this frequency table shows how many cousins a group of children have.

1 Draw a bar chart of this data, without grouping it.

2 Draw a bar chart of this data, putting it into two groups.

3 Why are these not good ways of displaying the data?

4 Draw a bar chart of this data putting it into six groups.

5 Compare your graphs. Which do you think displays the information best?

Number of cousins	Frequency
1	2
2	3
3	5
4	7
5	8
6	6
7	7
8	9
9	6
10	5
11	4
12	3
13	2
14	3
15	3
16	1
17	0
18	1
19	2
20	1
21	0
22	1
23	1
24	1

Assignment 3 Pocket money survey

- Carry out a survey of how much pocket money the pupils in your class are given.
- Display this information in a bar chart.
- Using a suitable group size, display this information as grouped data.
- Explain why you chose this group size.

Averages

Mode – the most common result

The mode is the measure of average which gives the most common or popular result.

Key fact

The mode is the most common result.

This is called the **modal result**.

It is easy to see which result is the mode when you have data displayed.

In a tally the largest number of tally marks shows the mode.

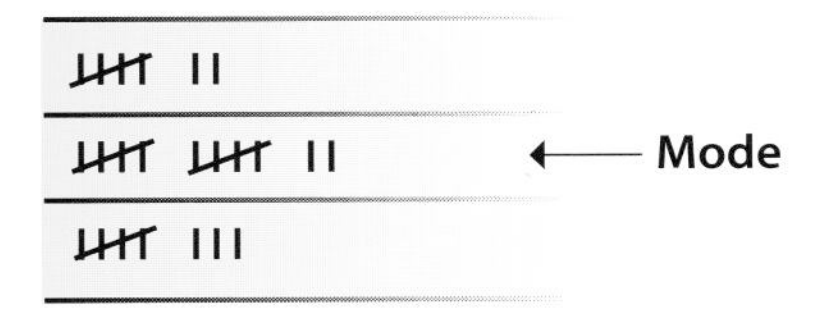

In a bar chart the tallest column or longest bar represents the most common result.

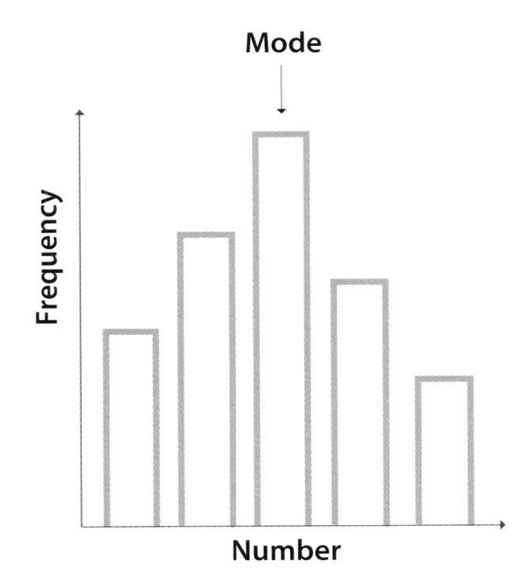

In a pictogram the most common result is represented by the largest number of pictures.

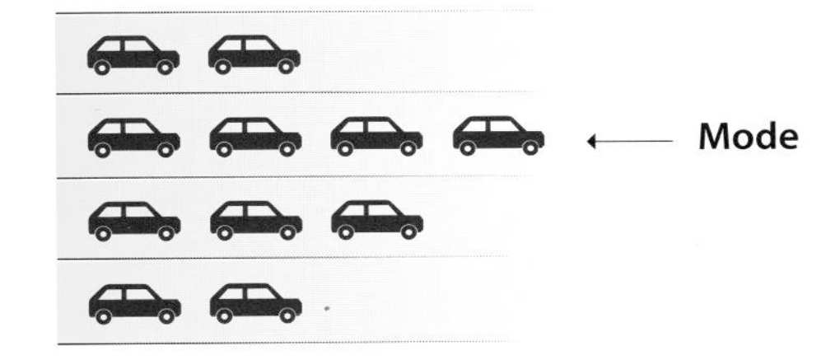

Exercise 12

1 Look at these sets of data. In each case, which is the mode or modal result?

(a)

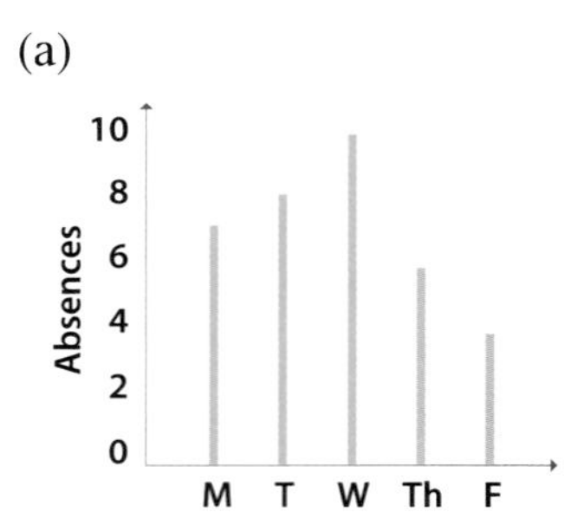

(b)

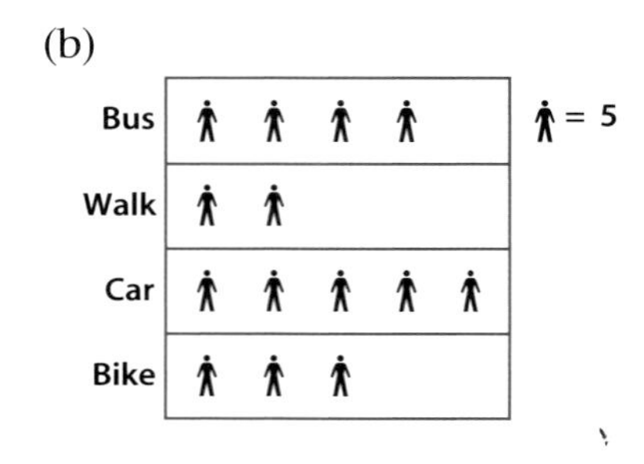

(c)

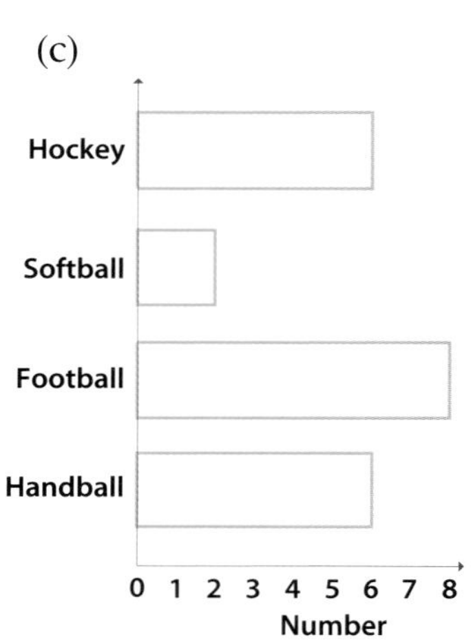

(d)

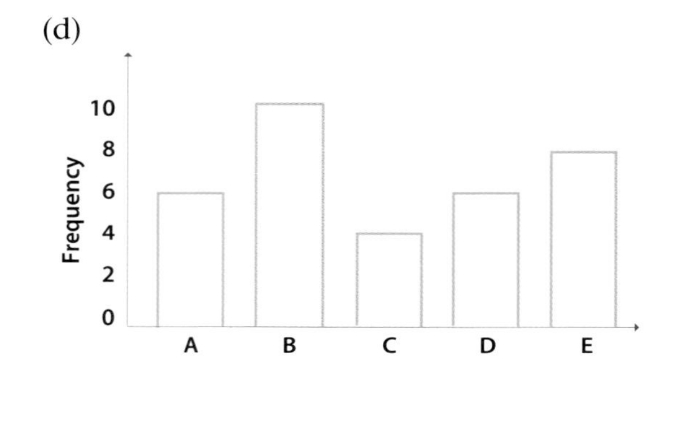

> **If results are not shown in a diagram, to find the mode tally the results first.**

2 Here are the shoe sizes of the pupils in 7M.

5	4	6	5	4	3	4	9	5	3
4	6	3	5	3	8	7	4	6	5
2	3	3	7	2	5	6	7	5	

(a) Put the results into a tally chart.

(b) Which is the most common shoe size?

3 These are the heights of a group of ten Year 7 pupils.

125 cm	124 cm	127 cm	127 cm	130 cm
125 cm	129 cm	128 cm	127 cm	

Which is the modal height?

Median – the middle result

The mode is not always the best measure of the average of a set of data. Look at this bar chart.

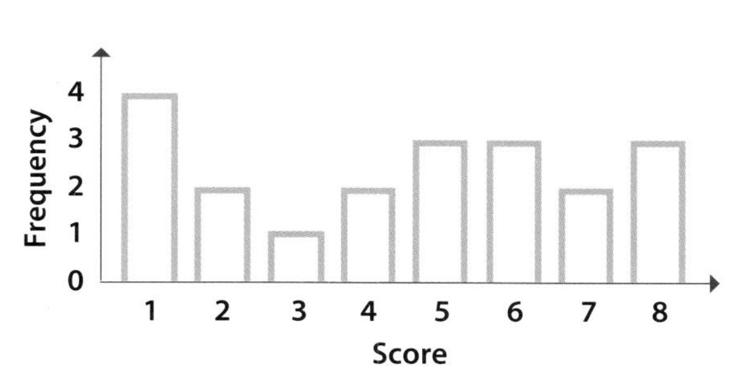

You can see that the modal value is represented by the tallest column. A better average for this data would be the middle value in the bar chart. This middle value is called the **median.**

Key fact
The median is the middle value in an ordered list, or the average of the middle two values.

The median is the middle value in a set of data, when the set of data is put into size order.

Example

1, 1, 1, <u>2</u>, 3, 5, 9

The median value is 2.

Example

1, 1, 7, 3, 6, 1, 4

Rewrite the data in size order.

1, 1, 1, <u>3</u>, 4, 6, 7

The median value is 3.

If there are two middle values you take the midpoint between the two values.

Example

1, 1, 1, 2, 3, 4, 4, 6

If you imagine the numbers in a strip you can see that there are two middle values, which are 2 and 3.

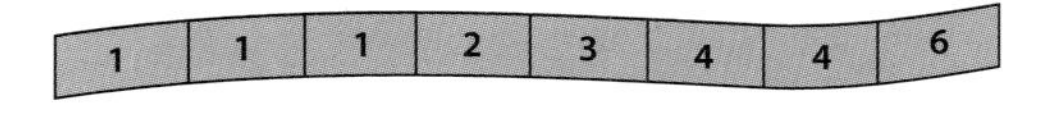

The median or mid-point value is 2.5

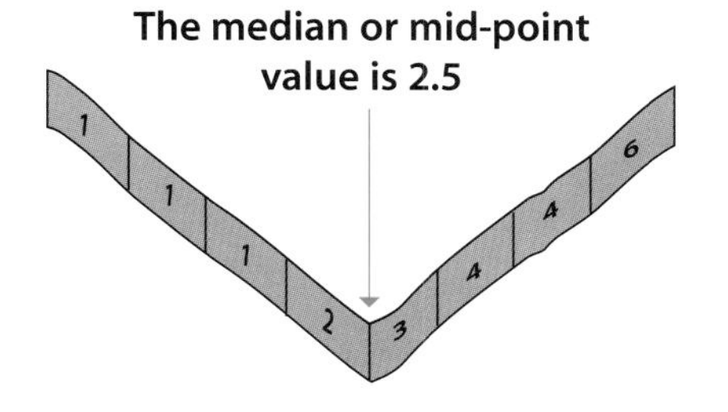

The midpoint between 2 and 3 is 2.5.

The median value is 2.5.

Key fact
The value of the median does not have to be a value from the original data.

Exercise 13

1 Write the following lists of numbers in order (smallest first). Find the median values from your ordered lists.

(a) 2, 4, 1, 8, 5

(b) 3, 6, 6, 5, 7, 1, 9

(c) 35, 38, 39, 36, 37, 39, 40, 34, 31

(d) 1, 2 , 3, 3, 4, 5

(e) 1, 2, 3, 4, 5, 6

Remember to put the data in order before you find the middle values.

2 Find the median and modal values in each of the following sets of data.

(a) 1, 4, 7, 1, 5, 2, 1, 3, 6, 3, 5

(b) £5, £4, £3.50, £5, £4.50, £3, £6

3 Two pupils in 7M have school dinners nine times and spend the following amounts.

Pupil A **£1.35, £1.55, £1.20, £1.20, £1.75, £1.35, £1.35, £1.75, £1.35**

Pupil B **£1.35, £1.55, £1.55, £1, £1.75, £1.35, £1.20, £1.20, £1.20**

(a) Calculate the median and modal amounts spent by Pupil A.

(b) Calculate the median and modal amounts spent by Pupil B.

(c) Calculate the median and modal amounts spent by both pupils together.

4 This data shows how many sisters and brothers the pupils in 7M have.

1, 1, 4, 1, 3, 1, 0, 3, 2, 3, 3, 2, 5, 2, 2, 2, 3, 5, 0, 4, 1, 6, 1, 2, 0, 2, 3, 4, 3

(a) By tallying, or some other method, put this data into an ungrouped frequency table.

(b) Use the table to find the modal number of brothers and sisters of the pupils in 7M.

(c) Use the frequency table to find the median of the set of data.

To find the middle value in a frequency table with an odd number of values, add one to the total number of values and divide by two.

5 Fifty pupils take part in a school play. The year group of each pupil is given below.

7, 10, 7, 9, 9, 10, 9, 11, 8, 9, 11, 9, 10, 8, 9, 11, 8, 7, 8, 7, 11, 9, 11, 8, 11, 7, 11, 9, 9, 9, 8, 7, 8, 11, 10, 8, 7, 10, 8, 8, 10, 7, 9, 10, 7, 8, 9, 9, 7, 9

(a) Draw a frequency table to show the year groups of these pupils.

(b) Use your table to identify the modal year group.

(c) Use your table to find the year group of the median pupil.

(d) Which provides the more useful average for this group of pupils? Explain your answer.

Assignment 4 Brothers and sisters

- Carry out a survey of how many sisters and brothers pupils in your class have.
- Display this information in a frequency table or bar chart.
- Identify the modal number of brothers and sisters for the pupils in your class.
- What is the median number of sisters and brothers of the pupils in your class?

Review Exercise

1 (a) Look at this chart. What name is given to this method of recording data?

(b) What does HHT represent?

(c) Copy this table and complete it by filling in the frequencies.

(d) What is the modal colour?

(e) What is the total frequency?

Favourite colour	?	Frequency
Red	HHT HHTI II	?
Blue	HHT HHT IIII	?
Green	HHT II	?
Yellow	HHTI HHTI	?
Black	HHTI HHTI I	?
White	HHTI II	?
	Total	?

2 (a) Display the data from question 1, using a bar line chart.

(b) What does the height of each line represent?

3 This bar chart displays the makes of the cars in the school car park.

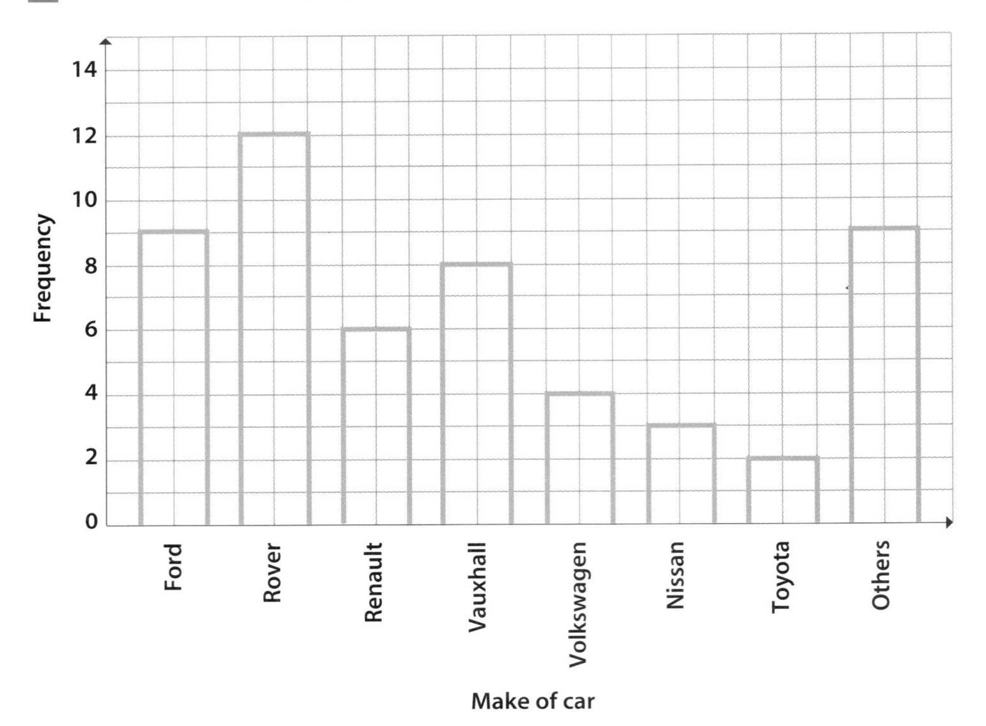

(a) What do the numbers on the vertical axis represent?

(b) Which two groups of car have the same frequency?

(c) Which is the modal make of car?

(d) How many cars are there in the modal group?

(e) What is the total number of cars in the car park?

4 The following 30 cards are picked from a pack.

♥ hearts
♦ diamonds
♣ clubs
♠ spades

(a) Tally this data.

(b) Use the data recorded in the tally chart to complete a pictogram.

5 The following marks out of 20 were recorded for two Year 7 classes.

Class A: **20, 13, 13, 14, 15, 11, 12 , 10 5, 7, 8, 16, 10, 12, 15 20, 15, 8, 14, 19, 3, 15, 16, 4, 19, 18, 17, 9, 12**

Class B: **20, 8, 19, 10, 13, 19, 5, 17, 18, 20, 2, 6, 14, 13, 12, 18, 15, 4, 19, 11, 4, 16, 13, 20, 5, 9, 19, 7, 13, 16**

(a) By dividing the marks from 1 to 20 into a suitable number of groups, display the marks of each class as grouped data.

(b) Using your displays, and the values for the mode and median of each group, decide which class has achieved the best marks.

6 The following are the shoe sizes of ten pupils.

4, 7, 4, 8, 9, 6, 5, 8, 5, 4

(a) What is the modal size for this group?

(b) Why is this not a good representative of the average size of the shoes?

(c) What value will the median take?

Assignment 5 More about our class

Using either the data sheet for class 7M, or data gathered from your own class, carry out one or more of the following.

- Find the mode and median amounts of pocket money received by pupils in your class per week. Use a suitable method for gathering and displaying this data.
- Find the modal and median numbers of sisters and brothers the pupils in your class have. Use a suitable method for gathering and displaying this data.
- Find the mode and median of the heights of the pupils in your class. Use a suitable method for gathering and displaying this data.

Class 7M				Data sheet					
Number	Gender	Birth month	Number of sisters	Number of brothers	Number of pets	Junior school	Favourite subject	How you get to school	Pocket money
1	G	February	1	0	3	Whitegates	Maths	Walk	£2
2	B	February	0	1	0	Oldham St.	English	Bus	£1
3	B	March	2	2	6	Park Junior	History	Walk	£5
4	B	May	1	0	20	Town Walls	IT	Bus	£3
5	B	April	3	0	15	Whitegates	Design	Car	£2
6	G	February	0	1	2	Town Walls	History	Bike	£2
7	B	September	0	0	13	Hadley	Maths	Walk	£5
8	G	October	1	2	8	Oldham St.	PE	Car	£3
9	G	January	1	1	1	Whitegates	English	Walk	£4
10	B	November	2	1	13	Whitegates	Info Tec	Walk	£7
11	B	September	1	2	4	Hadley	History	Walk	£4
12	G	May	1	1	4	London Rd.	PE	Bus	£5
13	G	March	1	4	21	Park Junior	Maths	Bus	£6
14	G	December	2	0	18	Town Walls	English	Walk	£20
15	B	April	0	3	9	Park Junior	English	Car	£4
16	G	September	4	1	3	Park Junior	Maths	Bike	£1
17	G	June	0	0	0	Park Junior	IT	Walk	£3
18	B	November	1	3	6	Oldham St.	Maths	Walk	£5
19	G	July	1	0	0	Hadley	Design	Walk	£2
20	G	December	1	5	1	Town Walls	PE	Bus	£10
21	G	January	0	1	12	Park Junior	IT	Walk	£4
22	B	October	2	0	0	Whitegates	French	Bus	0
23	B	December	0	0	16	Town Walls	English	Car	£2
24	G	May	1	1	2	Hadley	Maths	Bus	£3
25	G	October	1	2	6	Hadley	Design	Walk	£4
26	B	November	3	1	7	Oldham St.	History	Bike	£1
27	B	February	3	0	2	Park Junior	French	Car	£2
28	G	July	0	2	11	London Rd.	Maths	Walk	£3
29	B	April	2	0	0	Whitegates	Info Tec	Walk	£2

Four rules of number (+, −, ×, ÷)

In this chapter you will learn:

- → about place value
- → ways to add and subtract in your head
- → how to +, −, × and ÷ without using a calculator
- → how to multiply and divide by 10 and 100
- → to solve problems

Starting points

Before you start this chapter you will need to:

- read, write and order numbers up to 1000
- practise adding, taking away, multiplying and dividing single-digit numbers.

Exercise 1

Three hundred and sixty-two is written: 362.
You can show this with a mapping.

Copy the following words and numbers into your book. Draw mappings between the words and numbers which have the same meaning. One pair is done for you.

1

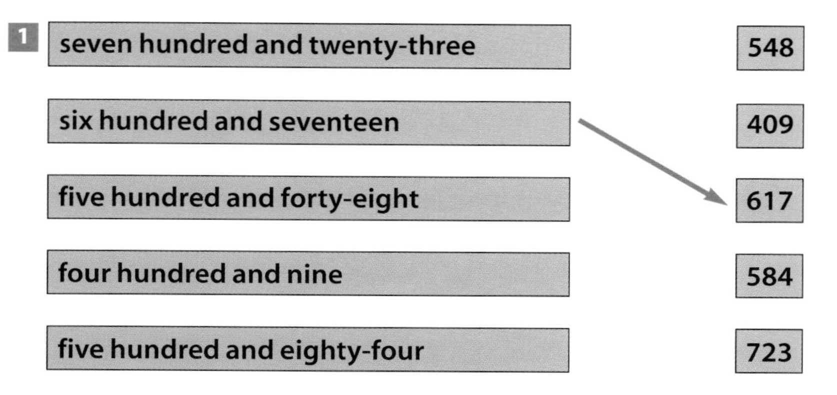

2 Copy these boxes and draw mappings from the numbers to the words that have the same meaning.

605	one hundred and five
120	two hundred and ten
210	one hundred and fifty
105	six hundred and five
150	one hundred and twenty

3 Write the following numbers in words.
(a) 625 (b) 789 (c) 210 (d) 105

Key fact
A digit is a single number, e.g. 6.

4 Write down in digits the number that is:
(a) six more than four hundred and twenty-one
(b) three less than nine hundred and thirty-eight
(c) eight more than five hundred and nineteen

5 Write down in words the number that is:
(a) 5 more than 873 (b) 3 less than 296
(c) 10 more than 356 (d) 6 more than 589

6 Write the following numbers in order, largest first.
505, 550, 515, 50, 500, 150, 510

Assignment 1 Three-digit numbers

This three-digit number 3 6 1 has been made by picking three cards from this group. 9 1 3 5 6

- What is the largest three-digit number that you can make from this group?
- What is the second largest three-digit number that you can make?
- What is the smallest three-digit number you can make?
- What is the largest three-digit *even* number that you can make?
- What is the smallest three-digit *odd* number you can make?

Key fact
An even number can be divided exactly by 2.

Now use only the cards 3 5 and 6 .

- List all the three-digit numbers you can make.
- Place your numbers in order, starting with the largest.

How numbers are related

The underlined digit in **$2\underline{7}9$** represents **70**.

Exercise 2

1 What is the value of the underlined number in each of the following?

(a) $3\underline{6}4$ (b) $\underline{8}70$ (c) $\underline{5}$ (d) $\underline{8}95$

(e) $\underline{3}4$ (f) $76\underline{4}$ (g) $4\underline{6}$ (h) $5\underline{0}0$

(i) $50\underline{5}$ (j) $\underline{6}23$ (k) $\underline{8}7$ (l) $8\underline{2}$

Use the numbers in question 1 for questions 2–12.

2 Which of the numbers in the list is 100 times as big as (c)?

3 Which of the numbers in the list is 10 times as big as (k)?

4 Which of the numbers in the list is 400 more than (a)?

5 Which of the numbers in the list is 5 less than (i)?

6 Which of the numbers in the list is 272 more than (j)?

7 How many tens are there in (b)?

8 How many hundreds are there in (h)?

9 Which number is ten times as small as (b)?

10 What is the value of the underlined number in (e)?
If you multiply (e) by 10, what will the value of the underlined number become?

11 What is the value of the underlined number in (g)?
If you multiply (g) by 10, what will the value of the underlined number become?

12 What is the value of the underlined number in (l)?
If you multiply (l) by 100, what will the value of the underlined number become?

Exercise 3

1 In this triangle the number at each corner is made by adding the numbers on the sides next to it.

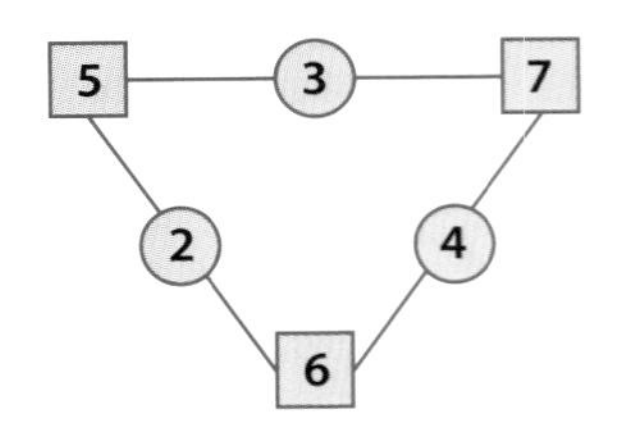

3 + 4 = 7
2 + 4 = 6

Copy these triangles and complete them.

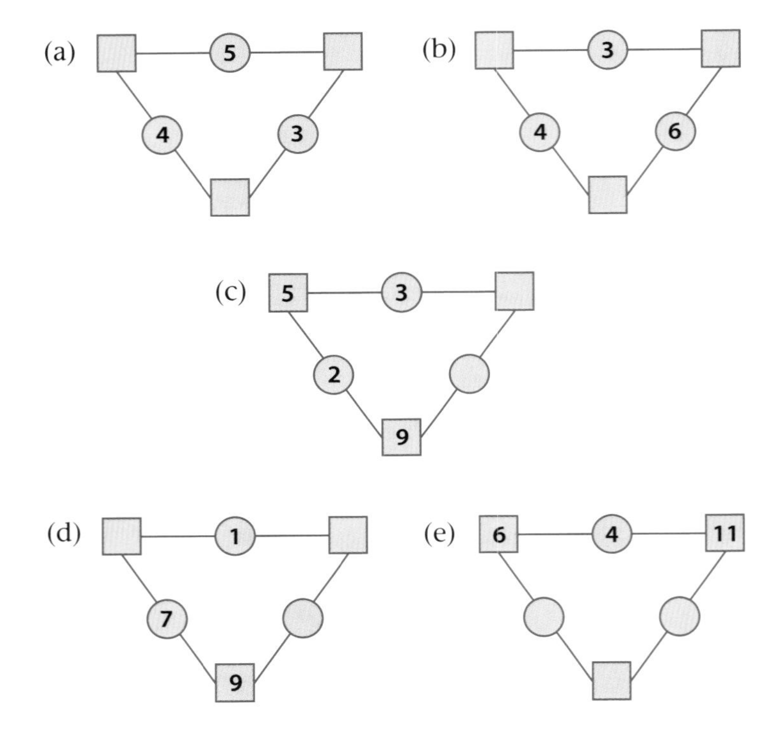

2 Using the numbers 1 to 10, find as many solutions to this number triangle as you can.

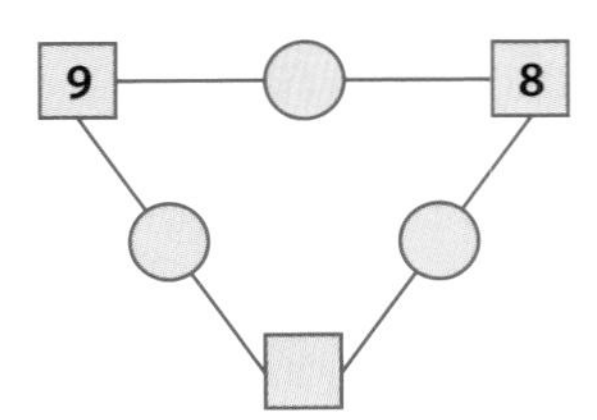

Assignment 2 A number maze

Look at this maze. You enter at **Start** and you leave at **Exit**.

As you travel through the maze, add the numbers written in the squares you cross, to find your total score.

Do not cross the same square more than once.

Do not use a calculator when you see this sign.

- What is the smallest total you can score?
- What is the largest total you can score?

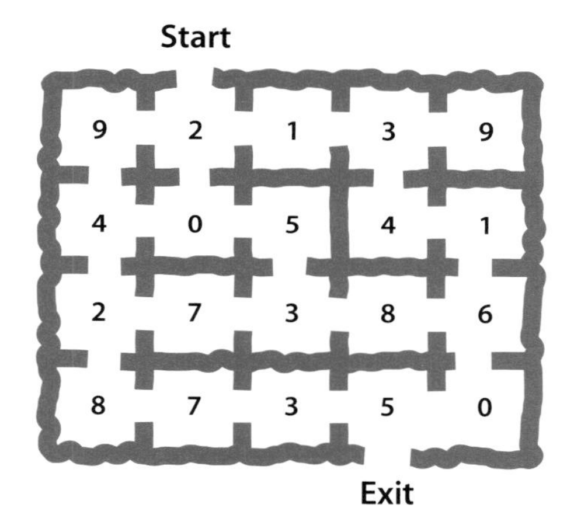

Adding and subtracting in your head

Looking for an adding pattern

Think about **28 + 43**.

There are several ways to work out the answer in your head, without writing down your working.

Example

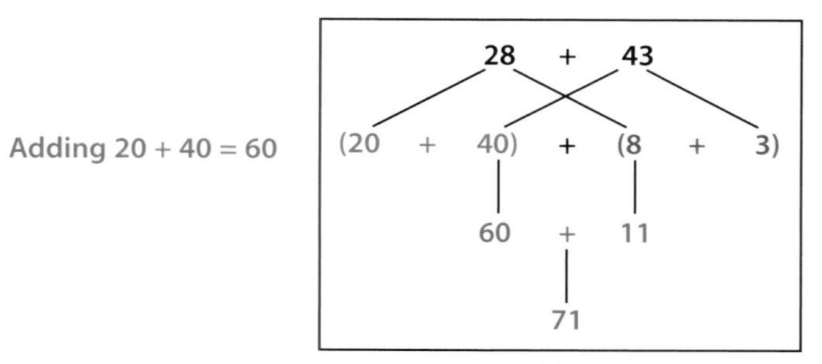

Adding 20 + 40 = 60

Adding 8 + 3 = 11

Adding 60 + 11 = 71

Exercise 4

1 Write down some other ways of adding the numbers in your head.

2 Add these in your head. Write down the answers.

(a) 34 + 12 = (b) 47+ 42 = (c) 38 + 15 = (d) 37 + 15 =

(e) 79 + 17 = (f) 35 + 29 = (g) 28 + 19 = (h) 64 + 18 =

(i) 46 + 39 = (j) 31 + 99 = (k) 74 + 28 = (l) 34 + 89 =

Now use a calculator to check your answers.

3 There are 27 girls and 35 boys taking part in an athletics meeting. Work out in your head how many pupils will be taking part.

4 You have 49p in one pocket and 32p in another. Work out in your head how much you have altogether.

5 You have 26 posters of your favourite pop-group. Your friend has 17 posters of the same group. Work out in your head how many you have together.

6 Your maths homework takes you 35 minutes. Your history homework takes you 45 minutes. Work out in your head how many minutes you spend on your homework.

7 Yesterday there were 81 pencils in the stock cupboard. Today there are 58. How many pencils have been given out?

8 In these magic squares, the rows, columns and diagonals each add up to give the target number.

Copy and complete the squares.

(a) Target 34

16	3		
5		11	8
	6		12
	15		1

(b) Target 40

	17		14
18	4		
12	6		
		2	16

Looking for a pattern for taking away

Think about **68 – 25**.

This type of problem might be solved like this.

Taking away 60 – 20 = 40

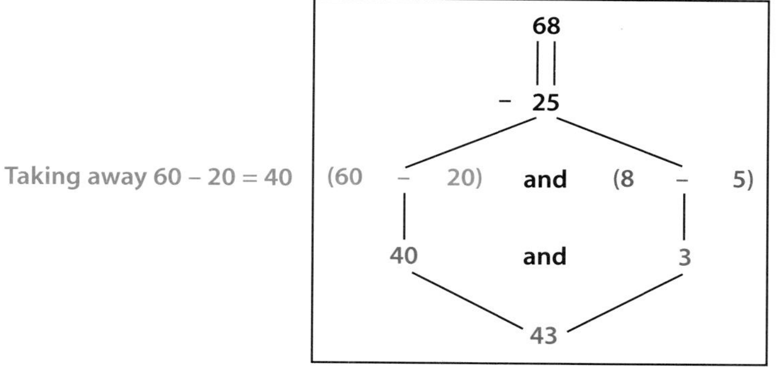

Taking away 8 – 5 = 3

Putting these together 40 + 3 = 43

Exercise 5

1 The example above is not the only way to take 25 away from 68 in your head. Explain another method.

2 Using any method you choose, work out the answers to the following in your head. Write down your answers.

(a) 27 – 4 =	(b) 38 – 6 =	(c) 28 – 15 =	(d) 87 – 76 =
(e) 42 – 7 =	(f) 53 – 6 =	(g) 27 – 18 =	(h) 43 – 25 =
(i) 72 – 59 =	(j) 91 – 19 =	(k) 45 – 27 =	(l) 63 – 36 =

Key fact
To find the difference you take away.

3 Work out the difference between 68 and 35.

4 You had 94p and then you spent 36p at break-time.
How much should you have left?

5 You have collected 46 tokens and your friend has collected 29.
How many more tokens do you have than your friend?

6 There are 32 pupils in your class and 29 in the class in the next room.

(a) How many pupils are there in both rooms?

(b) How many more pupils are there in your room than in the next room?

Adding larger numbers

When you add larger numbers together it is easier if you keep the units in columns.

```
   6 8 5
 + 1 3 6
   8 2 1
   1 1
```

Keep the numbers in their columns.
Remember to add in all the numbers you have carried.

Exercise 6

Calculate the following. Show all your working.

1 (a) 336 + 123 (b) 425 + 372 (c) 656 + 138

2 (a) 473 + 695 (b) 721 + 897 (c) 675 + 594

3 (a) 274 + 657 (b) 179 + 953 (c) 472 + 135 + 678

4 (a) 794 + 674 + 532 (b) 343 + 765 + 826 (c) 659 + 844 + 283

Remember to keep the units in line.

5 Write the following in columns and then add.

(a) 413 + 234 + 563 =
(b) 234 + 76 + 982 + 107 =
(c) 14 + 210 + 5 + 154 + 29 =
(d) 296 + 473 + 1 + 2 + 143 =
(e) 247 + 3 + 275 + 34 + 7 =
(f) 210 + 15 + 976 + 4 + 124 =

There may be more than one way to do this.

6 This calculation is wrong.
Change one digit to make it right.

```
   2 3 4
 + 1 8 6
   4 3 0   ✗
```

7 In a school there are 192 pupils in Year 7, 211 in Year 8 and 175 in Year 9. How many pupils are there altogether in these three years?

8 A charity receives the following donations.

Monday	£243
Tuesday	£167
Wednesday	£94
Thursday	£182
Friday	£207
Saturday	£193

What was the total amount donated in this week?

Assignment 3 Digits 0–9

This calculation uses each of the digits from 0 to 9 just once.

$$\begin{array}{r} 623 \\ +\ 475 \\ \hline 1098 \end{array}$$

- Investigate how many other pairs of three-digit numbers and their sums use each of the digits from 0 to 9 just once.

Assignment 4 Palindromic numbers

A palindromic number is one which has the same value when the digits are written in the reverse order. For example,

353 67576

are both palindromic numbers.

$$\begin{array}{r} 134 \\ +\ 431 \\ \hline 565 \end{array}$$

This is a one-step palindrome.

$$\begin{array}{r} 67 \\ +\ 76 \\ \hline 143 \\ +\ 341 \\ \hline 484 \end{array}$$

This is a two-step palindrome.

$$\begin{array}{r} 745 \\ +\ 547 \\ \hline 1292 \\ +\ 2921 \\ \hline 4213 \\ +\ 3124 \\ \hline 7337 \end{array}$$

This is a three-step palindrome.

You can make palindromic numbers by adding a number to its reverse.

- Using your calculator, find other numbers which become palindromic.
- Do all two- and three-digit numbers eventually become palindromes?

Taking away three-digit numbers

Look at the example 432 – 165.

Exercise 7

Remember to show all your working.

Calculate the following.

	(a)	(b)	(c)
1	$\begin{array}{r} 219 \\ -\ 108 \\ \hline \end{array}$	$\begin{array}{r} 379 \\ -\ 261 \\ \hline \end{array}$	$\begin{array}{r} 456 \\ -\ 310 \\ \hline \end{array}$
2	$\begin{array}{r} 587 \\ -\ 258 \\ \hline \end{array}$	$\begin{array}{r} 247 \\ -\ 108 \\ \hline \end{array}$	$\begin{array}{r} 425 \\ -\ 217 \\ \hline \end{array}$
3	$\begin{array}{r} 721 \\ -\ 346 \\ \hline \end{array}$	$\begin{array}{r} 567 \\ -\ 258 \\ \hline \end{array}$	$\begin{array}{r} 643 \\ -\ 449 \\ \hline \end{array}$
4	$\begin{array}{r} 800 \\ -\ \ 86 \\ \hline \end{array}$	$\begin{array}{r} 600 \\ -\ 152 \\ \hline \end{array}$	$\begin{array}{r} 700 \\ -\ 538 \\ \hline \end{array}$
5	$\begin{array}{r} 502 \\ -\ 309 \\ \hline \end{array}$	$\begin{array}{r} 301 \\ -\ 123 \\ \hline \end{array}$	$\begin{array}{r} 501 \\ -\ 274 \\ \hline \end{array}$
6	$\begin{array}{r} 713 \\ -\ 468 \\ \hline \end{array}$	$\begin{array}{r} 822 \\ -\ 618 \\ \hline \end{array}$	$\begin{array}{r} 976 \\ -\ 499 \\ \hline \end{array}$

7 Replace the boxes with the numbers 1, 2, 3 and 4.

$$\begin{array}{r} 7\,\square\,\square \\ -\ 5\,\square\,\square \\ \hline \\ \hline \end{array}$$

(a) Investigate which combination will give you the largest answer.

(b) Which combination will give you the smallest possible answer?

8 Two darts players are playing 501.
Player 1 scores 20, 36 and 1.
His total score is 57.

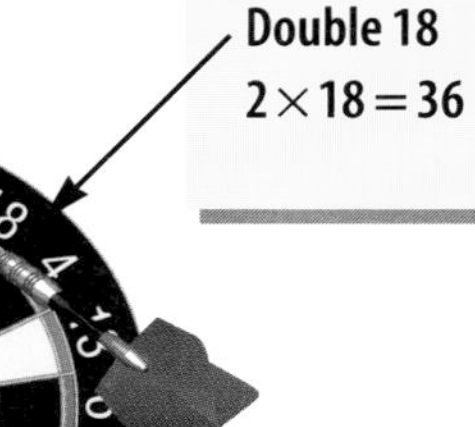

To work out the remaining score, 57 is taken away from 501.

$$\begin{array}{r} 5\ 0\ 1 \\ -\quad 5\ 7 \\ \hline 4\ 4\ 4 \\ \hline \end{array}$$

(a) Copy this table and complete it by filling in the remaining score after each round for both players.

Player 1		Player 2	
Score	**Remaining score**	**Score**	**Remaining score**
	501		501
57	444	80	
96		72	
101		99	
124		131	

(b) What is the remaining score for each player after four throws each?

Multiplying

Exercise 8

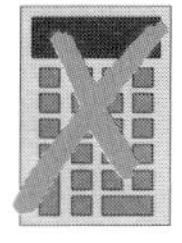

1 Copy the following multiplication tables and complete them. The first one is started for you.

(a)

×	2	3	4
1	2		
2			
3	6		12

(b)

×	5	3	1
4			
2			
1			

2 Now copy these tables and complete them. You need to work out what numbers are being multiplied.

(a)

×			
	6	10	8
	3	5	4
	15	25	20

(b)

×			
	5		4
		12	16
			8

3 In this table each number is multiplied by 2 across the rows and by 3 down the columns.

→ ×2, ↓ ×3

1	2	4
3	6	12
9	18	36

Copy these tables and complete them using the rules shown for across and down.

(a)

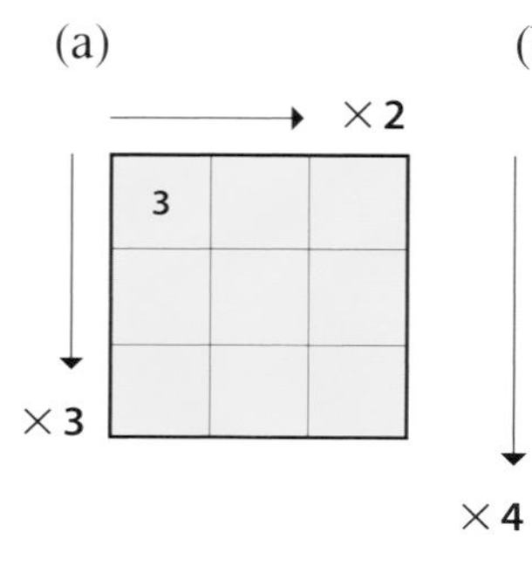

→ ×2, ↓ ×3

3		

(b)

→ ×3, ↓ ×4

2			

Assignment 5 Multiplying fingers

If you want to multiply numbers up to 10×10 together, you need to know your tables.

This may help. Number your fingers like this.

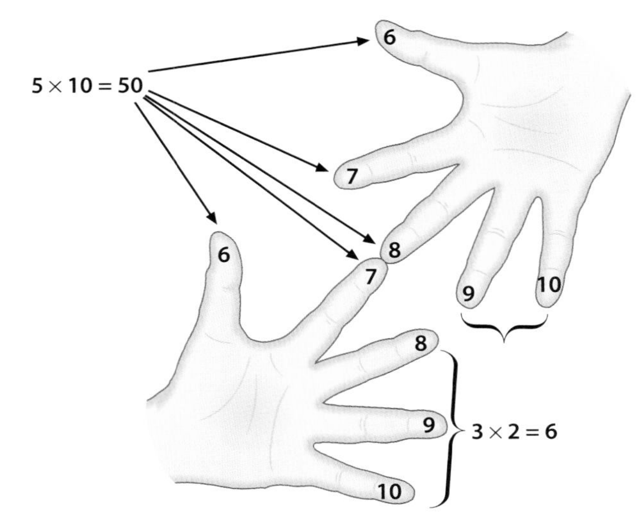

- If you want to multiply 7×8 put the 7 and the 8 together, as shown. Fingers touching and those above them are worth 10.

- Two fingers are touching, and there are three above, making 5.

  $\mathbf{5 \times 10 = 50}$

- Fingers below are worth 1 each. Count how many '1-fingers' there are on each hand.

  Multiply these two numbers together.

  3 (fingers) × 2 (fingers) = 3 × 2 = 6

- Now add the results.
  This gives $7 \times 8 = 50 + 6 = 56$.

- Try some of these.

7×9	7×10	9×9	7×7
6×8	9×6	8×10	8×9
8×8	10×6	7×6	6×6

  Check your answers using a calculator.

Multiplying by 10

Here are three examples of multiplying by 10.

6 × 10 = 60

8 × 10 = 80

23 × 10 = 230

Exercise 9

1 Copy the following and use your 10 times table to help you complete them.

(a) $5 \times 10 =$ (b) $3 \times 10 =$ (c) $7 \times 10 =$ (d) $8 \times 10 =$

(e) $10 \times 6 =$ (f) $10 \times 2 =$ (g) $15 \times 10 =$ (h) $27 \times 10 =$

(i) $10 \times 20 =$ (j) $4 \times 10 =$ (k) $14 \times 10 =$ (l) $10 \times 38 =$

Multiplying by 10 moves the digits one place to the left.

2 Copy the following and complete them.

(a) $10 \times 18 =$ (b) $10 \times 26 =$ (c) $10 \times 42 =$ (d) $10 \times 62 =$

(e) $16 \times 10 =$ (f) $29 \times 10 =$ (g) $35 \times 10 =$ (h) $87 \times 10 =$

Use a calculator to check your answers.

3 (a) What is the cost of a book of ten stamps at 28p each?

(b) Crayons cost 35 pence each. How much will ten crayons cost?

(c) Slabs are 45 cm long. How long will a row of ten slabs be?

(d) Pens come in boxes of ten. How many pens will there be in 36 boxes?

4 (a) A bakery uses 34 eggs to make a batch of cakes. How many eggs will they need for ten batches?

(b) The cakes made by the bakery cost £1.20 each. How much would ten cakes cost?

(c) The eggs come in trays of 48. How many eggs are there in ten trays?

5 In your own words, describe the effect on a number of multiplying it by 10.

Going backwards

The opposite or **inverse** of multiplying by 10 is dividing by 10.

Use your 10 times table.

Exercise 10

1 How many tens are there in the following numbers?
Copy these and complete them (the first one is done for you).

(a) $40 \div 10 = 4$ (b) $60 \div 10 =$ (c) $80 \div 10 =$

(d) $30 \div 10 =$ (e) $50 \div 10 =$ (f) $20 \div 10 =$

(g) $130 \div 10 =$ (h) $330 \div 10 =$ (i) $410 \div 10 =$

Explain what you notice about dividing by 10.

2 Copy this multiplication and division table and complete it.

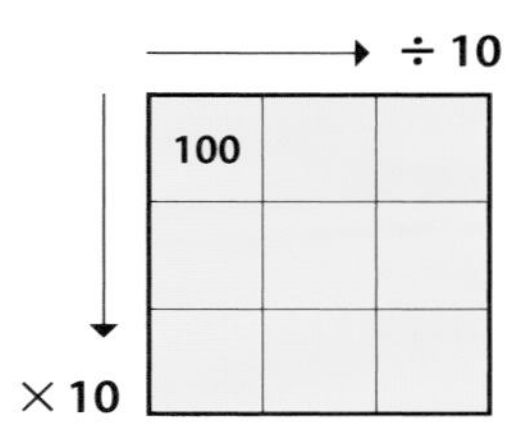

What is the effect of first multiplying by 10 and then dividing by 10?

Key fact
Input goes into the number chain.
Output comes out.

3 Copy the number chains and complete them for each of these inputs.

50	90	60	40
100	120	250	410

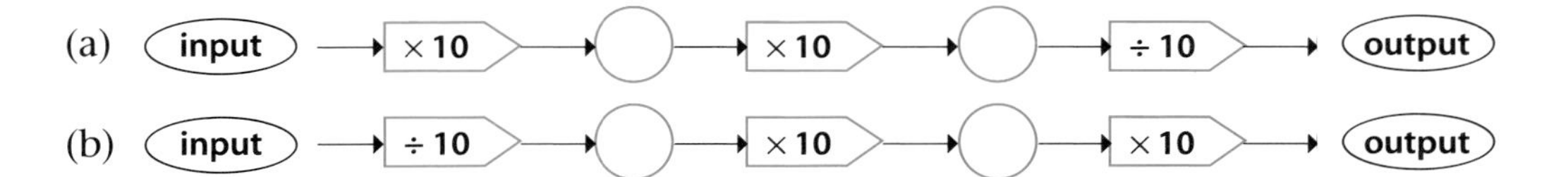

4 There are 260 pupils who are placed in ten equal classes.
How many would there be in each class?

5 If 480 fans have to be divided equally among ten coaches, how many fans will be on each coach?

6 In your own words, describe the effect on a number of dividing it by 10.

Beyond the ten times table

How do you multiply by numbers larger than 10?

Look at this example.

Example

If eight people pay £12 each to go on a trip, how much do they pay altogether?

$$\begin{array}{r} 12 \\ \times \quad 8 \\ \hline \\ \hline \end{array}$$

This can be shown as a rectangle that is 12 units long and 8 units high.

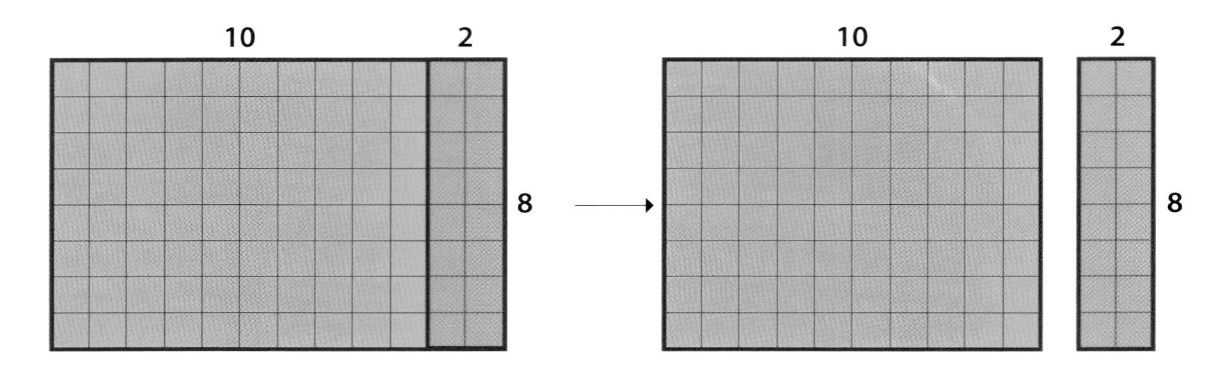

From the diagram you can see you have to find 8×10 and 8×2.

The calculation can be written like this.

$$\begin{array}{r} 12 \\ \times \quad 8 \\ \hline 16 \\ 80 \\ \hline 96 \\ \hline \end{array}$$

16 ← from 8×2

80 ← from 8×10

The number in the answer is the same as the number of small squares in the rectangle.

Exercise 11

1 Write down and complete the multiplication question which goes with each of the following diagrams.

(a)

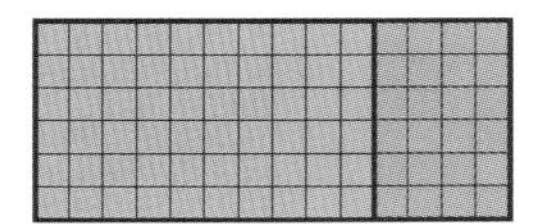

(b)

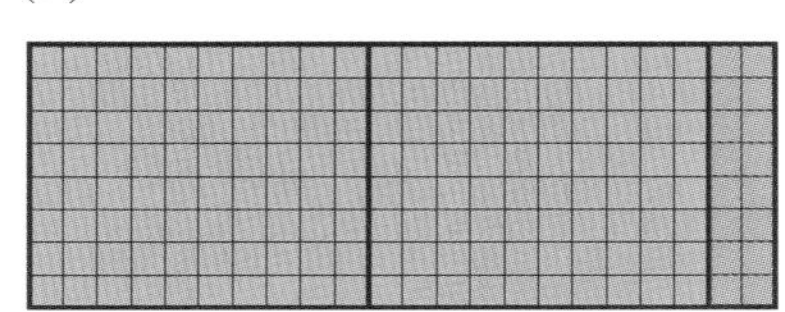

(c)

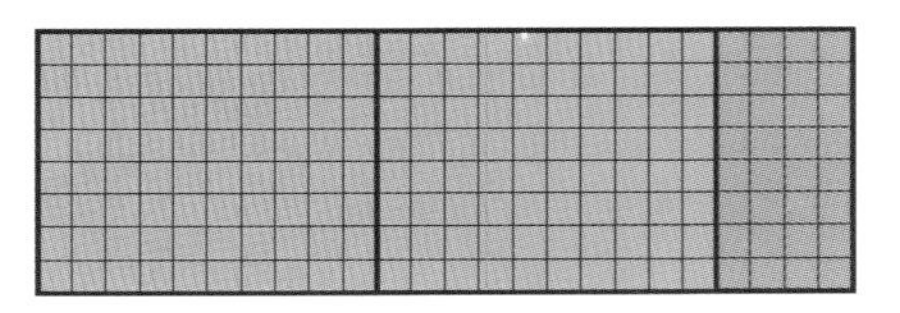

(d)

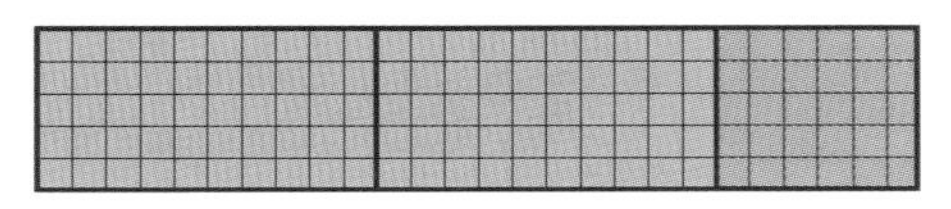

(e)

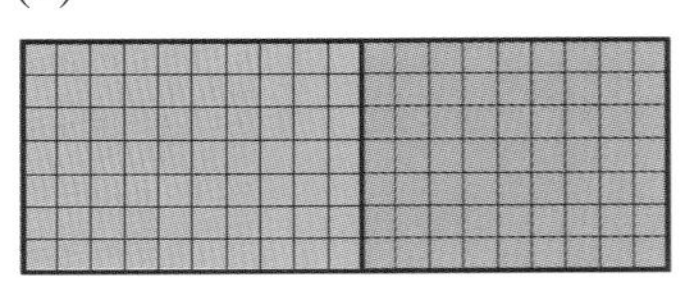

(f)

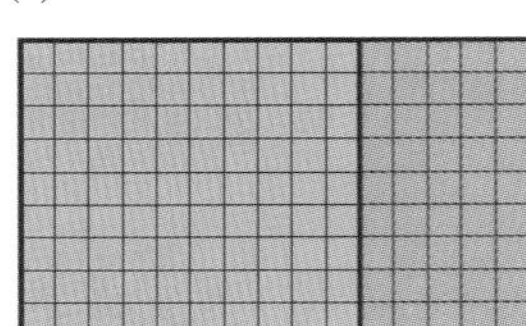

(g)

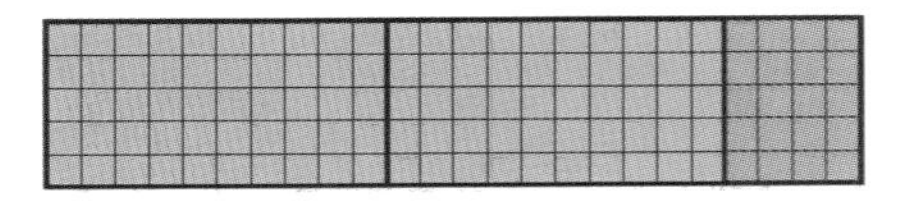

(h)

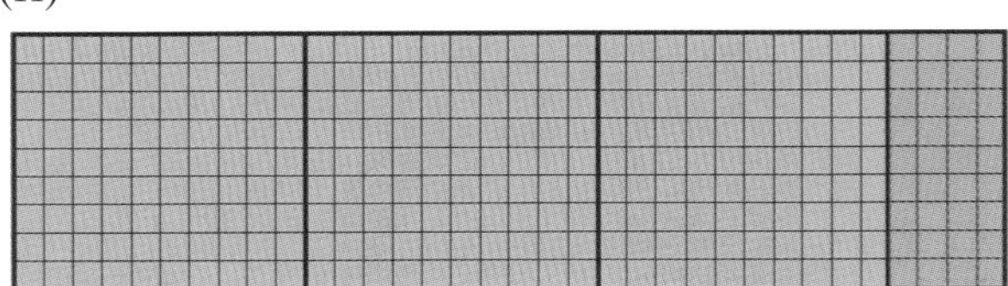

(i)

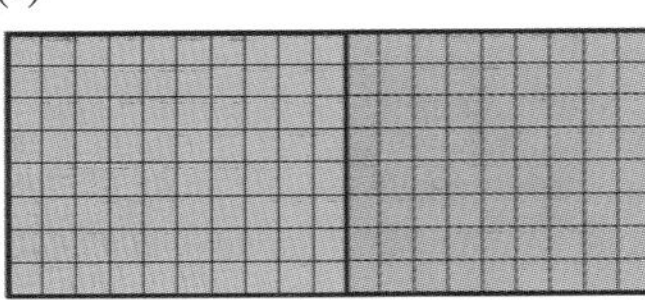

(j)

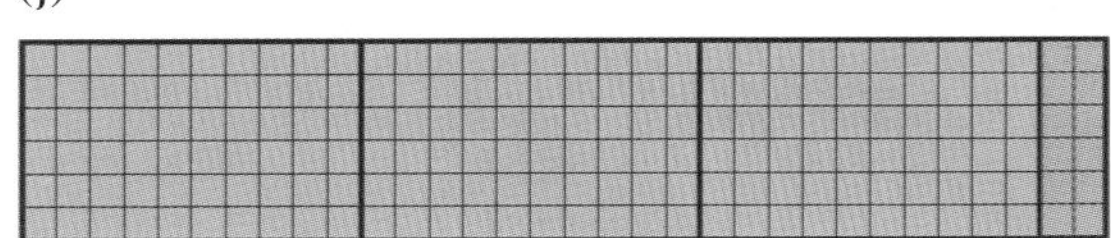

2 Copy the following and complete them, without using a calculator.

(a) $\begin{array}{r} 23 \\ \times\ 3 \\ \hline \end{array}$ (b) $\begin{array}{r} 32 \\ \times\ 4 \\ \hline \end{array}$ (c) $\begin{array}{r} 42 \\ \times\ 2 \\ \hline \end{array}$ (d) $\begin{array}{r} 83 \\ \times\ 2 \\ \hline \end{array}$

(e) $\begin{array}{r} 26 \\ \times\ 3 \\ \hline \end{array}$ (f) $\begin{array}{r} 34 \\ \times\ 4 \\ \hline \end{array}$ (g) $\begin{array}{r} 12 \\ \times\ 6 \\ \hline \end{array}$ (h) $\begin{array}{r} 13 \\ \times\ 7 \\ \hline \end{array}$

(i) $\begin{array}{r} 73 \\ \times\ 3 \\ \hline \end{array}$ (j) $\begin{array}{r} 62 \\ \times\ 4 \\ \hline \end{array}$ (k) $\begin{array}{r} 82 \\ \times\ 2 \\ \hline \end{array}$ (l) $\begin{array}{r} 93 \\ \times\ 3 \\ \hline \end{array}$

(m) $\begin{array}{r} 63 \\ \times\ 4 \\ \hline \end{array}$ (n) $\begin{array}{r} 75 \\ \times\ 8 \\ \hline \end{array}$ (o) $\begin{array}{r} 87 \\ \times\ 9 \\ \hline \end{array}$ (p) $\begin{array}{r} 96 \\ \times\ 7 \\ \hline \end{array}$

3 The following questions should be set out in the same way as those in question 2.

(a) $73 \times 2 =$ (b) $52 \times 3 =$ (c) $14 \times 6 =$ (d) $5 \times 31 =$

(e) $4 \times 42 =$ (f) $56 \times 7 =$ (g) $89 \times 7 =$ (h) $75 \times 8 =$

Use a calculator to check your answers before you continue.

4 (a) Seven people each pay £68 for a weekend in France. What is the total amount they pay?

(b) The minibus travels 27 miles on 1 gallon of petrol. How far will it travel if it has 8 gallons in the tank?

(c) What is the cost of 9 cards at 75p each?

5 Sam has £4.50 to spend. She wants to buy 8 books which cost 55p each. Can she afford to pay for them?

Assignment 6 Simply amazing

Try to get through the maze, from your chosen starting point to the hexagon of the same colour.

You can start on the red, blue or yellow hexagon.

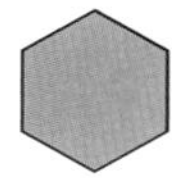

If you start on the red hexagon you must find calculations which include a 5 in the answer.

Example

$$\begin{array}{r} 6\ 3 \\ \times \quad 4 \\ \hline 2\ 5\ 2 \\ \hline \end{array}$$

There is a 5 in the tens column. It is on the red path.

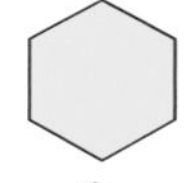

If you start on the yellow hexagon you must find calculations which include an 8 in the answer.

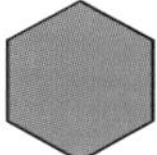

If you start on the blue hexagon you must find calculations which include a 3 in the answer.

Which is the shortest route through the maze?

Use tracing paper over the diagram to record your path.

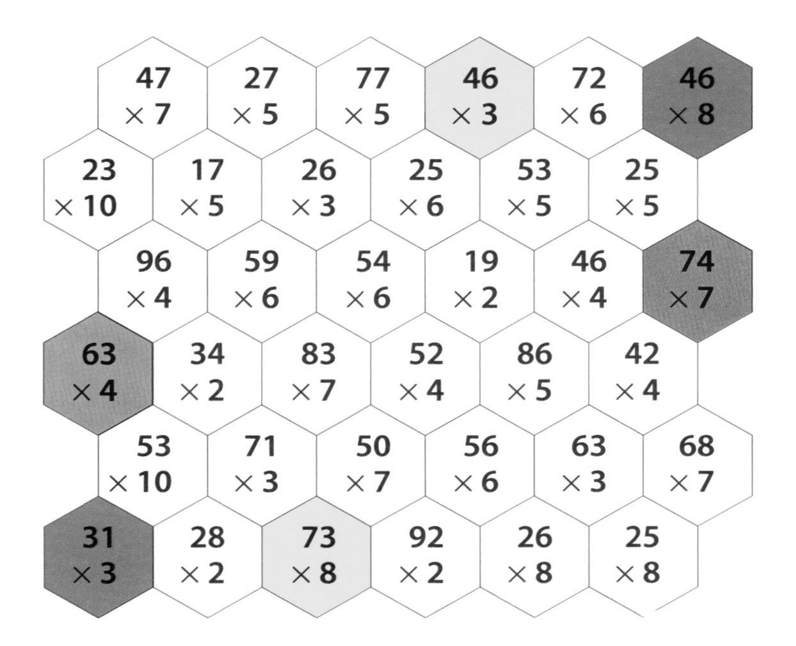

Multiplying by 100

You already have a quick method for multiplying by ten.

You can use that to help you to multiply by 100 because 100 is equal to 10×10.

Example

You have to work out: 5×100

You can think of that as: $5 \times 10 \times 10$

$5 \times 10 = 50$ and $50 \times 10 = 500$

Here are some more.

$8 \times 100 = 800$

$14 \times 100 = 1400$

Exercise 12

1 Complete the following without using a calculator.

(a) $100 \times 8 =$ (b) $100 \times 3 =$ (c) $100 \times 5 =$ (d) $100 \times 4 =$

(e) $6 \times 100 =$ (f) $2 \times 100 =$ (g) $100 \times 15 =$ (h) $27 \times 100 =$

(i) $100 \times 12 =$ (j) $4 \times 100 =$ (k) $14 \times 100 =$ (l) $100 \times 38 =$

Check your answers before you continue.

2 (a) £24 in pennies is collected for charity.
How many pennies is this?

(b) How many centimetres are there in 15 metres?

(c) A dollar ($1) is worth 100 cents.
You have $42; how many cents is this?

(d) Drawing pins come in boxes of 100.
How many pins will there be in nine boxes?
How many pins will there be in a pack of 12 boxes?
How many pins will there be in a jumbo pack of 36 boxes?

3 In your own words, describe the effect on a number of multiplying it by 100.

Dividing by 100

Dividing by 100 is the opposite or inverse of multiplying by 100.

Exercise 13

1 Copy the following and complete them (the first one is done for you).

(a) $500 \div 100 = 5$ (b) $800 \div 100 =$ (c) $600 \div 100 =$
(d) $200 \div 100 =$ (e) $1000 \div 100 =$ (f) $1500 \div 100 =$

2 If you have 800 pence, how many pounds (£) will you have?

3 A work surface is 300 cm long. How many metres is that?

4 In your own words, describe the effect of dividing by 100.

Dividing

Dividing without a remainder

$68 \div 2$ is usually written:

2)6 8

Divide the digit 6 by 2:

  3 ← Put the answer here.
2)6 8

Now divide the 8 by 2:

  3 4 ← The answer goes above the 8.
2)6 8

That means:

2)6 0 + 8

  3 0
2)6 0 + 8

  3 0 + 4
2)6 0 + 8

Numbers where you have to carry are more difficult, for example:

$34 \div 2$ is usually written as: 2)3 4

Divide the digit 3 by 2:

  1
2)3 4

The 1 lot of 10 left over is added to the 4 to give you 14.

Divide 14 by 2:

  1 7
2)3 ¹4

The answer goes above the 4.

Exercise 14

1 (a) $3\overline{)6\,3}$ (b) $2\overline{)8\,2}$ (c) $4\overline{)4\,0}$ (d) $2\overline{)6\,0}$

Write (e) to (p) in the same way as (a) to (d) and work out the answers.

(e) $42 \div 2 =$ (f) $36 \div 3 =$ (g) $48 \div 4 =$ (h) $25 \div 5 =$

(i) $54 \div 6 =$ (j) $63 \div 7 =$ (k) $72 \div 3 =$ (l) $65 \div 5 =$

(m) $84 \div 7 =$ (n) $90 \div 3 =$ (o) $57 \div 3 =$ (p) $85 \div 5 =$

2 (a) $3\overline{)4\,2}$ (b) $2\overline{)3\,6}$ (c) $4\overline{)5\,6}$ (d) $5\overline{)3\,5}$

(e) $6\overline{)3\,6}$ (f) $7\overline{)8\,4}$ (g) $5\overline{)6\,5}$ (h) $4\overline{)6\,4}$

(i) $8\overline{)9\,6}$ (j) $6\overline{)7\,2}$ (k) $3\overline{)8\,1}$ (l) $7\overline{)7\,7}$

(m) $4\overline{)8\,4}$ (n) $7\overline{)9\,1}$ (o) $6\overline{)8\,4}$ (p) $5\overline{)5\,5}$

3 Write these the way the division calculations were written in question 2 and then work out the answers.

(a) $45 \div 3 =$ (b) $28 \div 2 =$ (c) $68 \div 4 =$ (d) $95 \div 5 =$

(e) $52 \div 2 =$ (f) $48 \div 3 =$ (g) $28 \div 4 =$ (h) $75 \div 5 =$

(i) $44 \div 4 =$ (j) $96 \div 6 =$ (k) $80 \div 5 =$ (l) $93 \div 3 =$

4 (a) $3\overline{)7\,8}$ (b) $4\overline{)9\,2}$ (c) $3\overline{)5\,1}$ (d) $9\overline{)4\,5}$

(e) $3\overline{)7\,5}$ (f) $7\overline{)9\,8}$ (g) $4\overline{)8\,8}$ (h) $7\overline{)4\,9}$

(i) $8\overline{)6\,4}$ (j) $9\overline{)9\,9}$ (k) $6\overline{)9\,0}$ (l) $3\overline{)8\,7}$

(m) $7\overline{)7\,0}$ (n) $3\overline{)6\,9}$ (o) $4\overline{)7\,2}$ (p) $2\overline{)9\,8}$

Assignment 7 Division bugs

Each of the following numbers can be divided by other whole numbers.

- Copy and complete these division bugs. The first one is done for you.

(a)

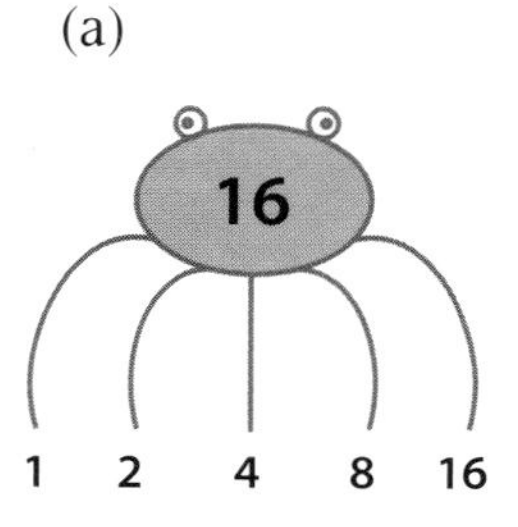

(b)

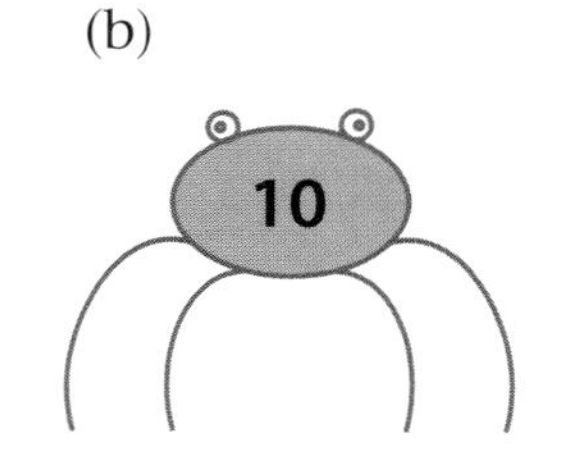

(c)

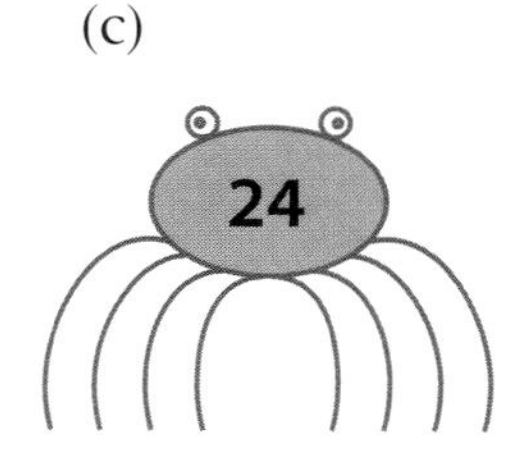

(d)

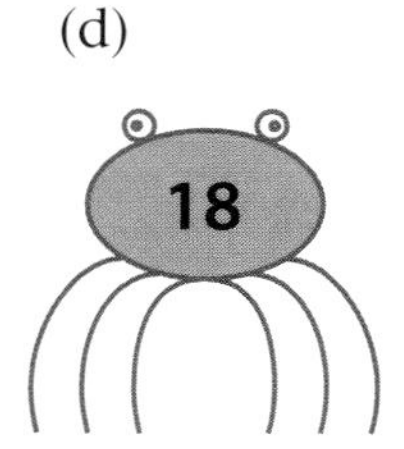

The division bug for 24 has eight legs. Investigate the number of legs for numbers less than 100.

- Which division bugs have eight legs?
- What is special about division bugs with odd numbers of legs?
- Which division bugs have the most legs?

Dividing with a remainder

In all the questions you have done so far the numbers have divided exactly. This will not always happen.
In some cases you may have a number left over or remaining.
You write it like this.

$$\begin{array}{r} 1\ 4\ \text{r}2 \\ 3\overline{)4^{1}4} \end{array}$$

3 into 44 goes 14 remainder 2

Exercise 15

1 Copy these and complete them in your book. (Some of the answers have remainders.)

(a) $3\overline{)5\ 5}$ (b) $2\overline{)3\ 5}$ (c) $4\overline{)3\ 6}$ (d) $5\overline{)3\ 6}$

(e) $6\overline{)4\ 0}$ (f) $7\overline{)8\ 0}$ (g) $5\overline{)6\ 3}$ (h) $4\overline{)8\ 4}$

(i) $27 \div 2 =$ (j) $49 \div 3 =$ (k) $68 \div 4 =$ (l) $77 \div 5 =$

2 (a) At break-time 87 pupils are divided equally into three rooms. How many pupils will go into each room?

(b) The bus fare for five friends costs a total of 95p. How much does it cost for each one?

3 Two friends are going camping. They have to carry six items. The weights of the items are shown.

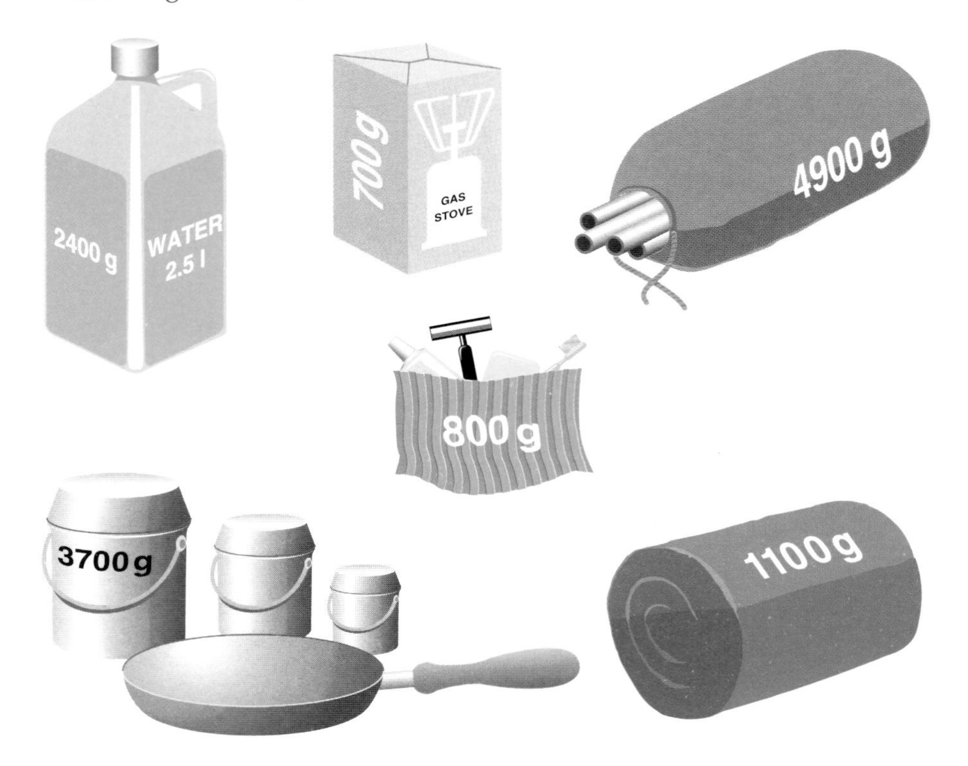

Divide these between the two friends, so they both have the same weight to carry.

Assignment 8 Remainders

The digits 3, 6, 8, 9 can be placed in any of the positions shown in this diagram. Each digit can be used only once.

- Investigate which division calculations you can do which give you no remainders.
- Choose four numbers of your own and repeat the process.

Review Exercise

Copy these calculations and fill in the missing numbers.

1 (a) $\begin{array}{r} 2\,\square \\ +\ \square\,4 \\ \hline 6\,6 \end{array}$ (b) $\begin{array}{r} 3\,\square \\ +\ \square\,5 \\ \hline 4\,8 \end{array}$ (c) $\begin{array}{r} 1\,\square \\ +\ \square\,6 \\ \hline 7\,8 \end{array}$ (d) $\begin{array}{r} \square\,3 \\ +\ 6\,\square \\ \hline 9\,3 \end{array}$

2 (a) $\begin{array}{r} \square\,7 \\ +\ 4\,\square \\ \hline 5\,9 \end{array}$ (b) $\begin{array}{r} \square\,2 \\ +\ 8\,\square \\ \hline 9\,7 \end{array}$ (c) $\begin{array}{r} 7\,\square \\ +\ \square\,9 \\ \hline 1\,2\,6 \end{array}$ (d) $\begin{array}{r} \square\,9 \\ +\ 6\,\square \\ \hline 1\,0\,6 \end{array}$

3 (a) $\begin{array}{r} 3\,\square \\ +\ \square\,9 \\ \hline 1\,2\,3 \end{array}$ (b) $\begin{array}{r} \square\,3 \\ +\ 7\,\square \\ \hline 1\,2\,1 \end{array}$ (c) $\begin{array}{r} 7\,\square \\ +\ \square\,9 \\ \hline 9\,3 \end{array}$ (d) $\begin{array}{r} 7\,\square \\ +\ \square\,\square \\ \hline 1\,1\,3 \end{array}$

4 At their new school 96 pupils are divided equally into four classes.
How many will go into each class?

5 A pupil reads 246 pages in one novel, 365 in the next and 172 in the third.
How many pages has the pupil read altogether?

6 In Year 10 there are 163 pupils. In Year 9 there are 147.
How many more pupils are there in Year 10 than Year 9?

7 A car travels 12 miles on 1 litre of petrol.
How far will it travel if it has 6 litres?

8 (a) How many centilitres are there in 12 litres? (1 *l* = 100 c*l*)

(b) A dollar ($1) is worth 100 cents.
You have $89; how many cents is that?

9 Twenty-four people pay £10 each to see their favourite group.
How much do they pay altogether?

10 Use the following clues to complete the number puzzle.
All the answers are different.

Across
1 Divisible by 5
3 Divisible by 7

Down
1 Divisible by 2
2 Divisible by 13

1	2
3	

11 What is the second largest number you can make with the digits 3, 6, 8 and 7?

12 Six people each pay 78p bus fare.
What is the total amount they pay?

13 Which of these six numbers

2 3 4 7 8 9

have been used in this calculation to give the answer 237?

```
    ? ?
  × ?
  -----
  2 3 7
```

Assignment 9 Reverse it

In this magic square each row, column and diagonal adds up to 15.

8	1	6
4	5	7
3	9	2

Thinking of each row, column or diagonal as a three-digit number, you can make 16 different three-digit numbers, for example: 816, 276 and 653.

(a) What is the largest number that can be made by adding two three-digit numbers from the grid?

(b) What is the smallest number that can be made by taking one of the three-digit numbers from another one?

(c) Take any three-digit number from the grid and reverse it.
Take the smaller number from the bigger number,
for example: $816 - 618 = 198$.
Add up the digits of the answer: $1 + 9 + 8$.
Repeat this task several times. Write down what you notice.

2-D shapes

In this chapter you will learn:

- → **different ways to sort shapes**
- → **some special words for shapes and angles**
- → **to make tiling patterns with various shapes**

Starting points

You will have come across shapes like these before.

Squares belong to the rectangle family.
Squares and rectangles have four square corners.

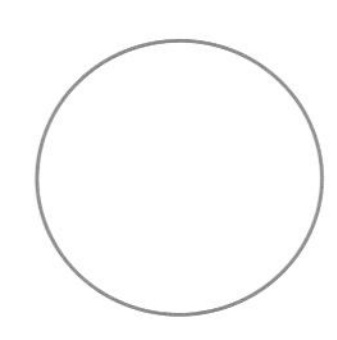

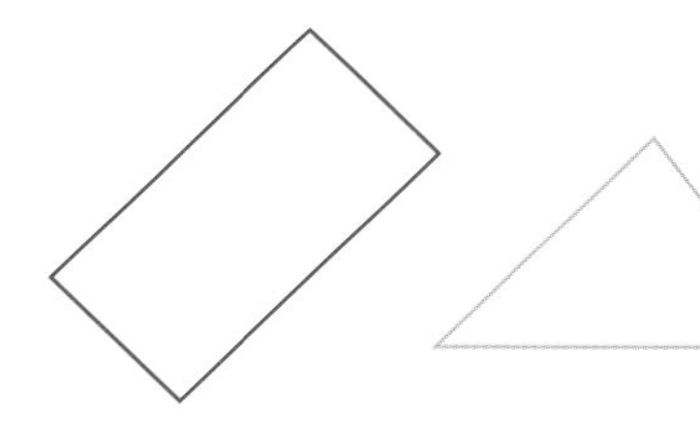

These are all flat shapes. They are called '2-dimensional' or '2-D' shapes.

A 2-D shape which is closed with all sides straight lines is a **polygon**. A shape that is not a polygon is called a non-polygon.

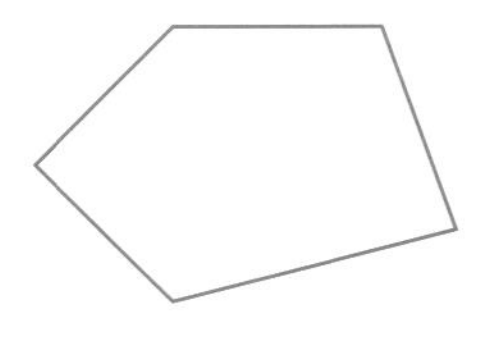

polygon

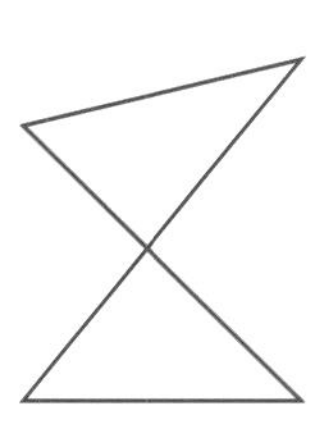

non-polygon

polygon

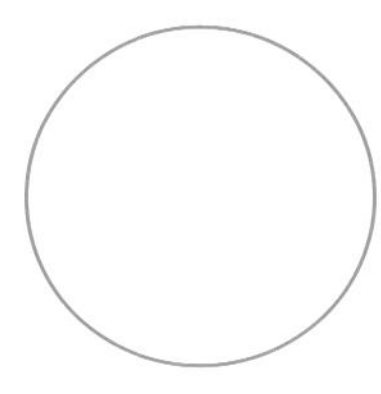

non-polygon

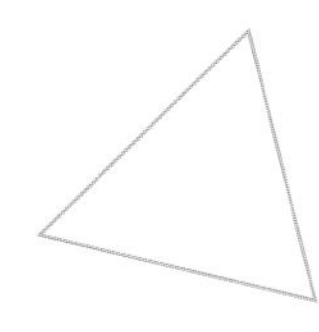

polygon

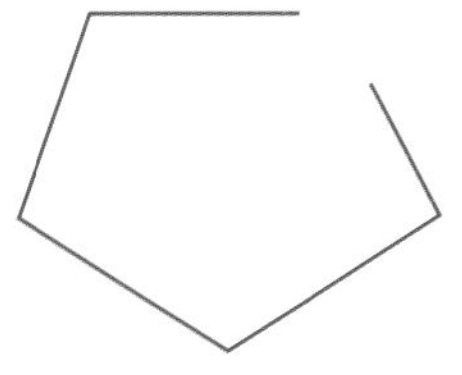

non-polygon

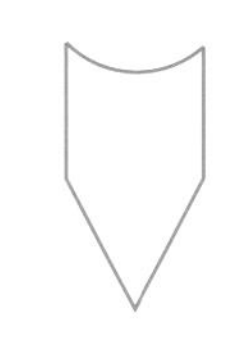

non-polygon

Exercise 1

1 Look carefully at the picture. What shapes can you see?

Make lists of the objects that are:
(a) squares (b) circles (c) rectangles (d) triangles.

2 The simplest polygon is a triangle.

It is possible to make different-shaped triangles on a 3 by 3 pinboard.

Key facts
A polygon is closed.
It has sides that are straight lines.

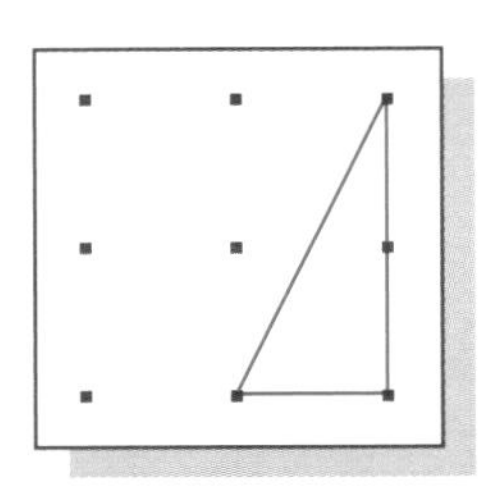

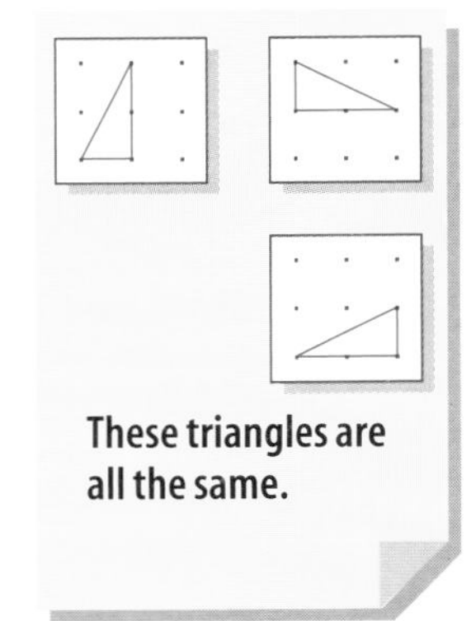

These triangles are all the same.

Using dotted paper, show how many different-shaped triangles you can make.

3 A polygon with four sides is called a **quadrilateral**. Squares and rectangles are quadrilaterals.

Key fact
A quadrilateral is a 4-sided polygon.

How many different-shaped quadrilaterals can you make?

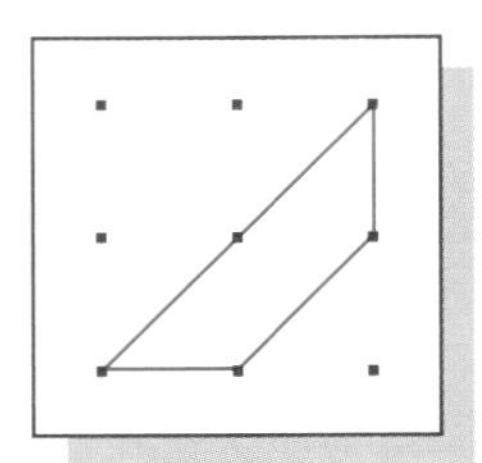

Use dotted paper to show your results.

4 Five-sided polygons are called **pentagons.**
Six-sided polygons are called **hexagons**.

It is possible to make pentagons and hexagons on a 3 by 3 pinboard.

How many different-shaped pentagons and hexagons can you make?

Use dotted paper to show your results.

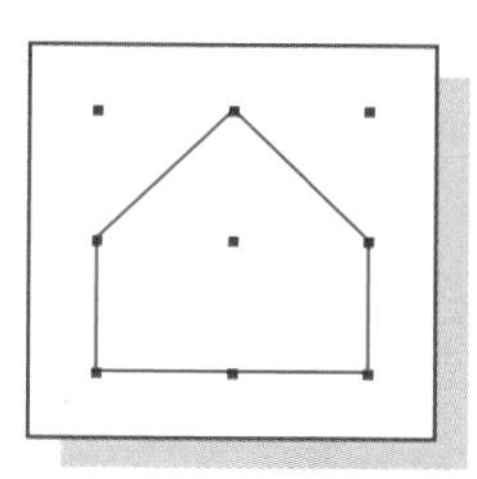

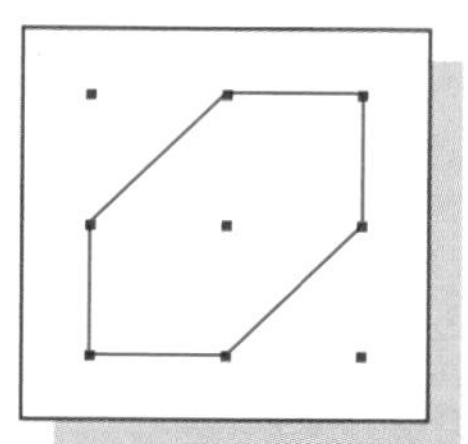

5 Is it possible to make a polygon with more than seven sides on a 3 by 3 pinboard?

Show your results on dotted paper.

Key fact

A regular polygon has equal sides and equal angles.

6 A polygon that has all its sides the same length and all its angles the same size is called a **regular polygon**.
Which of these shapes are regular polygons?

(a)

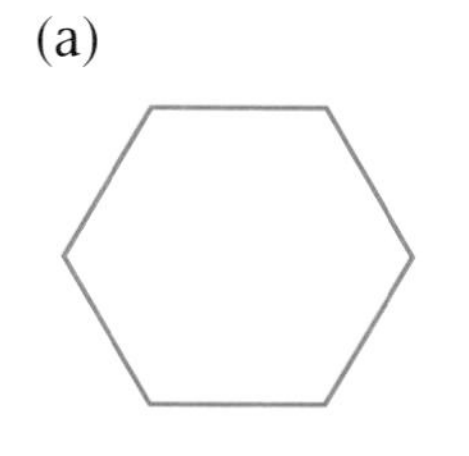

(b)

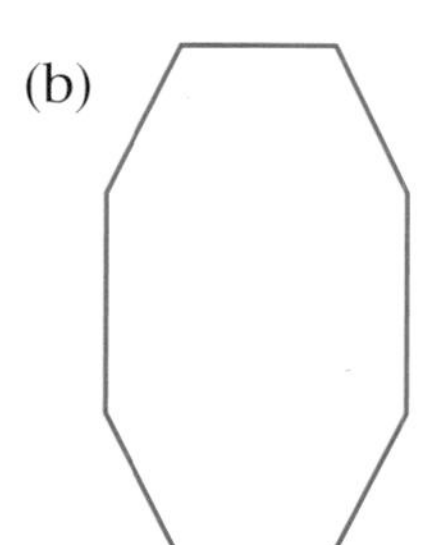

(c)

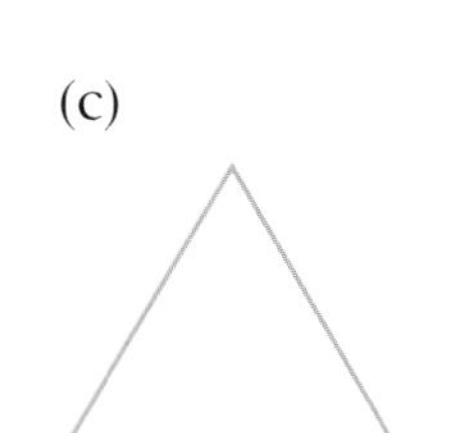

(d)

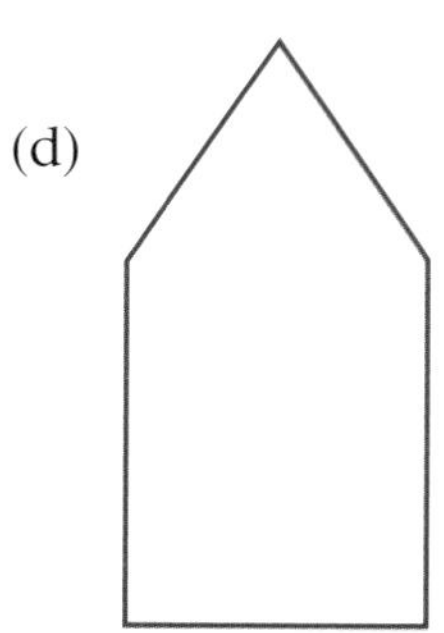

(e)

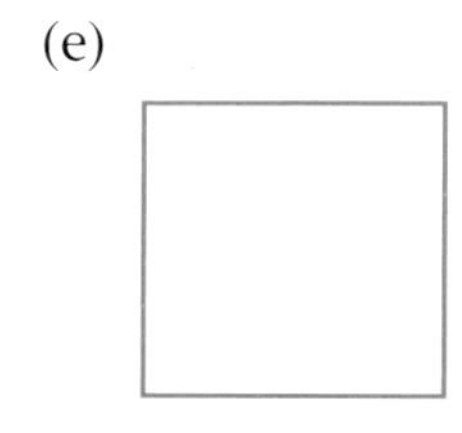

(f)

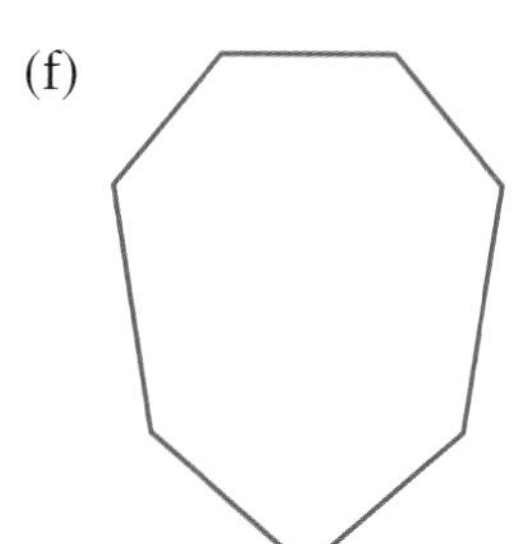

(g)

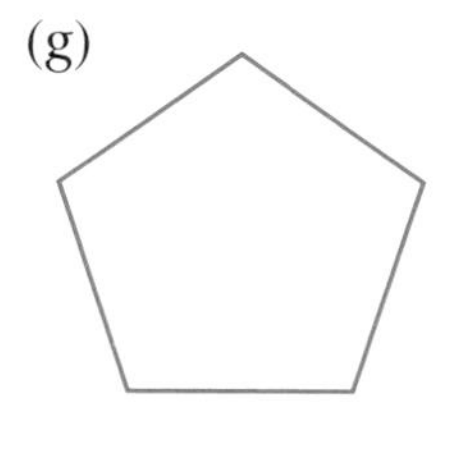

Angles

Right angles

You can make a square corner measure from a piece of paper.

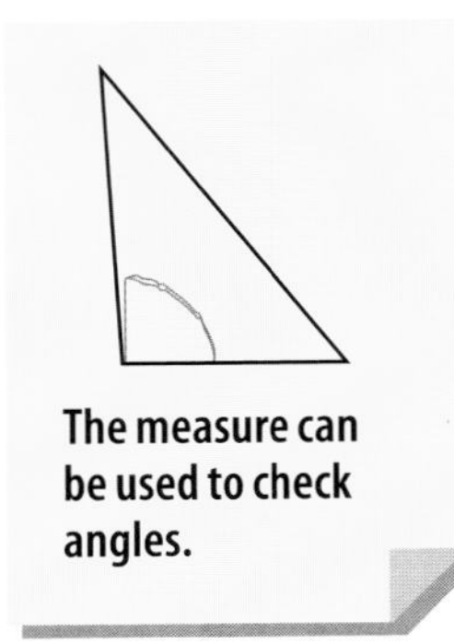

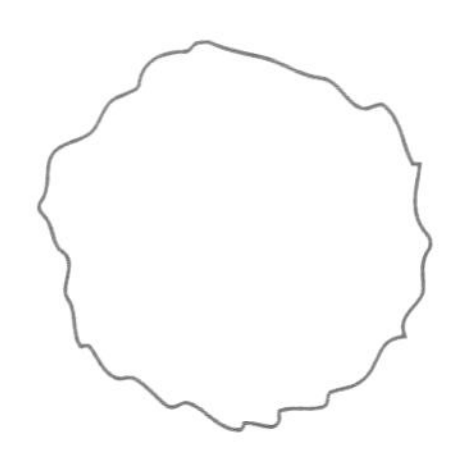

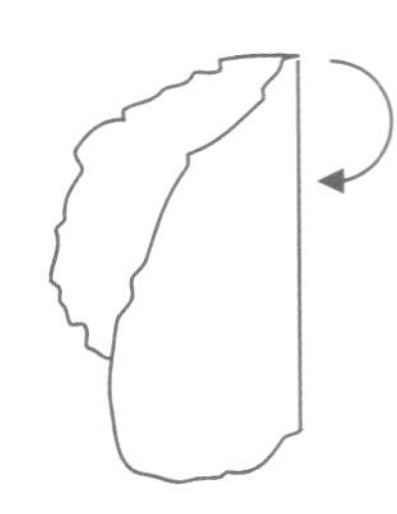

Fold it over to make a straight edge.

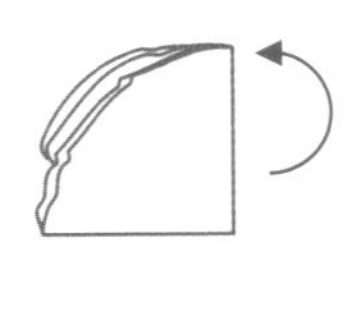

Fold it again across the straight edge to make a square corner.

A square corner is also called a **right angle**.

Exercise 2

1 Use your square corner measure to decide whether each labelled corner is less than a right angle, a right angle, or more than a right angle.

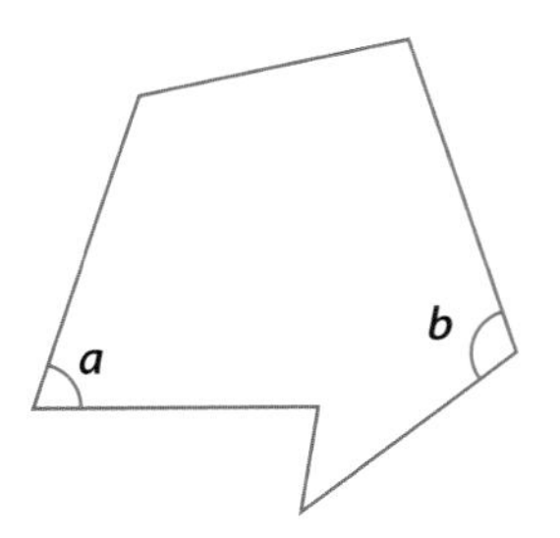

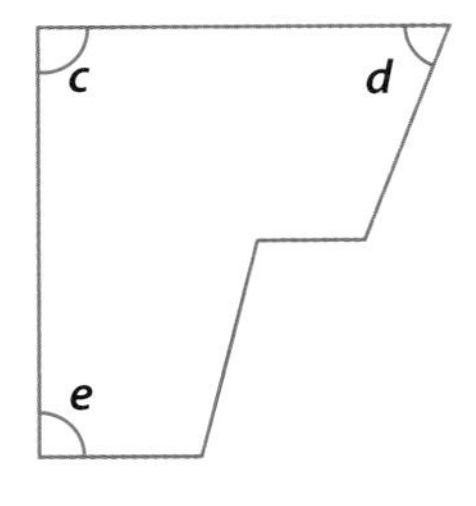

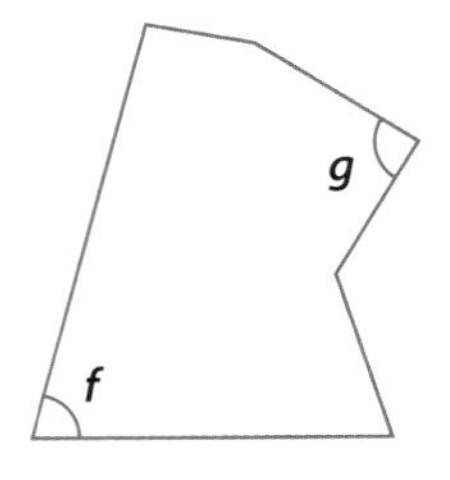

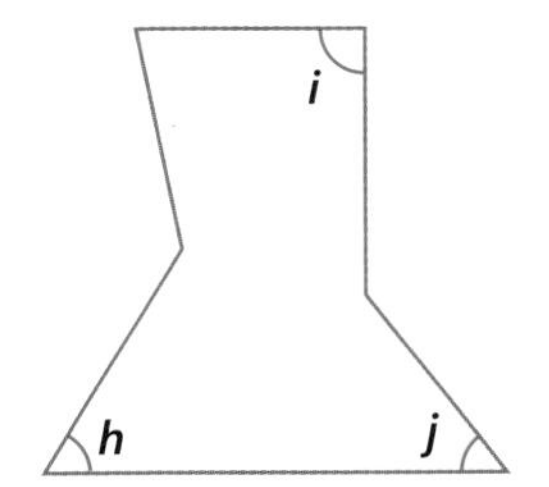

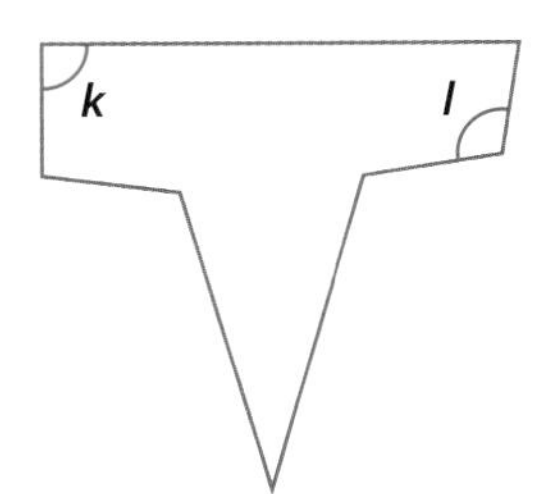

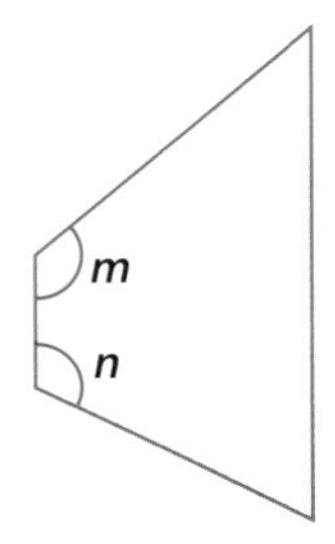

Copy the table below and put a tick in the column you think describes each corner.

Corner	Less than a right angle	A right angle	More than a right angle
a			
b			
c			
d			
e			
f			
g			
h			
i			
j			
k			
l			
m			
n			

Other types of angle

An angle that is less than a right angle is called an **acute angle.**

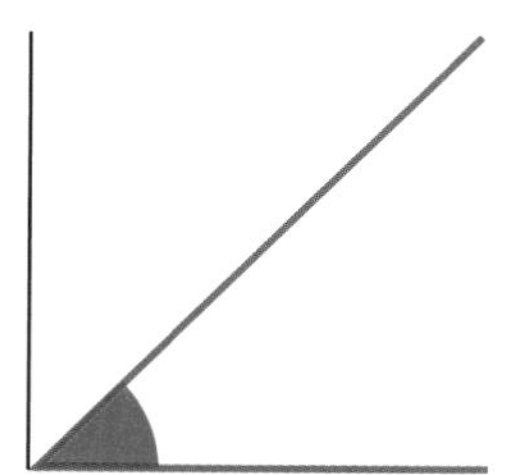

An angle that is more than a right angle is called an **obtuse angle.**

Key facts

An acute angle is smaller than a right angle.

An obtuse angle is bigger than a right angle.

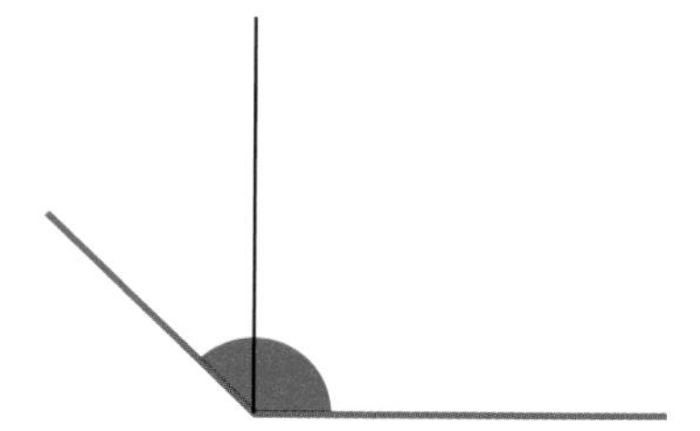

Exercise 3

Look at these polygons.

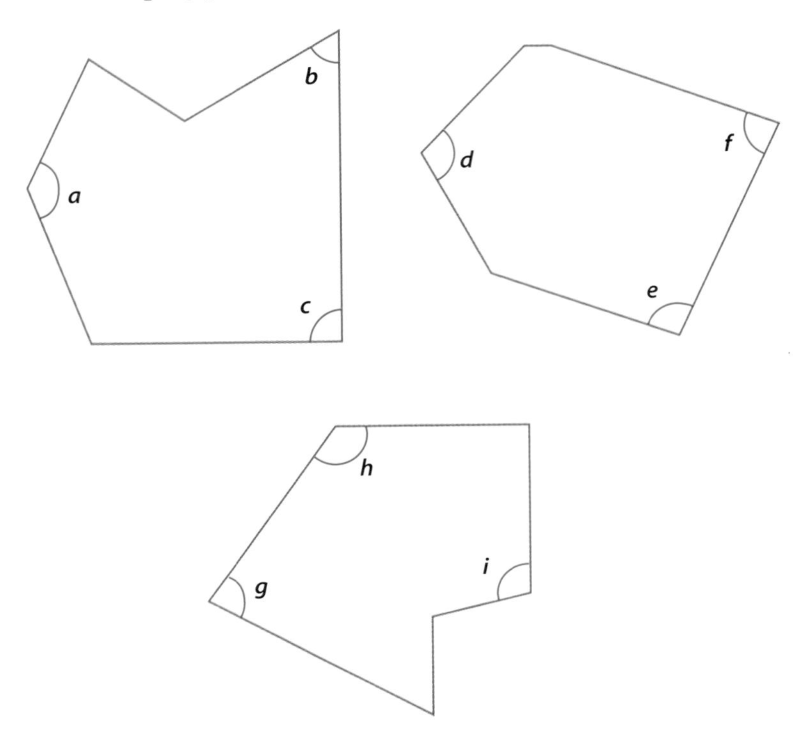

Use your square corner measure to help you.

1 Write down the letters of all the acute angles.

2 Write down the letters of all the obtuse angles.

3 Write down the letters of all the right angles.

Constructing simple polygons

You can draw simple polygons using a ruler, a pencil and your square corner measure to compare the angles with a right angle.

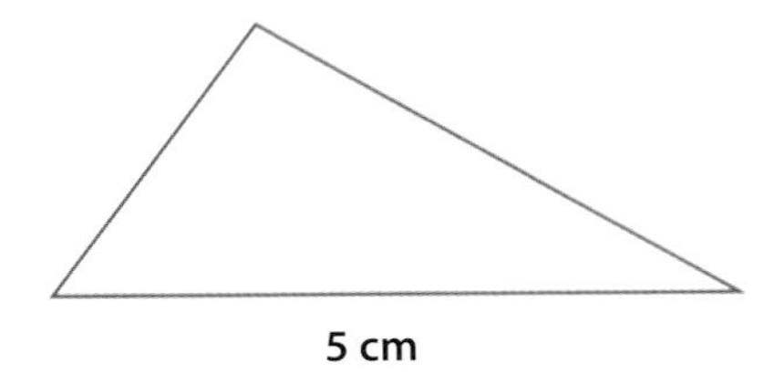

This triangle has a side of 5 cm and two acute angles.

Exercise 4

1 Construct a triangle with one side 7 cm long, one right angle and one angle acute.

2 Construct a triangle with one side 5 cm long, another side 7 cm long and with a right angle between them. Complete the shape by drawing the third side. Use your square corner measure to check the size of each corner, and write down what you find.

3 Construct a square with sides of 5 cm using only your ruler and the square corner measure. Write down clearly how you constructed the square.

4 Construct a rectangle with a long side of 8 cm and a short side of 5 cm.

5 Construct a triangle with sides of 3 cm, 4 cm and 5 cm. Use your square corner measure to check the size of each corner. Write down what you notice.

Circles

Assignment 1 Round a dot

Follow these steps.

- Mark a dot on a piece of paper.
- Place your ruler on the paper so that one side touches the dot.
- Draw a line along the other side of the ruler.
- Keeping the same side against the point, move your ruler round a little and draw a second line.
- Keep repeating this action until your ruler is back to where it started.

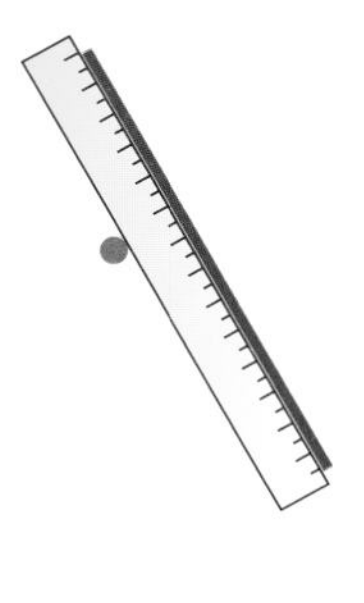

Write down what you think the shape you have constructed looks like.

Parts of a circle

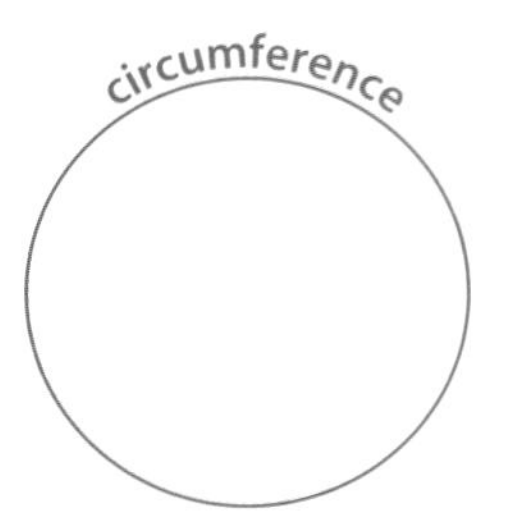

The **circumference** is the distance around the circle. It is the perimeter of the circle.

Key fact
diameter = 2 × radius

The **diameter** is the distance across the widest part of the circle. It passes through the centre of the circle.

The **radius** is the distance from the centre to the edge of the circle. It is half the length of the diameter.

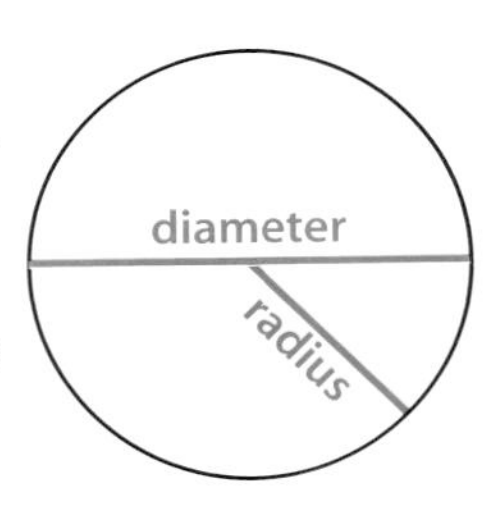

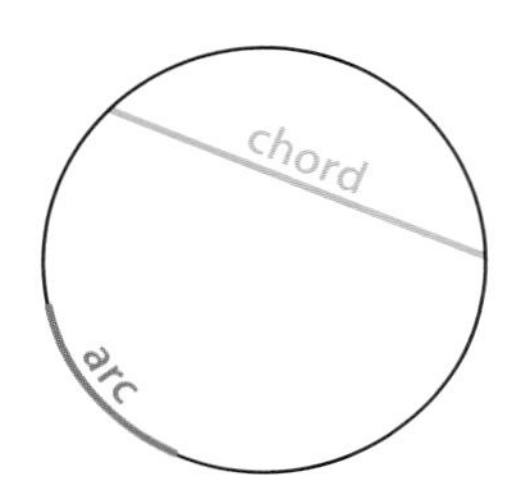

A **chord** is a line which cuts the circle into two parts. The diameter is a special chord.

An **arc** is a part of the circumference.

Exercise 5

1 Use your compasses to draw a circle which has a radius of 3 cm.

2 Use your compasses to draw a circle which has a diameter of 8 cm.

3 Use your compasses to copy these patterns. Is there a relation between the radii of the circles?

(a)

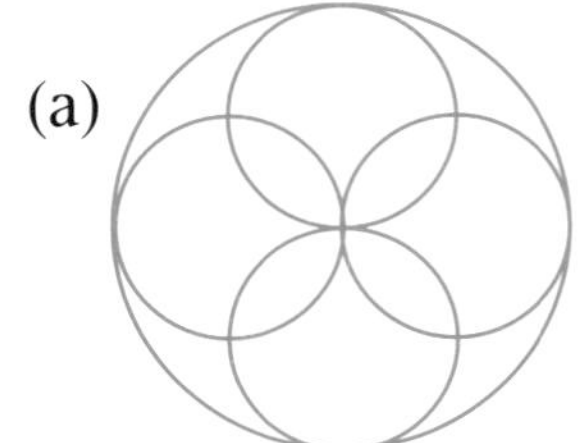

(b)

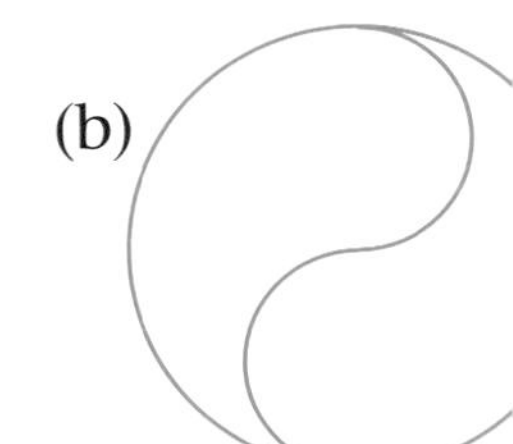

4 Now try these.
What do you notice about the centres of the circles?

(a)

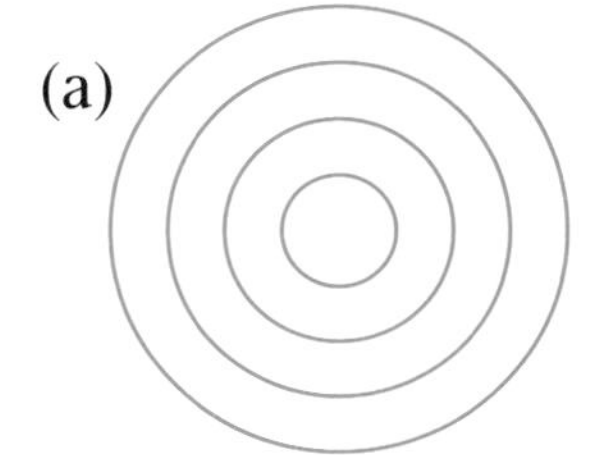

(b)

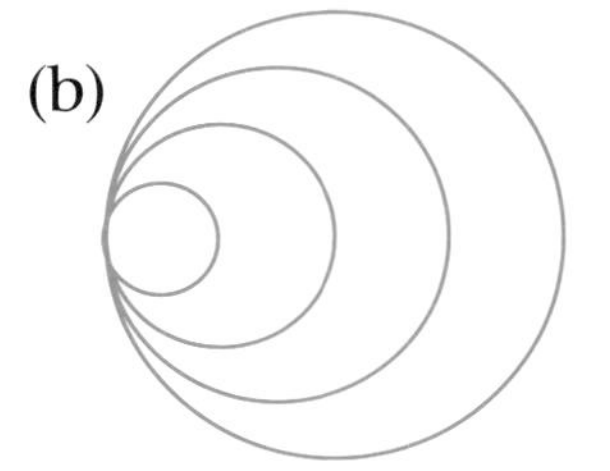

Accurate constructions

Exercise 6

You will need a ruler and a pair of compasses. Make sure your pencil is sharp. This work should be done on plain paper. Use a new page or sheet of paper for each question.

1 Draw a triangle

An arc is part of the circumference.

(a) Draw a line 4 cm long in the middle of the page.
Label one end of the line A and the other end B.

(b) Open the compasses to 4 cm, the same length as the line.

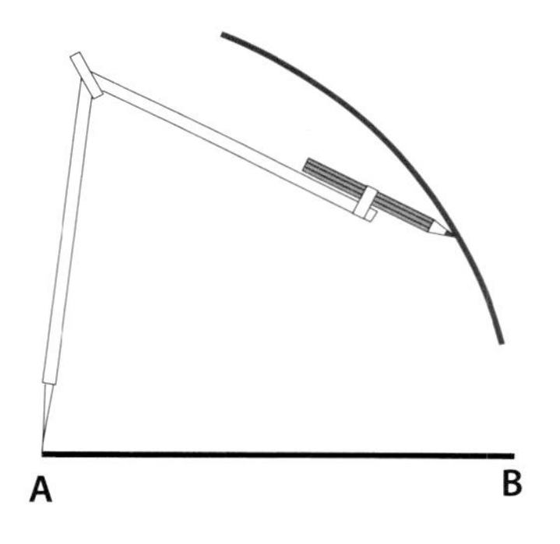

(c) Put your compass point carefully on point A and draw an arc above the line.

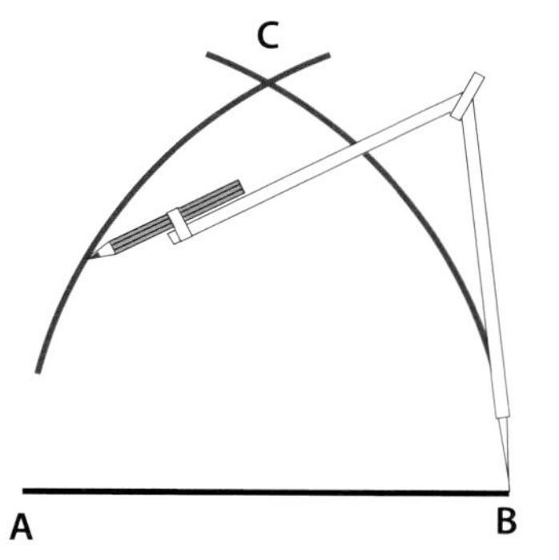

(d) **Do not adjust your compasses.**
Put your compass point on point B.
Draw another arc which crosses the first arc. Label the point where the arcs cross C.

(e) Use your ruler to draw lines joining A to C and B to C.

Write down what each of the angles in the triangle you have drawn is called. Use your square corner measure to help you.

2 Draw more triangles

Choose two different lengths for your lines. Construct two triangles in the same way as before using the lengths you have chosen.
Cut out the two triangles.
Compare the angles of these two triangles and your triangle from question 1. Write down as much as you can about them.

A semi-circle is half a circle.

3 Draw an angle

(a) Draw a straight line about 10 cm long. Mark a point X somewhere near the middle of the line.

(b) Open your compasses to about 4 cm. Place your compass point on point X. Check that, without moving the compasses, the pencil will touch the line on both sides of the point. If not, set the compasses a little smaller and check again.

Now that you have set the size for your compasses **do not adjust them again.**

(c) Draw a semi-circle which touches the line at both ends. Label these points A and B.

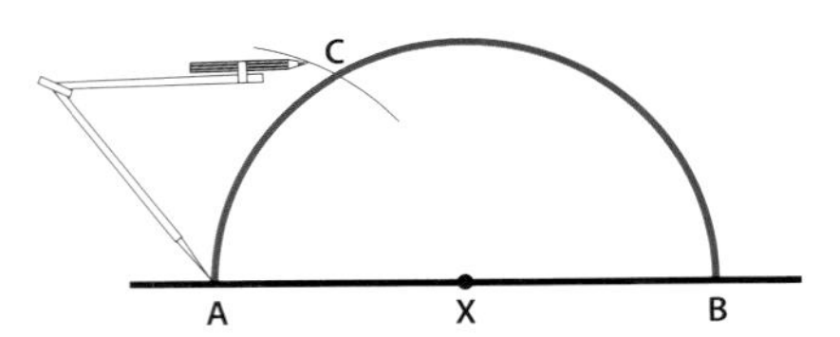

(d) Place your compass point on point A. Draw an arc which crosses the semi-circle. Label this point C.

(e) Repeat this at point B. Label the new point D.

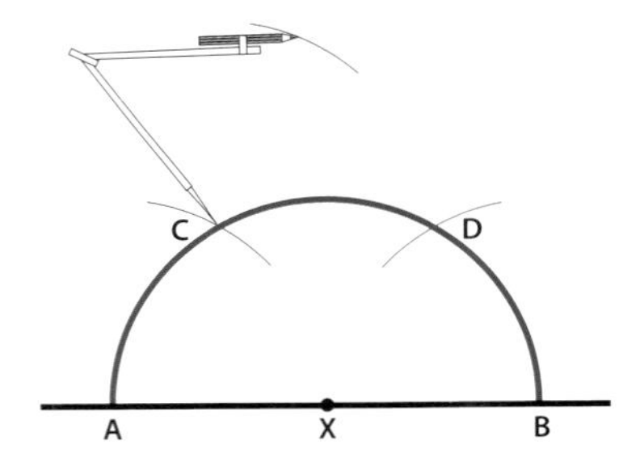

(f) Place your compass point on C. Draw another arc.

(g) Repeat this at point D so the two arcs cross each other. Label the point where the arcs cross Y.

(h) Join the points X and Y.

Use your square corner measure to check the angle. What type of angle have you constructed? What do you think it is called?

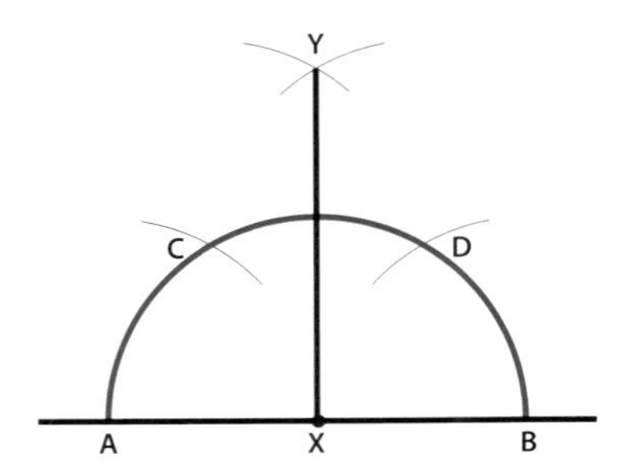

4 Use a ruler and compasses to draw a square with sides 5 cm long.

5 Use a ruler and compasses to draw a rectangle. Make the long sides 6 cm and the short sides 3 cm.

6 Inside a circle

(a) Using compasses, draw a circle with a radius of 4 cm. Now that you have set the size of your compasses, **do not adjust them.**

(b) Mark a point A anywhere on the circumference of the circle. Put your compass point on point A and draw a small arc which crosses the circle. Label this point B.

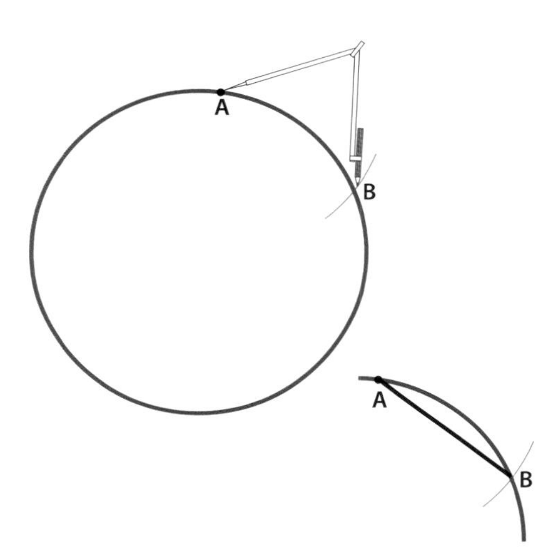

(c) Draw a straight line between the points A and B.

(d) Put your compass point on point B. Draw another arc which crosses the circle at point C. Draw a straight line joining points B and C.

(e) Repeat this all the way round the circle.

Write down the name of the shape you have drawn.

Assignment 2 Patterns with circles

- Using a pair of compasses, draw a circle with a radius of 2 cm. Now that you have set the size of your compasses, **do not adjust them.**

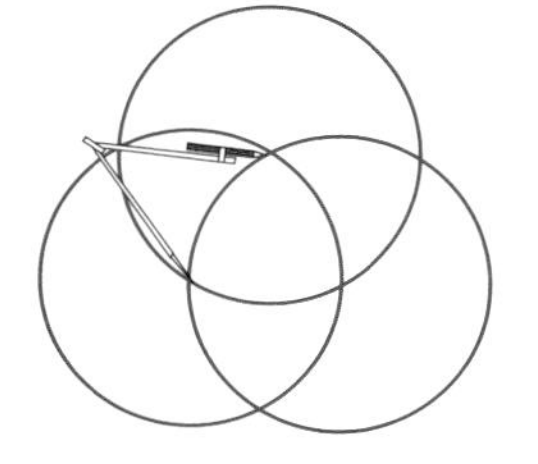

- Put your compass point anywhere on the circumference of the circle. Draw another circle.
- Put your compass point on the point where the two circles cross. Draw another circle.

- Repeat this to make a pattern of circles over the page.
- Draw straight lines joining the points where the circles cross, to make a repeating pattern.

- Draw the circle pattern again. Try to find how many different ways you can join the points, to form different repeating patterns.

Key fact

A tessellation is a pattern of repeated tiles.

Tessellations

Some shapes fit together to make a tiling pattern which could carry on for ever. There are no spaces between the shapes.
Here are two examples.

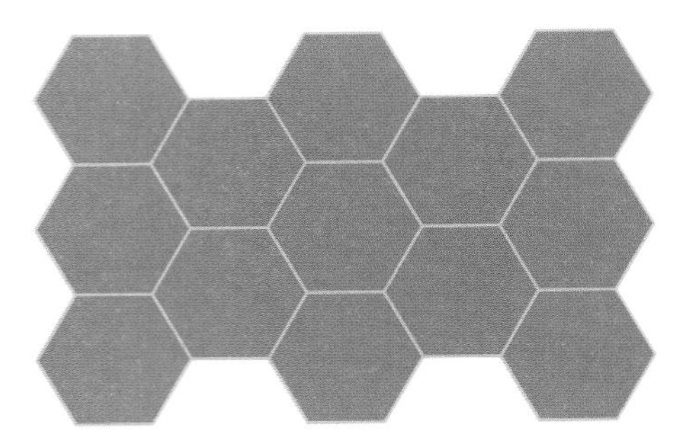

This pattern of repeating tiles is called a **tessellation**.

Circles do not tessellate because they do not fit together without leaving spaces between them.

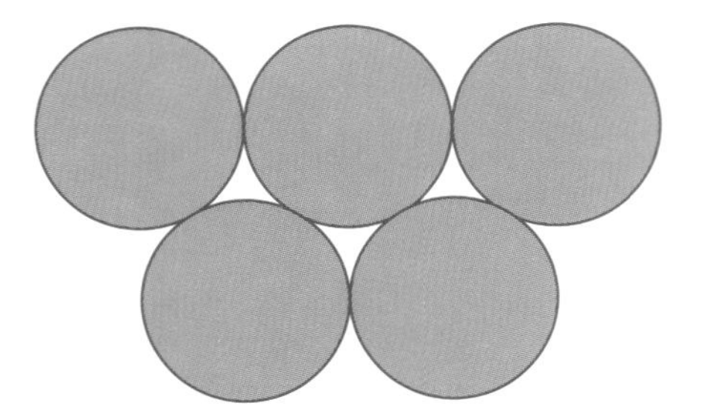

The shapes in a tessellation are called tiles.

Exercise 7

1 Here is the start of a tessellation using a rectangle. Copy it onto squared paper and continue it to cover half a page.

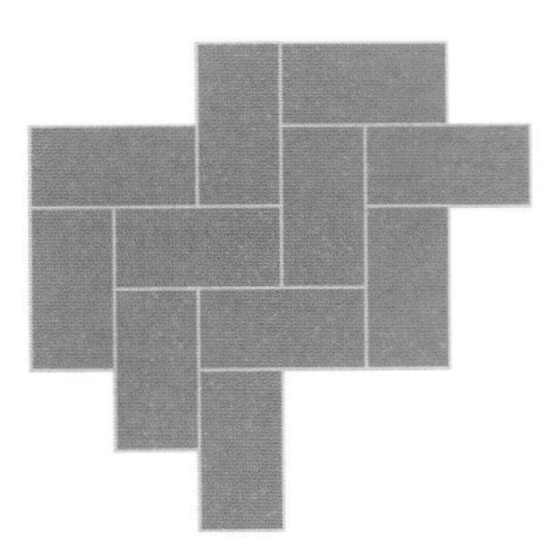

2 Draw a tessellation on squared paper starting like this.

3 Sometimes it is easier to draw tessellations using special paper.
Here is the start of a tessellation on squared dotted paper.
Copy the pattern and continue it to cover half a page.

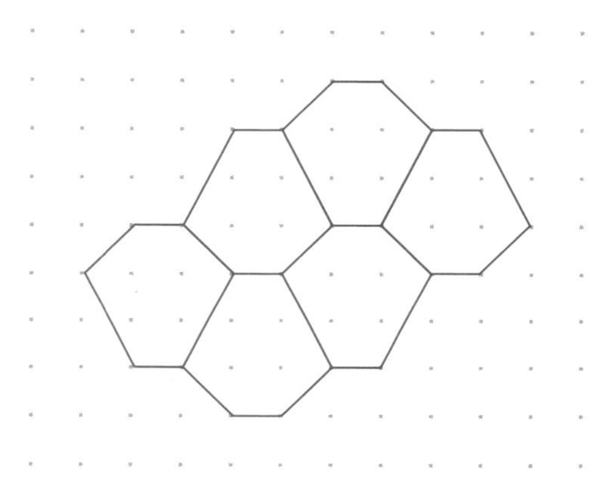

4 Tetrominoes

Draw tessellations of each of these shapes on squared dotted paper.

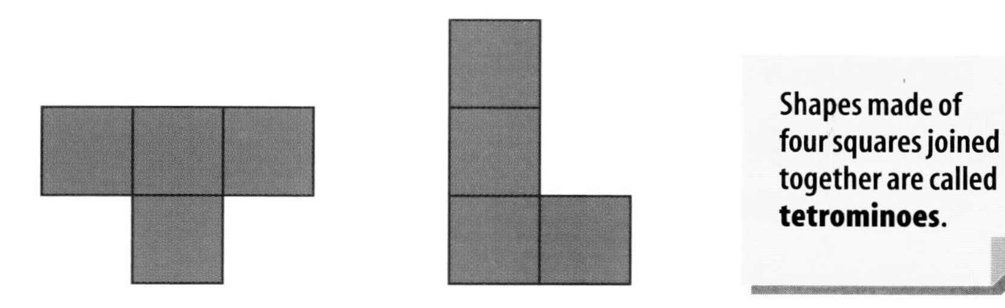

Shapes made of four squares joined together are called **tetrominoes.**

Draw the other shapes which can be made using four squares.
Use each of these to draw a tessellation.

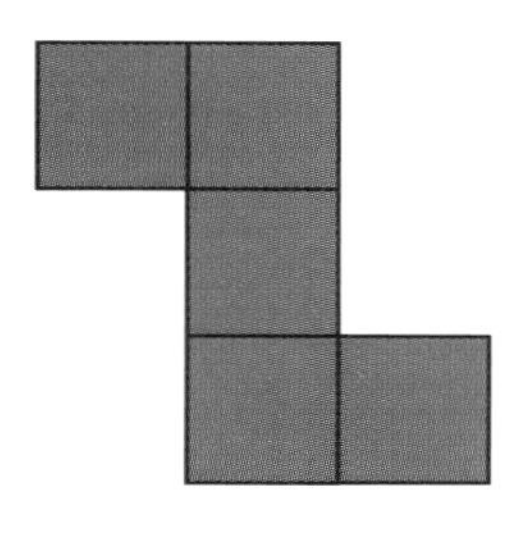

5 Pentominoes

Draw as many different pentominoes as you can on squared paper.
Which of your pentominoes will tessellate?

Shapes made of five squares joined together are called **pentominoes.**

6 Here is the start of a tessellation on triangular dotted paper. Copy it and draw more so that it fills half a page.

7 Another way to draw a tessellation is to use tracing paper. Continue the following tessellations to fill half a page. Use tracing paper and draw on plain paper.

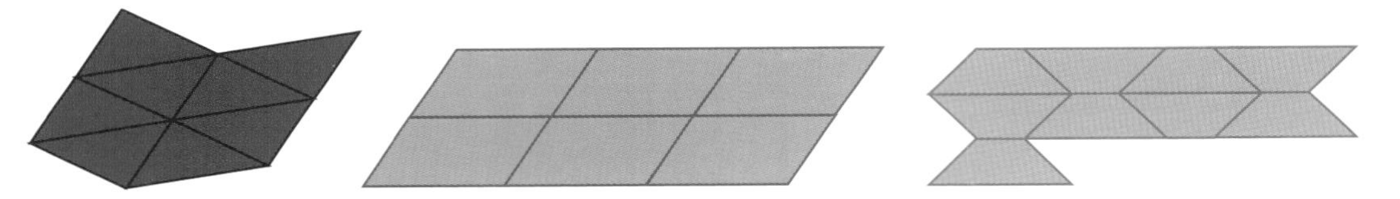

8 Tessellations can use more than one shape. This one uses an octagon and a square. Copy it and make it cover half a page.

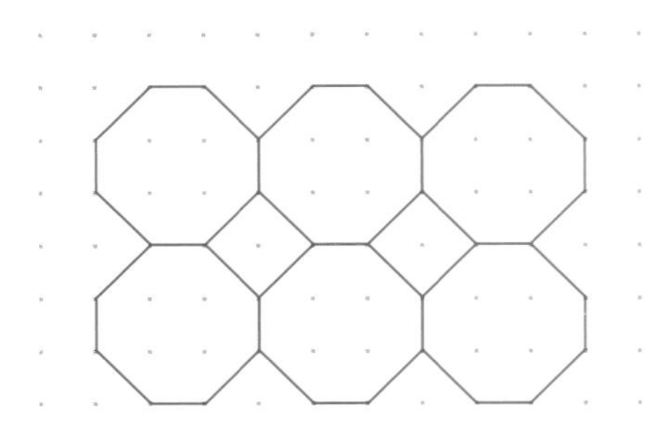

9 Draw another tessellation using an octagon and a square. Start with a square drawn like the one shown.

Assignment 3 Tessellations with two shapes

- Use squared dotted paper. Make up a tessellation using two different shapes.
- Use triangular dotted paper. Make up a tessellation using two different shapes.

Assignment 4 Tessellations in art

Tessellations have been used in art. A famous artist called Escher used them a lot. This is an example of his work.

You can make a tessellation look like a picture. You have to change the shape you are using but make sure it still fits together. This is an example.

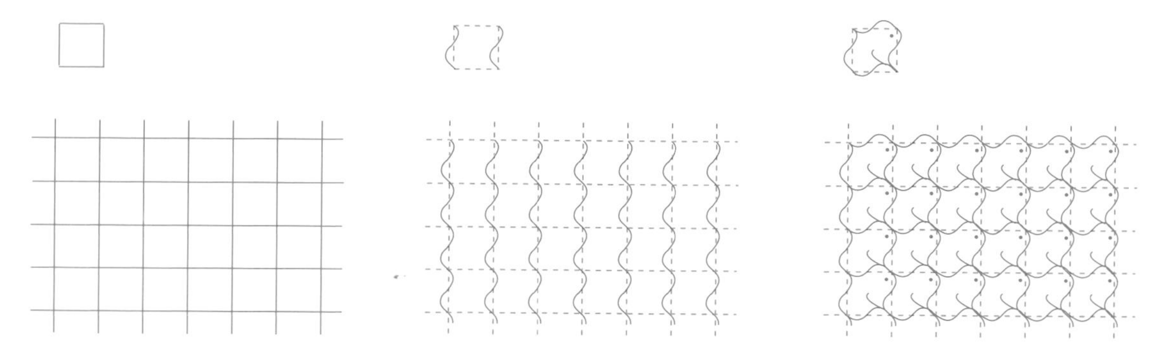

Using squared paper draw your own tessellation. Alter it to make a picture.

Review Exercise

1 Write down the names of these shapes.

(a) 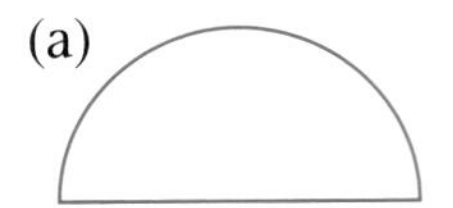 (b) 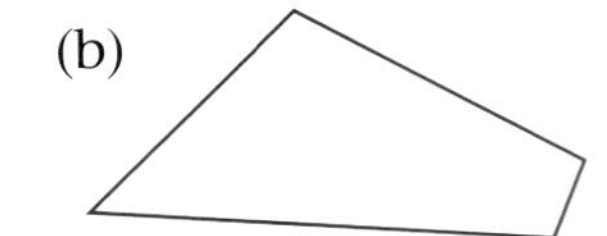 (c) 

2 Draw a circle using a pair of compasses. On your circle,
(a) draw and label a diameter (b) draw and label a radius
(c) label the circumference.

3 Write down the difference between a square and a rectangle.

4 Which of the following shapes are polygons?

(a) 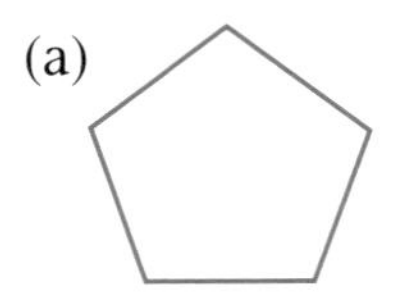 (b) 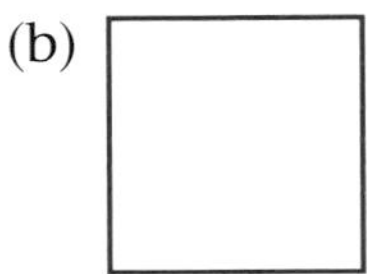 (c) 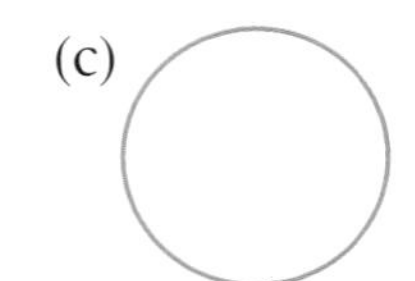 (d) 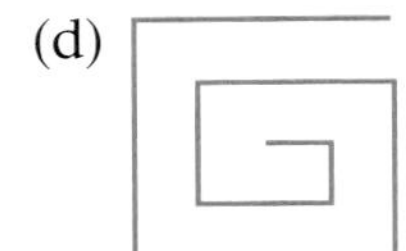

(e) 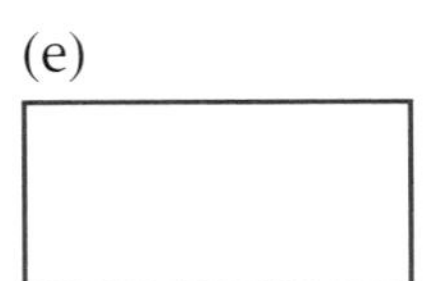 (f) 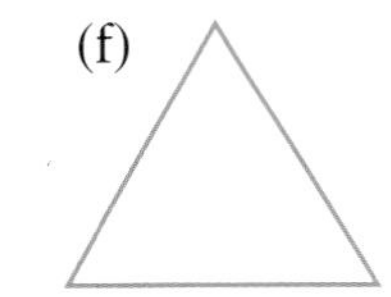 (g) 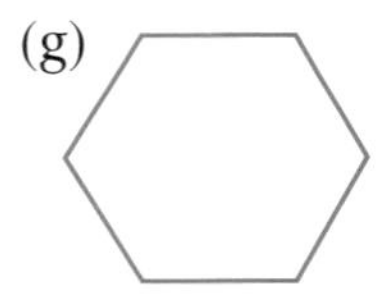

5 Draw a shape which is a non-polygon. Write down why it is not a polygon.

6 Construct a right-angled triangle using a ruler and compasses.

7 Draw a quadrilateral which has two obtuse angles, using a ruler and compasses.

Assignment 5 Tessellations with squares and triangles

- How many different-sized squares can you draw on a grid like this?
  The corners of the squares must be on the dots.
- How many different-sized triangles can you draw on the grid?
  How many of them are right-angled triangles?

Discovering patterns and sequences

In this chapter you will learn:

- ➜ **how to continue patterns and sequences**
- ➜ **how to describe and generate patterns**

Starting points

Before you start this chapter, you will need to know how to use patterns to find out what is missing.

Exercise 1

What is missing from each of the following patterns?
In your own words, describe how you worked it out.

1 1, 2, 3, 4, ?, 6, 7, 8, 9

2 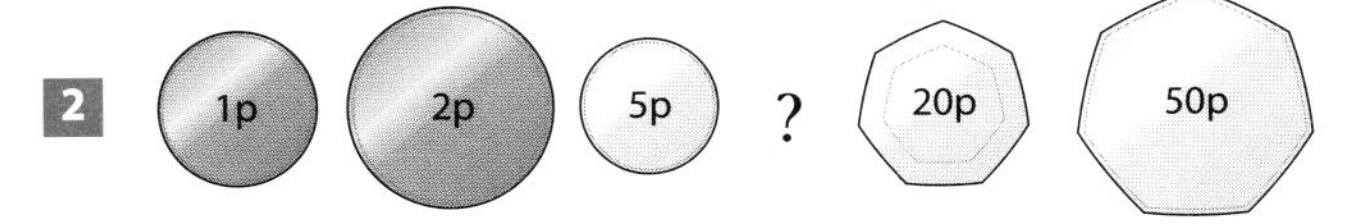

3 A, B, C, ?, E, F, G

4 

5 M A T ? E M A T I C S

6 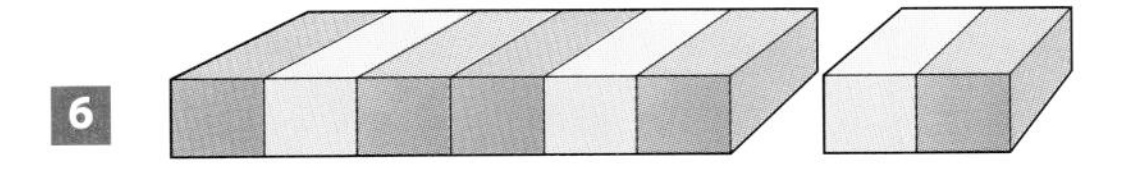

7 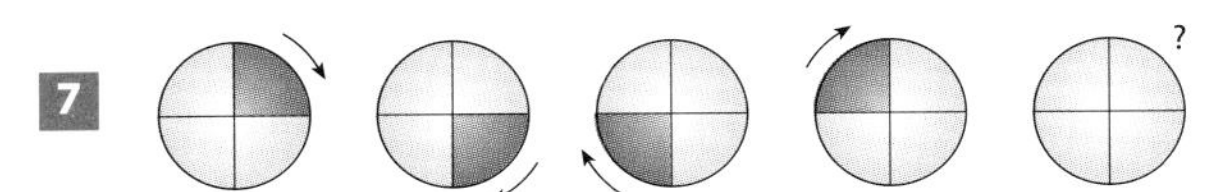

8 S C H O ? L

A sequence of events

Exercise 2

These pictures show things which happen in a certain sequence, but they are not arranged in the proper order.

For each set, write down the order in which they should be arranged.

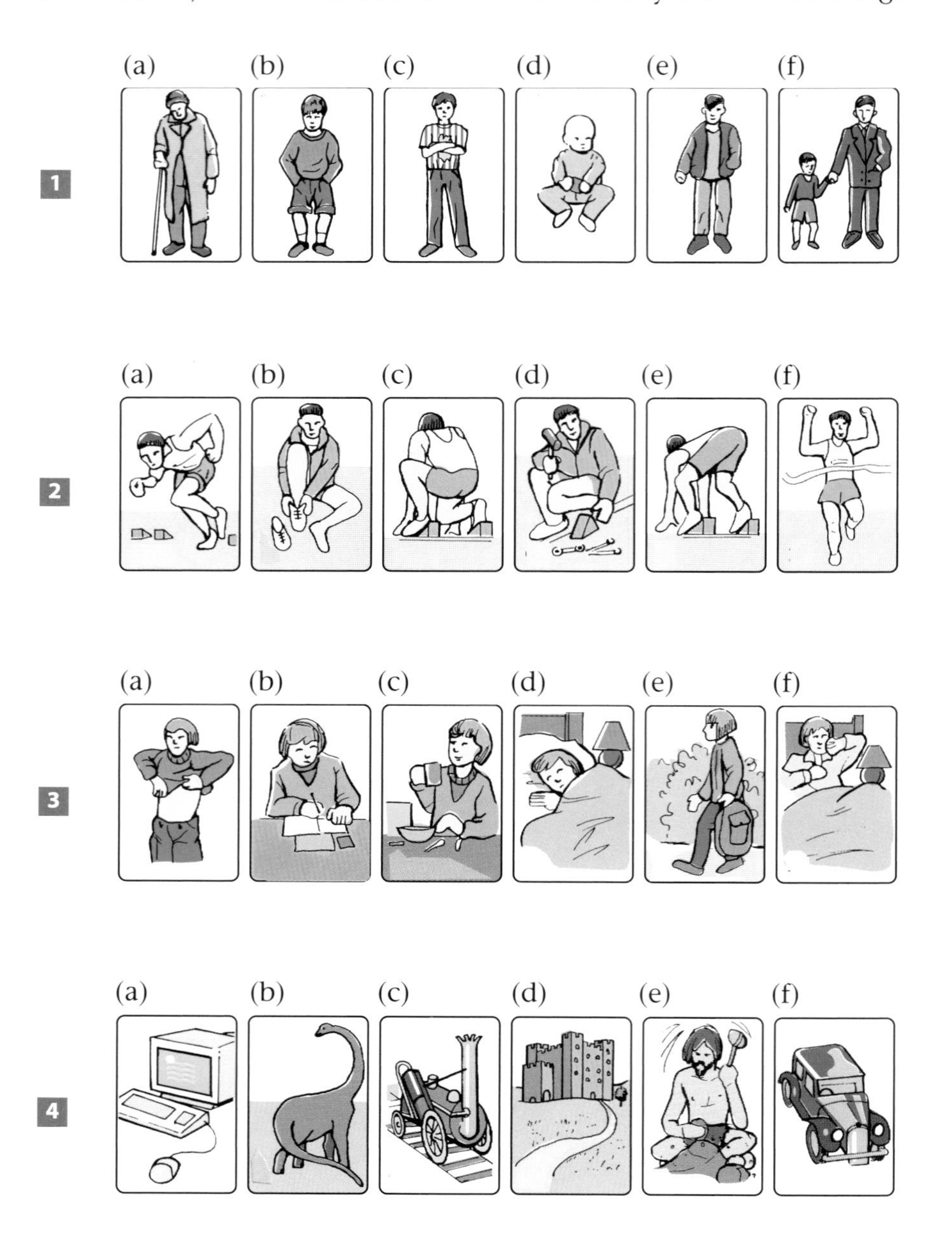

Making patterns

Exercise 3

Copy these patterns and continue them across your page. For each one, describe how the pattern is formed.

1 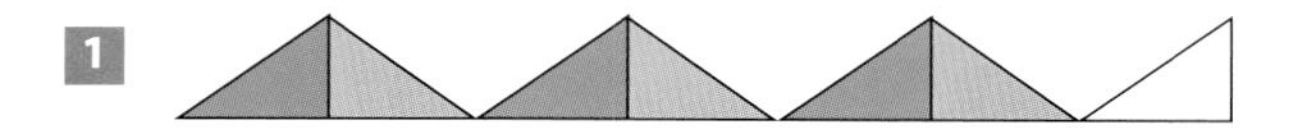

2 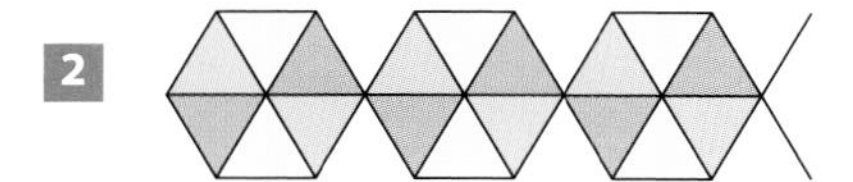

3 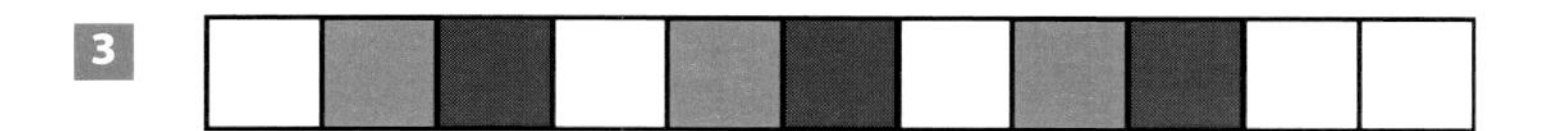

4 

5 

6 

7 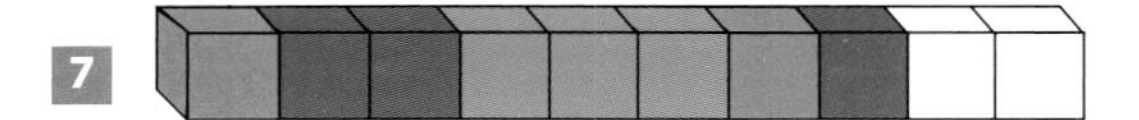

Growing patterns

Key fact

A term is one part of a pattern.

Exercise 4

The following patterns grow according to a rule. Continue these growing patterns for three more terms. Use squared paper.

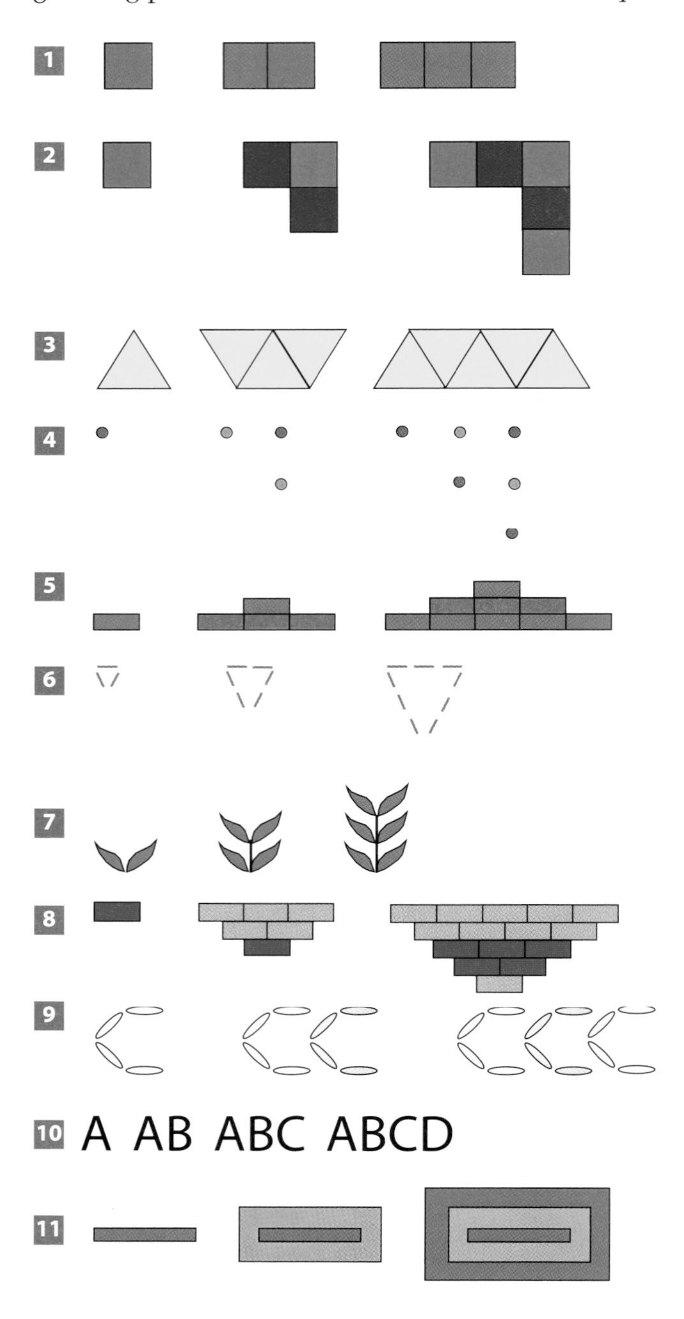

Following a pattern

Exercise 5

1 (a) Look at this flow diagram. Follow the instructions and write down the answers you get.

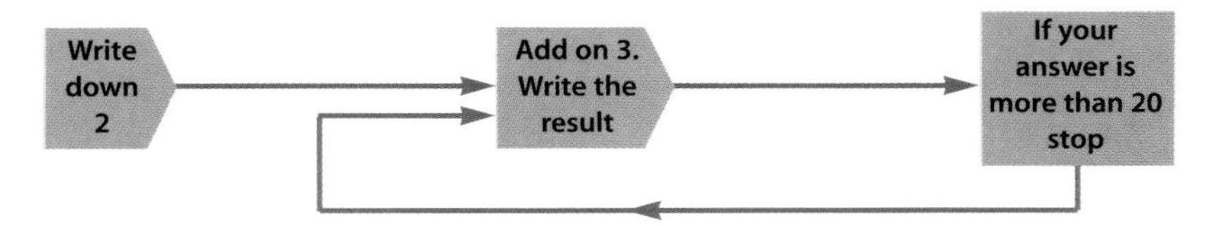

(b) In your own words, describe the pattern formed by following the instructions.

2 (a) Look at this flow diagram. Follow the instructions and write down the answers you get.

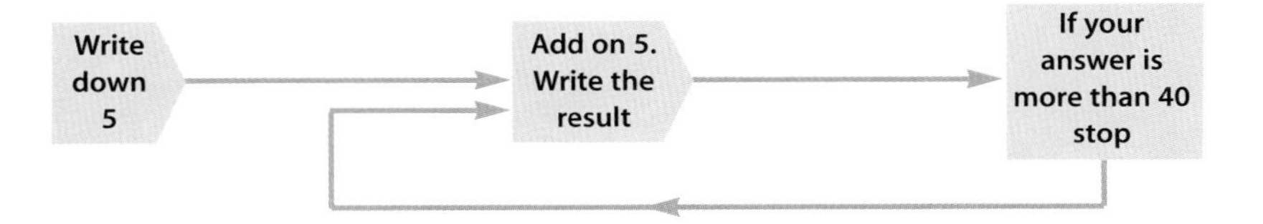

(b) In your own words, describe the pattern formed by following the instructions.

3 (a) Look at this flow diagram. Follow the instructions and write down the answers you get.

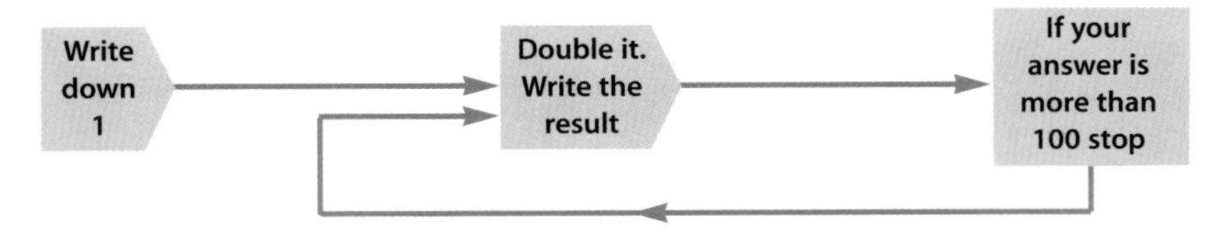

(b) In your own words, describe the pattern formed by following the instructions.

4 (a) Think of a pattern of numbers of your own.

(b) Make up a flow diagram that will give your pattern of numbers.

Exercise 6

Key fact

Numbers which follow a pattern are called a sequence.

1 Copy these growing patterns and continue them for three more terms.

In your own words, describe how you worked out the next terms in each of the sequences in (a) to (g).

(a) 2 , 4, 6, 8 , 10, _ , _ , _ (b) 1, 2, 3, 4, 5, _ , _ , _ ,

(c) 1, 3, 5, 7, 9, _ , _ , _ (d) 5, 10, 15, 20, 25, _ , _ , _

(e) $\frac{1}{2}$, 1, $1\frac{1}{2}$, 2, $2\frac{1}{2}$, _ , _ , _ (f) 1, 3, 6, 10, 15, _ , _ , _

(g) 1, 4, 9, 16, _ , _ , _

2 These sequences follow a pattern but do not grow bigger. Copy the sequences and continue them for three more terms.

In your own words, describe how you worked out the next terms.

(a) 10, 9, 8, 7, 6, _ , _ , _ (b) 30, 28, 26, 24, _ , _ , _

(c) 100, 90, 80, 70, _ , _ , _ (d) 64, 32, 16, 8, _ , _ , _

(e) 10, 11, 9, 10, 8, 9, 7, 8, _ , _ , _

Making sequences

To work out the terms of a sequence you need:
- the starting point for the sequence
- the pattern or rule which generates the sequence.

A sequence needs a starting value and a pattern to generate it.

Example

Start with 3, add on 4 each time.

This gives 3, 7, 11, 15, ...

Exercise 7

Write the first five terms of each of these sequences.

1 Start with 2, add on 3 each time.

2 Start with 4, add on 5 each time.

3 Start with 1, double it each time.

4 Start with 20, take away 2 each time.

5 Start with 120, divide by 2 each time.

6 Start with 1, add on 1 then two and then keep increasing the difference between the terms by 1 each time.

7 Make up a sequence of numbers. Write down the rule to give the numbers. Swap with a partner and write down the first five terms of each other's sequences.

Searching for a pattern

One way of looking for a pattern in a sequence of numbers is to look at the gaps between the terms.

Example

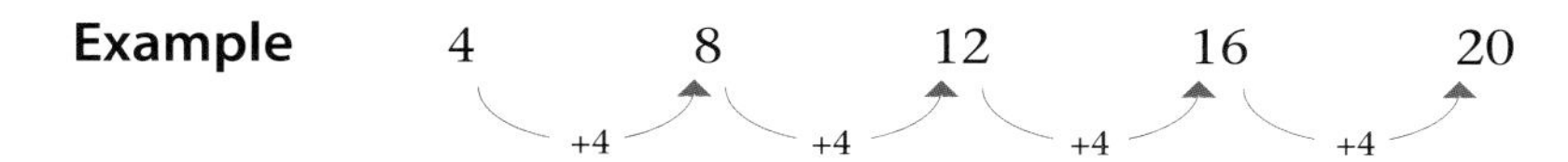

This sequence goes up in 4s.

Example

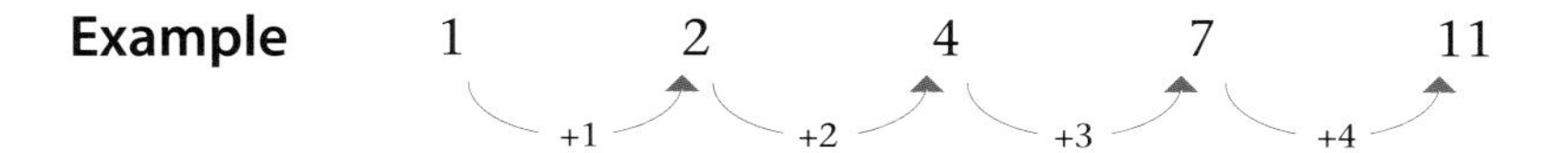

The difference between the terms increases by 1 each time.

Exercise 8

Use the differences between the terms in these sequences to continue them.

1 (a) 2, 5, 8, 11, 14, _, _, _,

(b) 19, 22, 25, 28, 31, _, _, _,

(c) 3, 10, 17, 24, _, _, _,

(d) 6, 12, 18, 24, _, _, _,

(e) 2, 5, 9, 14, _, _, _,

(f) 1, 1, 2, 4, 7, 11, _, _, _,

(g) 3, 4, 7, 12, 19, 28, _, _, _,

(h) 1, 1, 2, 3, 5, _, _, _,

(i) 50, 45, 40, 35, 30, _, _, _,

2 Look at each of the following sequences, then describe, in your own words, how the next term can be found.

(a) 3, 6, 9, 12, _ , _ , _ , (b) 4, 7, 10, 13, _ , _ , _ ,

(c) 4, 8, 12, 16, _ , _ , _ , (d) 50, 40, 30, 20, _ , _ , _ ,

(e) 1, 2, 4, 8, _ , _ , _ , (f) 64, 32, 16, 8, _ , _ , _ ,

(g) 0, 5, 11, 16, _ , _ , _ ,

3 Write down the 10th term for each sequence in question 2.

4 Write in words the starting point and pattern which generates each of the following sequences.

(a) 2, 7, 12, 17, 22, ... (b) 5, 9, 13, 17, 21, ...

(c) 12, 22, 32, 42, 52, ... (d) 13, 11, 9, 7, 5, ...

(e) 100, 90, 80, 70, 60, ... (f) 2, 3, 5, 8, 12, ...

5 There are five chairs around this desk.
Desks can be single or in pairs.

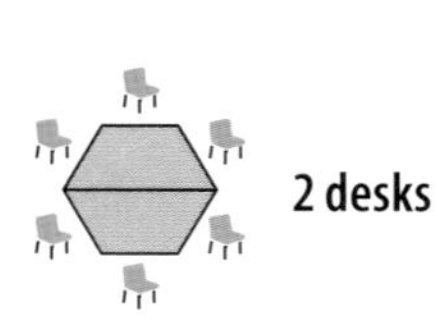

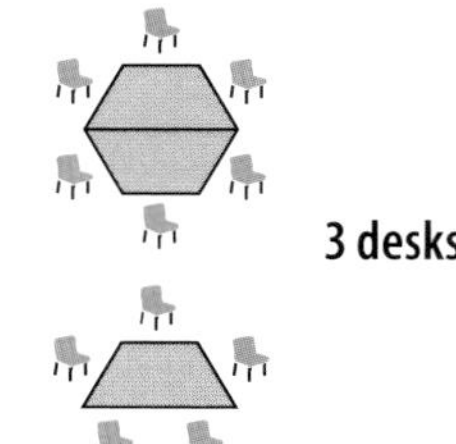

(a) Copy the table and complete it to show the numbers of chairs that can be put around the desks.

Number of desks	1	2	3	4	5	6	7	8
Number of chairs	5	6	11					

(b) Describe how you work out the next number of chairs.

Assignment 1 Hidden faces

The growth of these shapes follows a pattern.

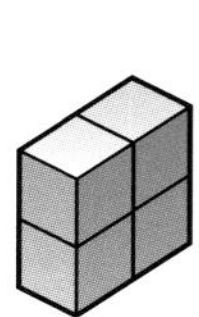

- Record the sequence generated by the numbers of cubes.
- Record the sequence generated by the size (in cubes) of the hole in the shape.

When two cubes are put together two of the faces are hidden.

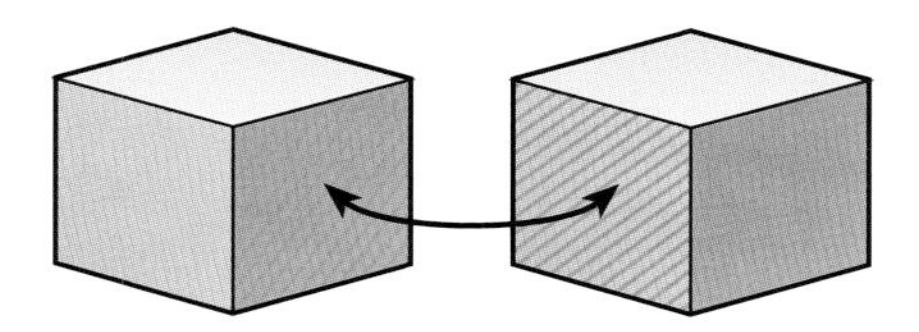

- How many hidden faces are there in each of the shapes in the sequence?
- Explain how the pattern develops.

- Investigate the number of cubes in this sequence.

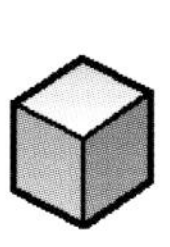
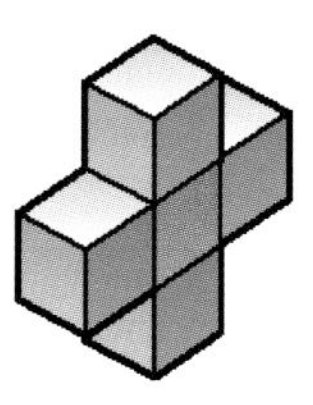
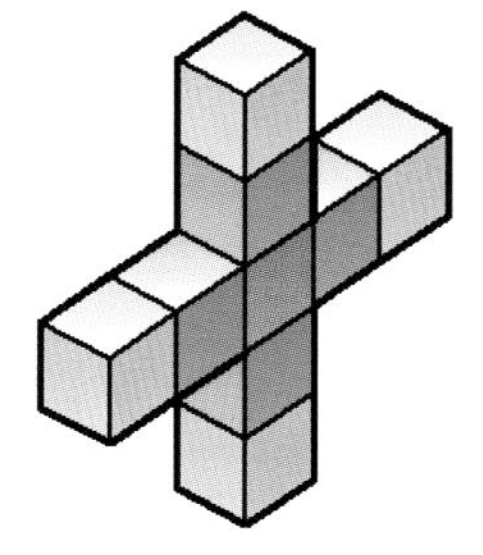

- Investigate the number of hidden faces.

The value of higher terms

To find the 10th term in the sequences in the last exercise you had to find all the values in the sequence. This is not too difficult if you are asked for the 10th term. What about the 100th term or the 1000th?

We need a different method to solve this type of problem.

Exercise 9

In this shape each side is one matchstick long.
The shape is made with three matchsticks.

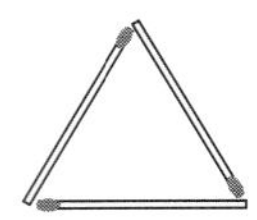

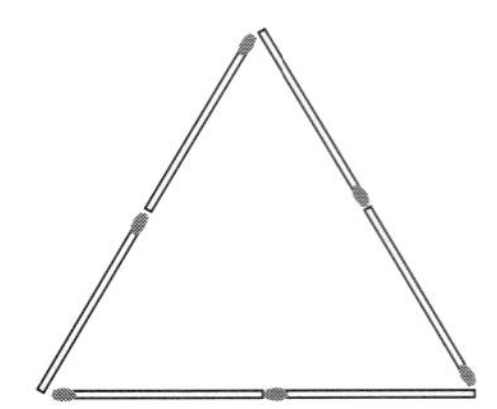

In this shape each side is two matchsticks long.
The shape is made with six matchsticks.

1 (a) Draw the next two shapes in this sequence.

(b) Copy this table and complete it for shapes made with up to eight matches per side.

(c) What is the difference between the numbers of matches per side?

Matches on each side	Number of matches
1	3
2	6
3	
4	
5	
6	
7	
8	

(d) Is there a pattern between the length of the side and the number of matches? What is it?

(e) Use your pattern to find the number of matches if the side length is 100.

2 (a) Copy this sequence and continue it for three more terms.

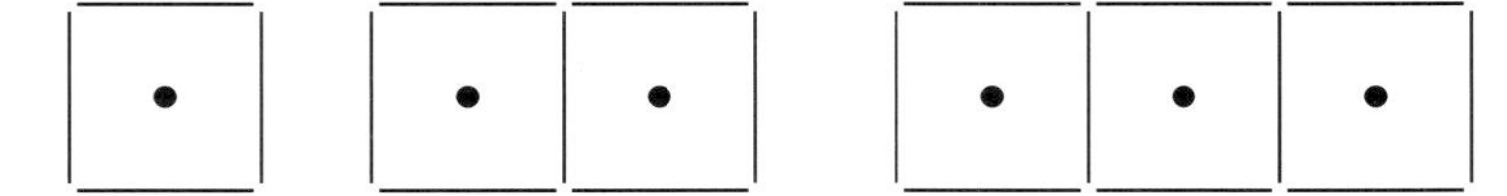

(b) Copy this table and complete it for the sequence of dots and lines, until there are eight dots.

Dots	Lines
1	4
2	7
3	10
4	
5	
6	
7	
8	

(c) Describe the growing pattern for this sequence.

(d) Is there a pattern between the number of dots and the number of lines? Try to write it down.

(e) How many lines would there be for 20 dots?

3 These circles have dots on the points where they cross.

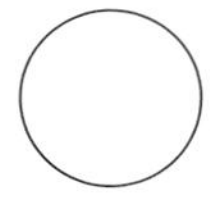

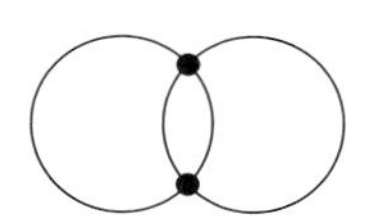

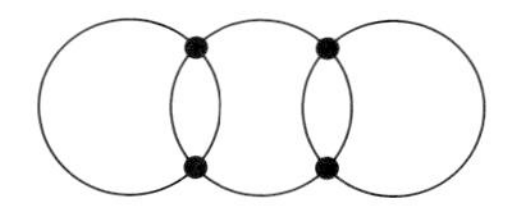

(a) Copy this table and complete it to show the numbers of circles and the numbers of dots, up to eight circles.

Circles	Dots
1	0
2	2
3	4
4	
5	
6	
7	
8	

(b) Describe the growing patterns in the sequences.

(c) Is there a pattern between the number of dots and the number of lines? Try to write it down.

(d) Use this to find the number of dots for 50 circles.

Assignment 2 Rectangular nails

This sequence of rectangles is formed by hammering nails into a piece of pegboard. The nails are 1 cm apart, up and across.

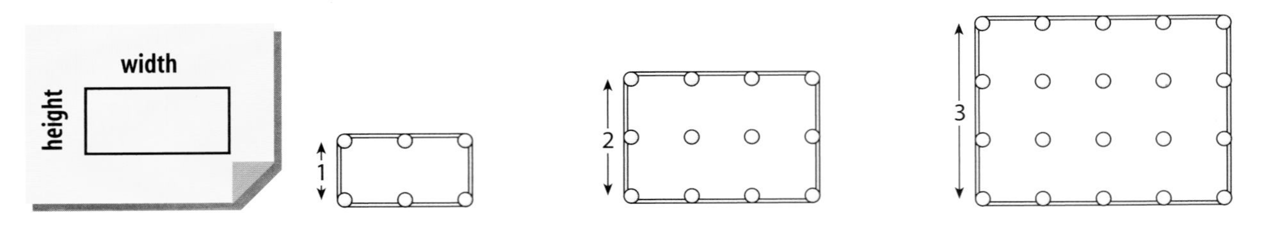

For each rectangle, its width is 1 cm more than its height.

- Write down what you notice about the number of nails along the side of a rectangle, and the length of that side.
- How many nails would there be up the side if the height was 100 cm?

The table shows the numbers of nails needed to make rectangles of different heights.

Height	Nails
1	6
2	12
3	
4	
5	
6	

- Copy the table and complete it.
- Investigate the relationship between the height of the rectangle in the sequence and the number of nails in the rectangle.
- What would be the height of a rectangle made from 132 nails?

Assignment 3 A number sequence

This sequence is generated by adding to each term the sum of the digits of the previous number.

13 17 25 32 37 47

1 + 3 = 4 1 + 7 = 8 2 + 5 = 7 3 + 2 = 5 3 + 7 = 10

- Find the next eight terms of the sequence.
- Can you find a value to go before 13 in the sequence?
- Investigate sequences like this with different starting points.
- Is it possible to find a value that would go before 20 in a sequence?

Assignment 4 Pascal's triangle

This triangle is named after a French mathematician called Blaise Pascal.

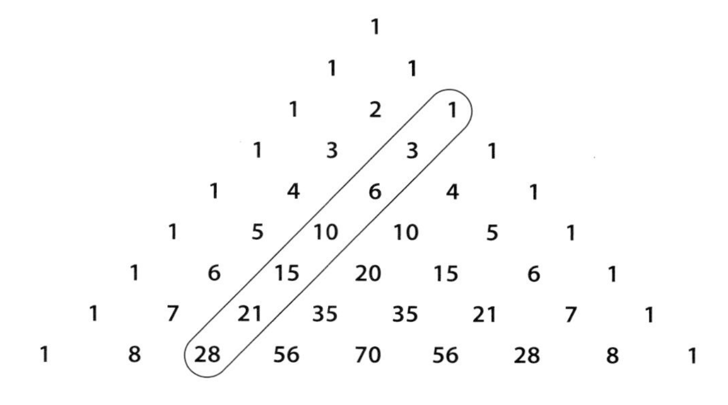

and so on.

On each diagonal there is a different sequence of numbers.
e.g. 1, 3, 6, 10, 15, 21, 28, ...

- Investigate sequences which appear in Pascal's triangle.
- Use your school library or information system to find out about Blaise Pascal.

Review Exercise

1 Copy each of the following sequences and continue it for three more terms.

(a) 

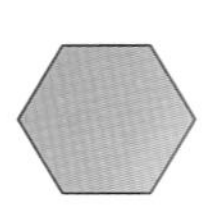 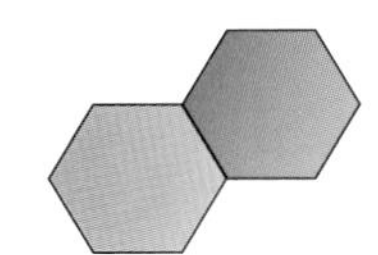 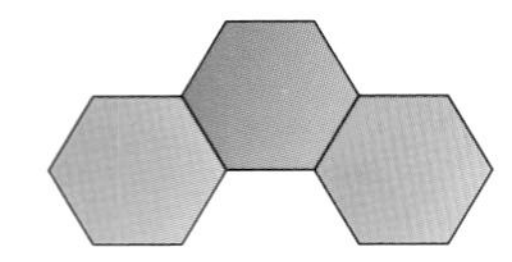

(b) 5, 9, 13, 17, _ , _ , _

(c) 13, 11, 9, 7, _ , _ , _

2 Write down the first five terms of the sequences generated by the following rules.

(a) Start with 5 , add on 5 each time.

(b) Start with 2 , add on 3 each time.

(c) Start with 3 , double it each time.

(b) Start with 20 , take off 3 each time.

(e) Start with 1, add on 2, then take away 1, repeat this pattern.

3 What are the next three terms in each of the following sequences?

(a) 1, 5, 9, 13, _ , _ , _

(b) 6, 12, 18, 24, _ , _ , _

(c) 14, 12, 10, 8, _ , _ , _

In your own words, explain how you found the terms.

4 This pentagon has five edges. 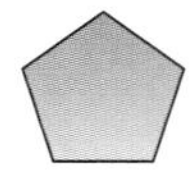

Two pentagons together have eight edges. 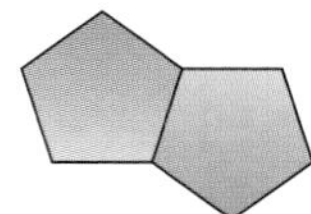

Three have eleven edges. 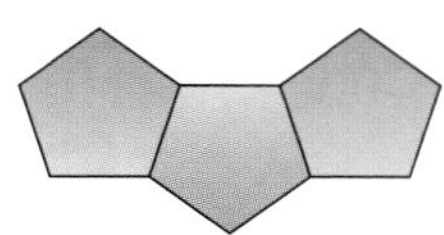

(a) Copy the sequence and continue it for three more terms.

(b) Copy this table and complete it to show the numbers of pentagons and the numbers of edges, up to eight pentagons.

Pentagons	Edges
1	5
2	8
3	11
4	
5	
6	
7	
8	

(c) Describe the growing pattern in the sequence of numbers of edges.

(d) Is there a pattern between the number of pentagons and the number of edges? Describe it in your own words.

(e) Use this pattern to find the number of edges for 20 pentagons.

Assignment 5 The Fibonacci sequence

Fibonacci discovered a mathematical sequence which seems to appear in many natural situations.

1, 1, 2, 3, 5, 8, 13, 21, 35, ...

- Work out how the sequence is formed.
- Continue the sequence for five more terms.
- Use your school library or information system to find out about the Fibonacci sequence.

CHAPTER 5

Measures

In this chapter you will learn:

- ➜ **more about measures for length, mass, capacity and time**
- ➜ **how to change measures into smaller or larger units**
- ➜ **how to read scales**
- ➜ **about numbers above and below zero.**

Starting points

Before starting this chapter you will need to know that the metric units are:

- **metre** for length
- **gram** for mass
- **litre** for capacity.

You should also know some imperial units such as **mile**, **pound** and **pint**.

Assignment 1 **Find out about measures**

Find out about other units of measure.

You can use a dictionary, or an encyclopaedia.

Look at the containers around your house or school and make a list or display of different ways the labels tell us how much is in a box or bottle.

Exercise 1

Using the lists shown, decide which are the most suitable units for describing the following.

1 Length

(a) the length of a playing field

(b) the distance of your journey to school

(c) the length of the classroom

(d) the length of a strand of hair

(e) the thickness of a book

(f) the length of a river

(g) the length of the Channel Tunnel

(h) your own height

2 Mass

(a) the mass of an apple

(b) the mass of a car

(c) the mass of a piece of paper

(d) the mass of an elephant

(e) your own mass

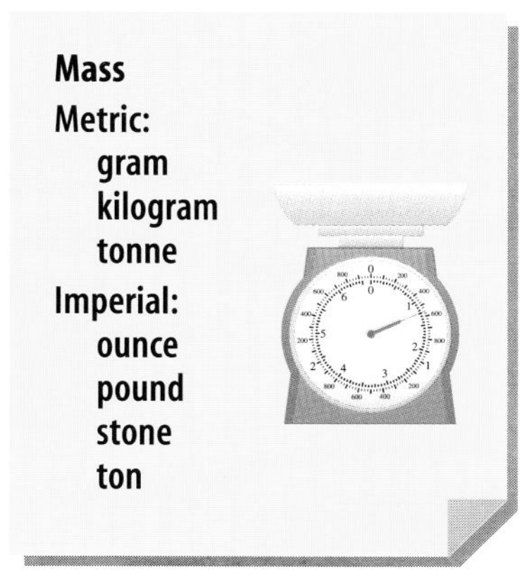

3 Capacity

(a) the amount a cup holds

(b) the amount of water in a swimming pool

(c) the amount of petrol in a car

(d) the amount of orange squash in a bottle

(e) the amount of medicine in a teaspoon

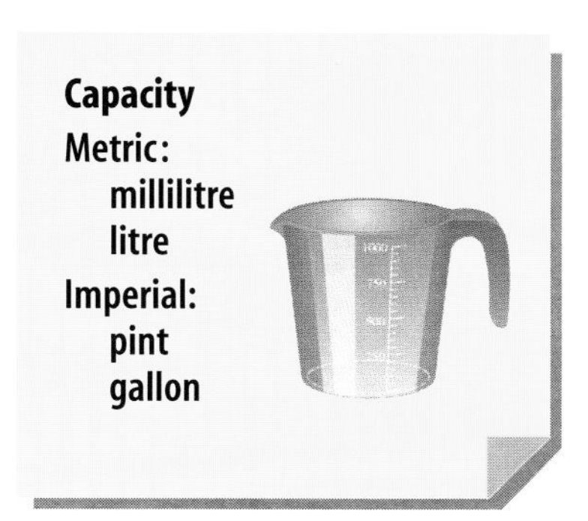

4 Time

(a) the time it will take you to get to your next lesson

(b) the length of a family summer holiday

(c) the age of a pet

(d) the time it would take you to run 100 m

(e) how long it would take you to run 1000 m

(f) how long it would take you to count to 1000

Time
second
minute
hour
year

Assignment 2 Alternative units

- Look again at your answers to question 1 of Exercise 1.
  If you used metric units, give them in imperial units.
  If you used imperial, give them in metric.
- Do the same for questions 2 and 3 of Exercise 1.

Useful approximations	
8 kilometres is about	5 miles
1 kilogram is just over	2 pounds
2.5 centimetres is about	1 inch
1 metre is just over	1 yard
1 tonne is almost	1 ton

Assignment 3 The history of a unit

Find out about the history of a unit of your choice.

- When was it first used?
- Who first used it?

Write some notes about your findings.
These questions may help you decide what to investigate.

- What are imperial units?
- How did the Romans measure length?
- Why is time based on sixty?
- How did calendars start?
- How do we know that there are 365 days in a year?
- When was the metric system introduced?
- How are clothes or shoes measured?
- How is temperature measured?

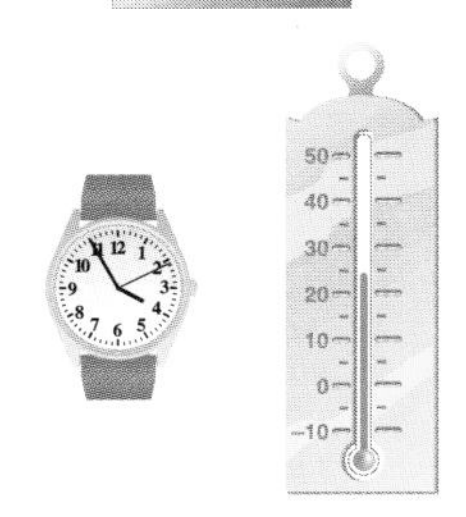

FEBRUARY

S	M	T	W	T	F	S
1	2	3	4	5	6	7
8	9	10	11	12	13	14
15	16	17	18	19	20	21
22	23	24	25	26	27	28

The metric system

The metric system is based on *ten*. It uses standard **prefixes**, which go in front of the units. This helps to give a better idea of the sizes they represent.

The *main* prefixes are as follows.

The highlighted prefixes are used most often.

milli	(m)	meaning	one thousandth
centi	(c)	meaning	one hundredth
deci	(d)	meaning	one tenth
deca	(da)	meaning	ten
hecta	(h)	meaning	one hundred
kilo	(k)	meaning	one thousand

Example: 1 kilogram of sugar means 'one thousand grams of sugar'.
1 kg = 1000 g

1 kg – say 'one kilogram'

card 3 millimetres thick is 'three thousandths of a metre' thick.
3 mm = 0.003 m

3 mm – say 'three millimetres'

We can use these prefixes as column headings like this. The decimal point separates the whole numbers from the fractional parts.

	kilo	hecta	deca	unit	•	deci	centi	milli	
1 km	1	0	0	0	•				1000 m
3 mm				0	•	0	0	3	0.003 m
14 m			1	4	•				14 m
18 mm				0	•	0	1	8	0.018 m
2.4 km		2	4	0	•	0			2400 m

Exercise 2

Units
metre (m)
gram (g)
litre (*l*)

1 Write each of these in the base unit – metre, gram or litre.

(a) 4 km (b) 4 cm (c) 60 cm (d) 3.5 km
(e) 300 mm (f) 24 km (g) 2 kg (h) 200 mg
(i) 0.5 kg (j) 300 c*l* (k) 500 m*l* (l) 2500 m*l*

Exercise 3

Place these measures in order of size, so that the largest is first.

Example

3 ml 4 l 3.2 l

First write them all in the same units.

3 ml 4000 ml 3200 ml

Now place them in order.
4000 ml, 3200 ml, 3 ml

Now put them back in the units they started with.
4 l, 3.2 l, 3 ml

The usual abbreviation for seconds is s, for hours is h.

1	7.2 cm	18 mm	5 m
2	13 m	2.4 km	140 cm
3	130 s	2 min 30 s	40 s
4	5 kg	250 g	70 mg
5	0.75 km	800 m	5000 cm
6	0.5 m	52 cm	528 mm
7	1 h 3 min	64 min	3660 s
8	180 l	4 l	0.8 l
9	1.4 g	150 mg	1.08 g
10	36 h	1 day	22 h
11	15 g	1.5 kg	1500 mg
12	7.2 km	7020 m	70 000 cm

Measuring and recording

In this section you will be measuring and recording the sizes of various objects.

Assignment 4 Measuring and recording length

You will need a variety of instruments.

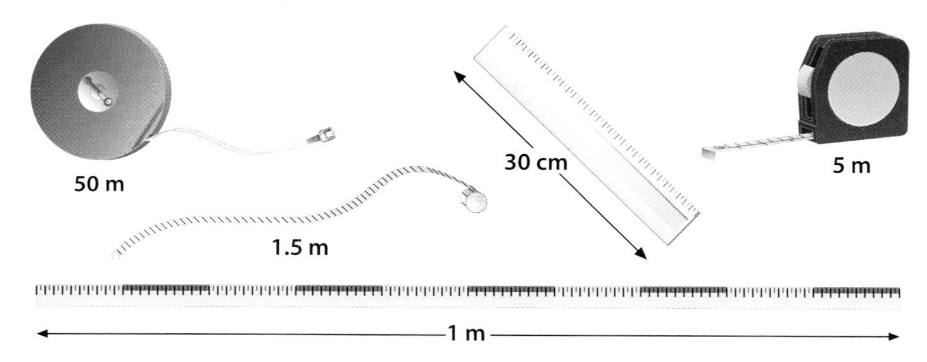

Copy this table into your book.

Always include your unit of measure.

Item	Length

Measure the lengths of these items and record them in your copy of the table.

- the length of your desk
- the height of your desk
- the length of the classroom
- the height of the classroom door
- the circumference of your wrist
- the length and width of this text book
- the width of the corridor nearest your room
- the height of the person next to you
- your hand span

Assignment 5 Measuring and recording mass

You will need:
bathroom scales
kitchen scales
balances.

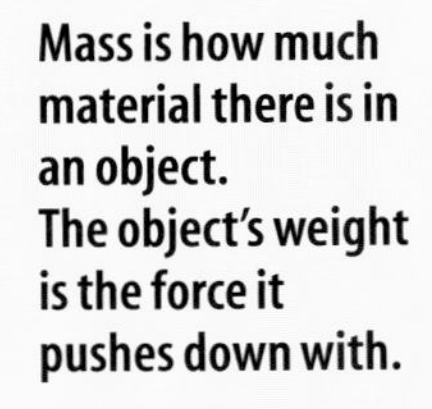

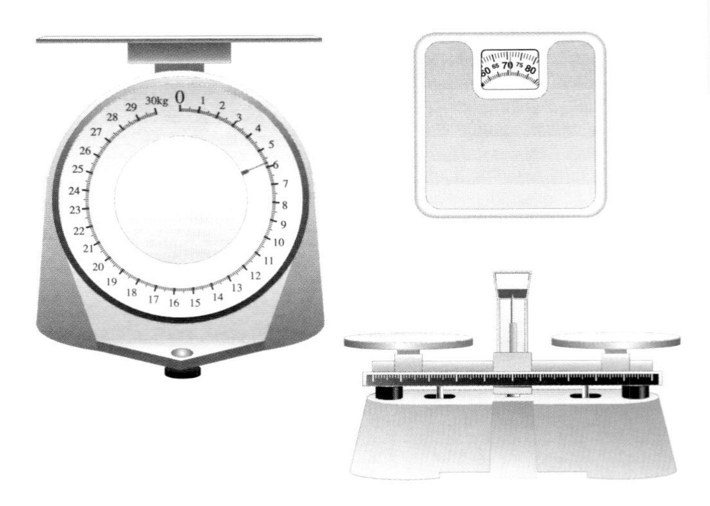

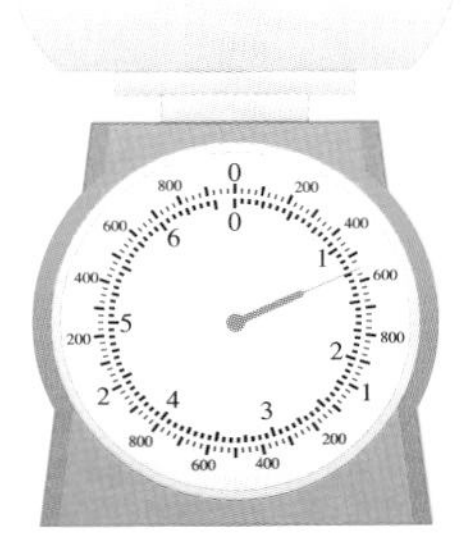

- Copy the following table into your book.

Item	Mass

- Measure the masses of these items and record them in your copy of the table.

this text book — your school bag
your pencil case — the board wipe
a ruler — a shoe
a school tie — a paper clip
a piece of paper — a chair

- Choose two or three items and record their masses in your table.

Assignment 6 Measuring and recording capacity

You will need:
a variety of measuring cylinders
a measuring jug
water to measure

- Copy the following table.

Item	Capacity

- Measure the capacity of water that fills each of these items. Record the capacities in your copy of the table.

  a plastic cup

  an ordinary cup

  a mug

  three different bottles

  a bowl

  a bucket

Key fact
Capacity is the amount of room a liquid takes up.

- Select three of the items you measured above and find out how much the contents *weigh* in each case. Write down your results carefully.
- Would the same *capacity* of different fluids or liquids have different *mass*? Explain your answer.

Assignment 7 Measuring and recording time

You will need:
a stopwatch or
a watch with a seconds hand or
a digital watch.

- Copy the following table into your book.

Item	Time

- Time the items below and record your results in your copy of the table.
  time to count to 50
  time to walk around the classroom once
  time to read a page in this book
  your estimate of 2 minutes
  time to say:
  'Peter Piper picked a peck of pickled peppers.
  Where's the peck of pickled peppers Peter Piper picked?'

- You can find your pulse rate using the carotid artery.
  Count how many pulses you feel in 15 seconds. Multiply this by four to give your pulse rate (beats per minute). Run on the spot for 1 minute. Then take your pulse again.

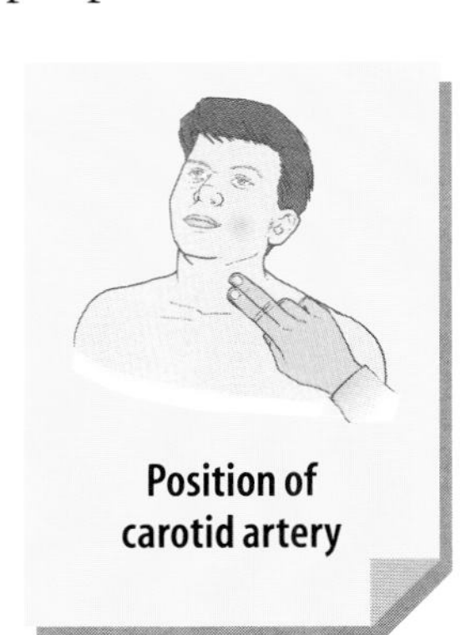
Position of carotid artery

- What is your pulse rate before exercise?
- What is your pulse rate after exercise?

Exercise 4

You will have practised reading several scales by now. Read the measures shown on these scales.

1

2 3 cm

2

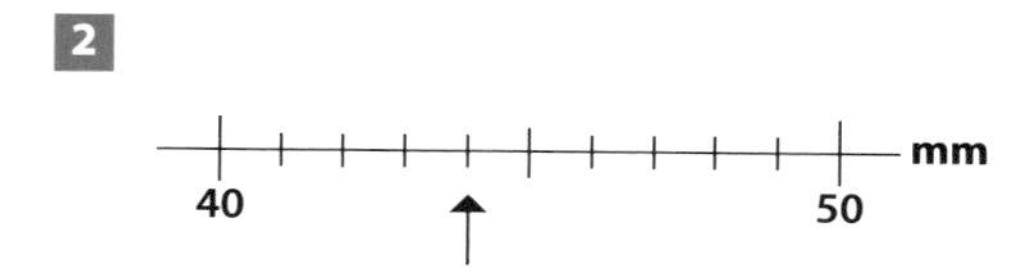

3

24 25 cm

4

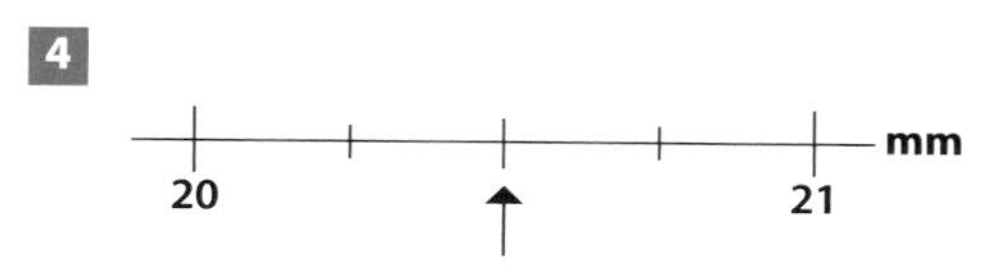

5

12.4 12.5 cm

6

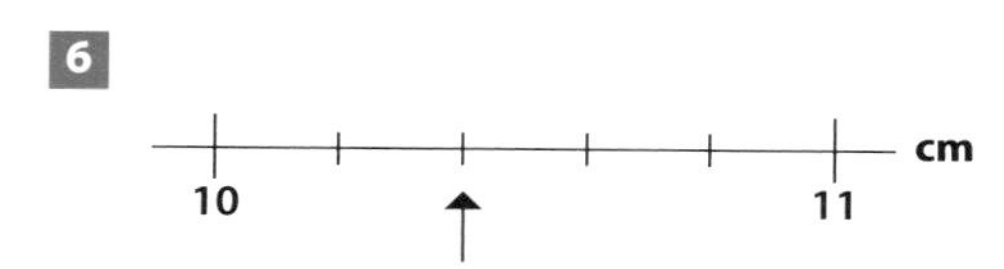

7

15 16 cm

8

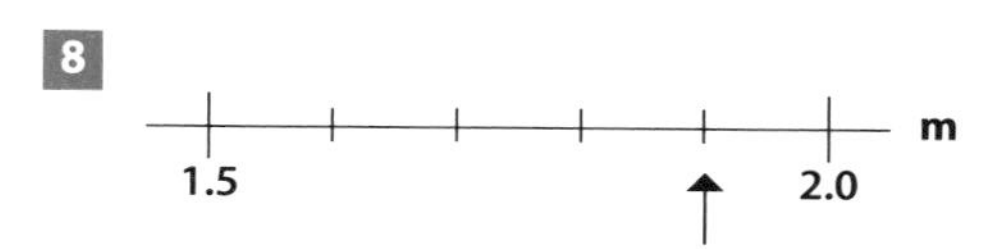

9

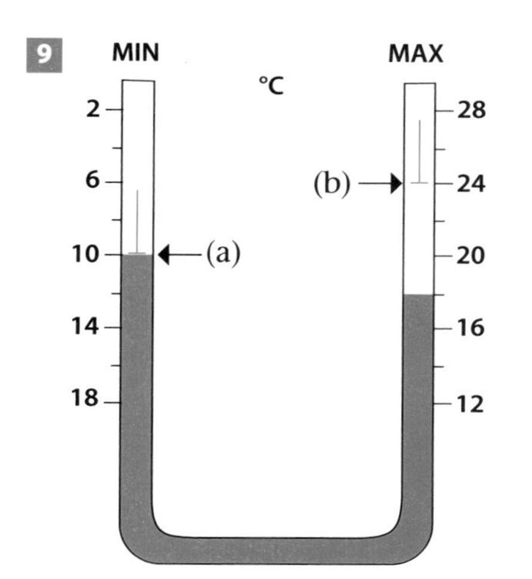

10

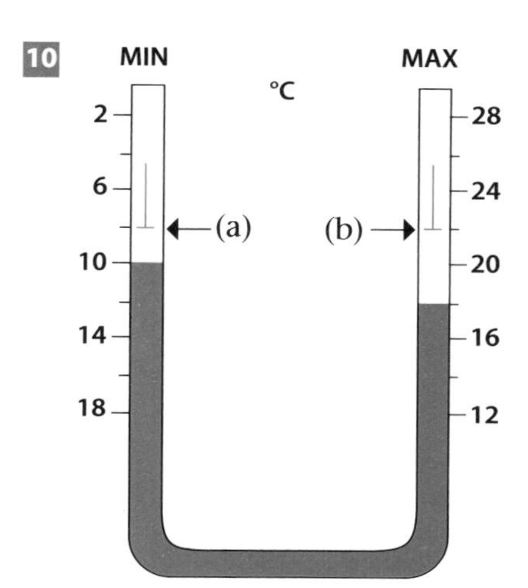

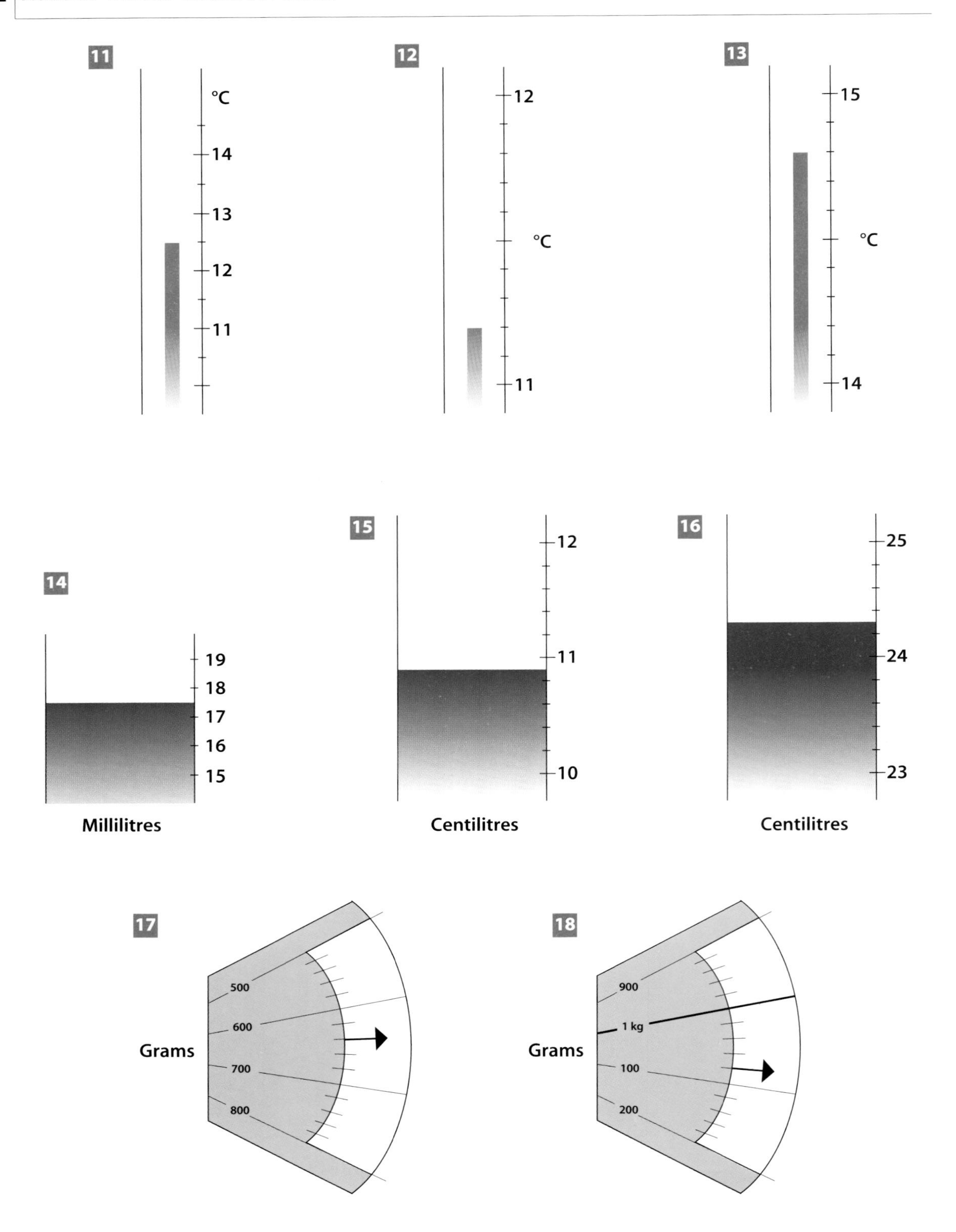
11
°C
14
13
12
11
12
12
°C
11
13
15
°C
14
14
19
18
17
16
15
Millilitres
15
12
11
10
Centilitres
16
25
24
23
Centilitres
17
Grams
500
600
700
800
18
Grams
900
1 kg
100
200

19

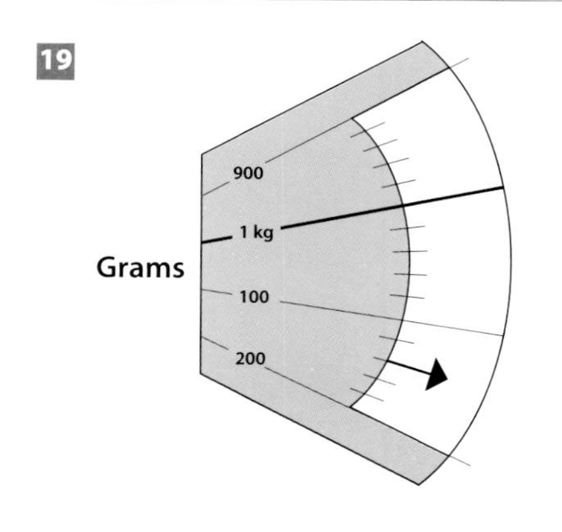

20

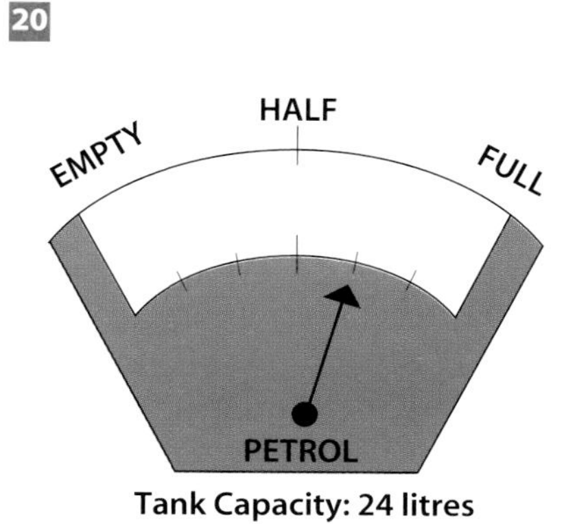

Tank Capacity: 24 litres

21

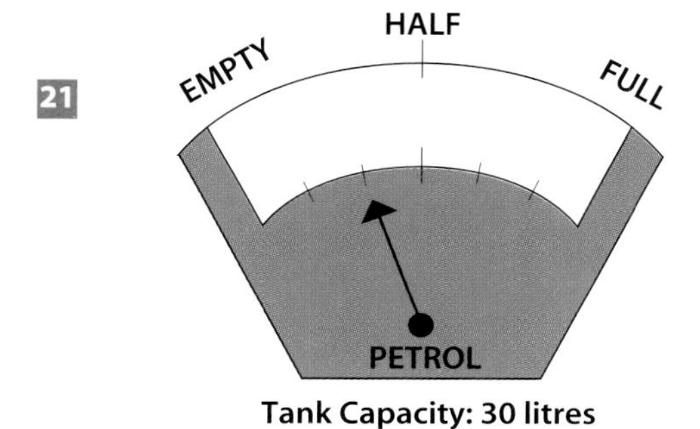

Tank Capacity: 30 litres

22

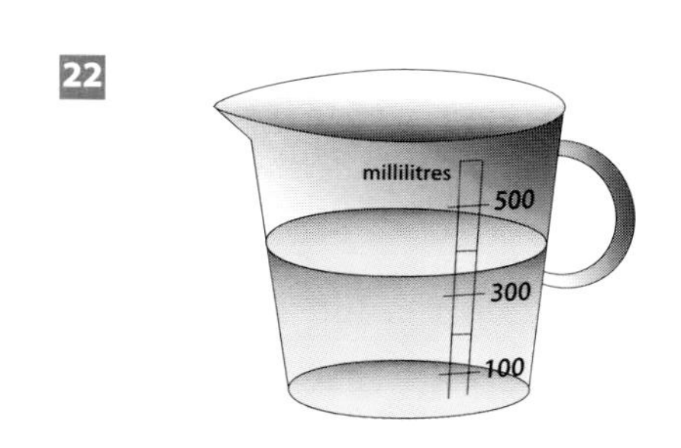

Assignment 8 Estimating and checking

0 10 20 30 40 50 60 70 80 90 100 cm

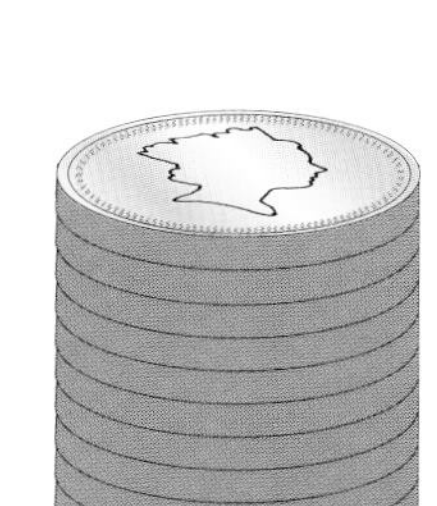

A metre of £1 coins is worth £44.

A pile of ten £1 coins is 3 cm high.

If you had to pack a million (1 000 000) £1 coins into a container with a square base measuring 1 m by 1 m, how high would the container need to be?

- Write down how high you estimate the container would need to be.
- Now work out the answer as accurately as you can.
  Remember to write down how you did it.

Temperature

The temperature scale we use today was developed by a Swedish astronomer called Anders Celsius who lived in the sixteenth century. At first he used the boiling point of water as 0°C and the freezing point as 100°C to set his scale. Later it was reversed, so that the freezing point of water is 0°C and the boiling point of water is 100°C.

Any temperatures can be given using this scale, even if they are lower than 0°C or higher than 100°C.

Exercise 5

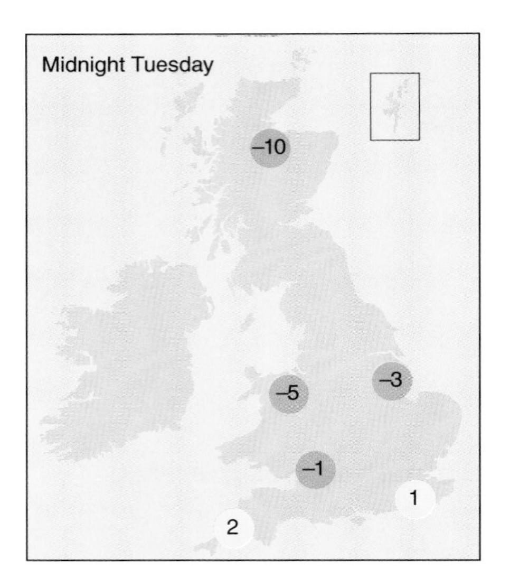

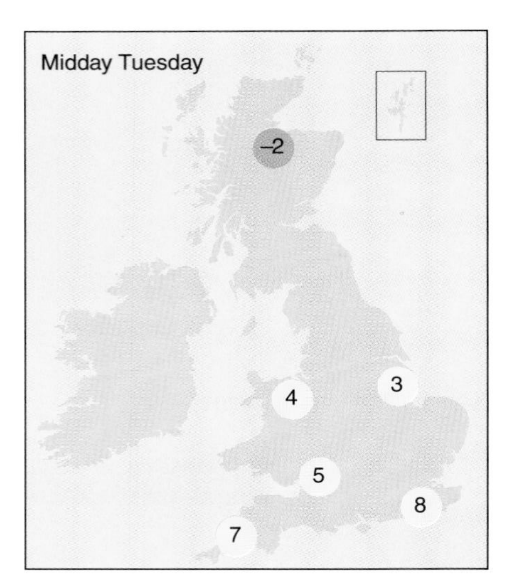

Here are two weather maps of the UK, showing different times on the same day.

Key fact
The symbol °C means 'degrees Celsius'.

1. What time of year do you think these maps were printed?
2. Which is the lowest temperature shown?
3. Which is the highest temperature shown?
4. Look at the chart for midday. What is the difference between the lowest and the highest temperatures shown?
5. Look at the chart for midnight. What is the biggest temperature difference shown?

Assignment 9 Temperatures around the world

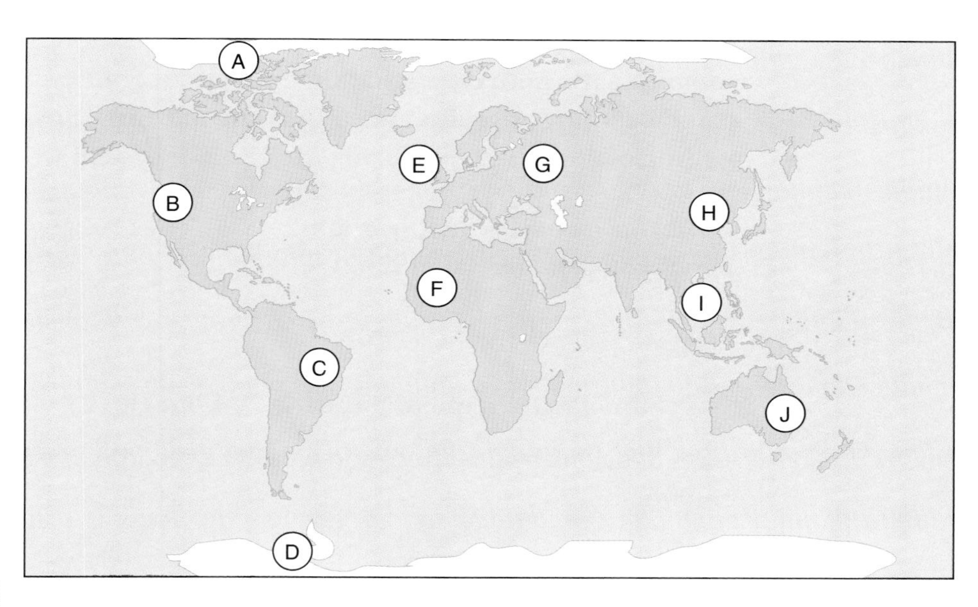

> **The temperature range is the difference between the highest temperature and the lowest.**

Using references such as a world atlas, find out the maximum and the minimum temperatures for the areas marked.

- In which area(s) is the temperature range greatest?
- In which area(s) is the temperature range smallest?

Exercise 6

1 Write down the temperatures marked A to E.

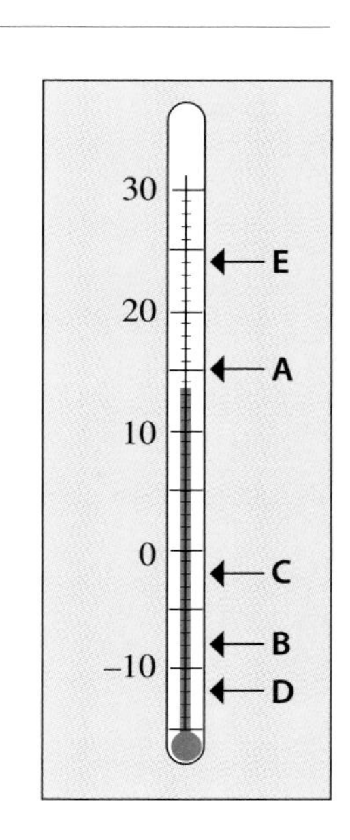

2 What is the difference between the highest temperature (E) and the lowest (D)?

3 Find the difference between each of the following pairs.

(a) E and A (b) A and C

(c) A and B (d) C and D

(e) A and D (f) C and E

(g) E and B (h) A and E

(i) B and C

Time

When we talk about time, we usually use the 12-hour system, but sometimes it is not clear what time we mean. For example 3 o'clock could be in the afternoon or early in the morning. So when we are writing or recording time using the 12-hour system we add:

a.m. (*ante meridian*) 'before noon'
p.m. (*post meridian*) 'after noon'

60 seconds = 1 minute
60 minutes = 1 hour
24 hours = 1 day

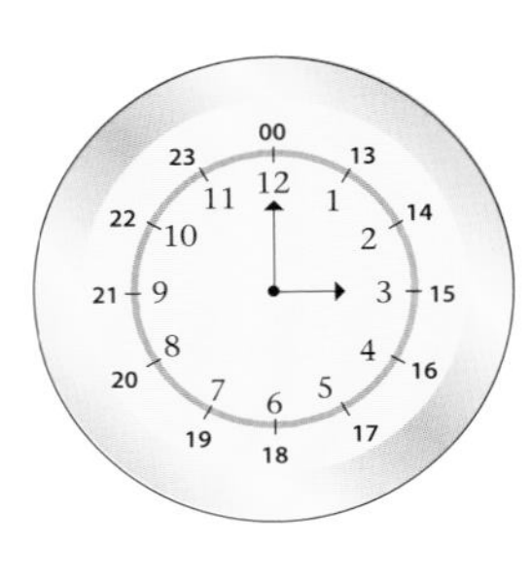

or we use the 24-hour system.

Travel timetables frequently use the 24-hour clock.

Thus 3 o'clock in the afternoon would be recorded as 3:00 p.m. or 15:00hrs.

Exercise 7

Copy this table into your book. On your copy, write down suitable times for the following activities in both the 12-hour and the 24-hour systems.

Event	12-hour	24 hour

1 breakfast

2 bed time

3 late night shopping

4 lunch time

5 delivering milk

6 doing homework

7 the dawn chorus

8 evening meal

Write the following times in both the 12-hour and 24-hour systems.

Time is written using a colon between hours and minutes. 02:15 means 'quarter past two'.

13 Copy the following table of equivalent times and complete it.

12-hour	24-hour
12:00 p.m.	
	15:45
2:30 p.m.	
11:10 a.m.	
	03:08
	21:40
	09:15
9:20 p.m.	
2:30 a.m.	
4:15 p.m.	
	00:00

Exercise 8

	BBC1		ITV
5:00	Newsround	4:45	Tiny Toon Adventures
5:10	Byker Grove	5:10	After 5
5:35	Neighbours	5:40	News
6:00	News	6:00	Home and Away
6:30	Regional News	6:30	Regional News
7:00	FILM: To Trap a Spy	7:00	Mork & Mindy
		7:30	Coronation Street
8:30	Commonwealth Games	8:00	Burke's Law
9:00	News	9:00	FILM: When Love Kills
9:30	Human Animal (Part 1)		
10:20	Commonwealth Games	10:00	News
2:00 a.m.	Close down	10:40	When Love Kills Part 2
		11:30	EuroMatch
		12.30	Local variations

1 How long is 'Neighbours'?

2 What is the length of the film 'When Love Kills'?

3 How much viewing time is given to sport here?

4 How much time altogether is spent on the news?

5 What are the lengths of:

(a) 'Byker Grove'
(b) 'Human Animal'
(c) 'Tiny Toon Adventures'?

6 Which tape would you use to record *all* the films?

Review Exercise

1 Write 273 cm in metres.

2 Write 15:35hrs in the 12-hour system.

3 A cyclist rode 10 miles. Approximately how far is this in kilometres?

4 Write these measures in order of size, largest first.

2.3 m   420 cm   1.95 m

5 How many minutes are there in 2 hours 40 minutes?

6 A journey began at 13:14 and ended at 15:36. How long did the journey take?

7 Write 3420 m*l* in litres.

8 What mass would I need to add to 834 g to make it up to a kilogram?

9 If you woke up in the night and looked at this clock, what time would it be? Write it in two ways.

10 Look at the label on the milk. How much fat would there be in 200 m*l* of milk?

11 Estimate the length of this line, in centimetres.

12 Write down the length marked by the arrow, in centimetres.

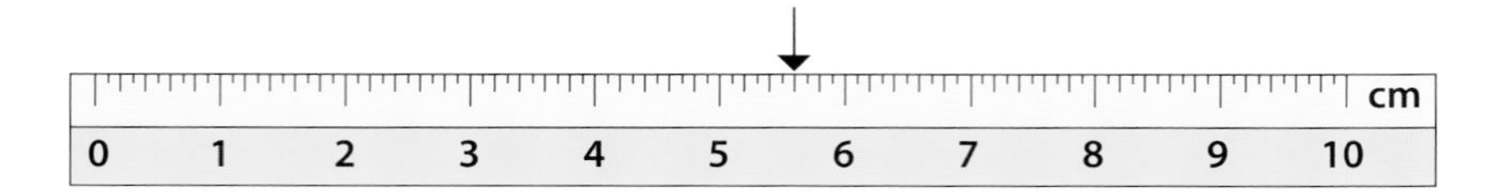

13 What is the difference between the maximum and the minimum temperatures shown?

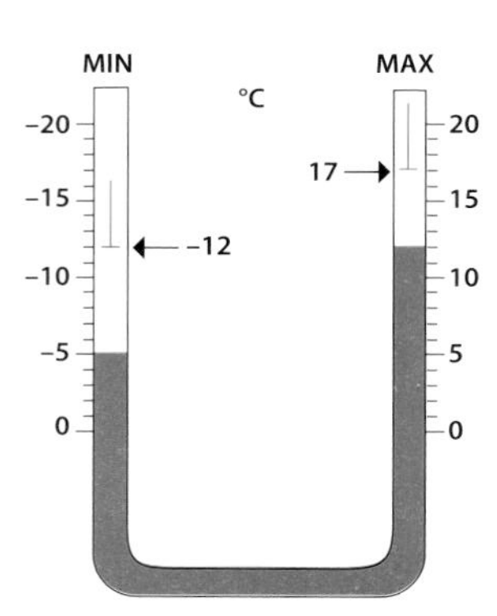

14

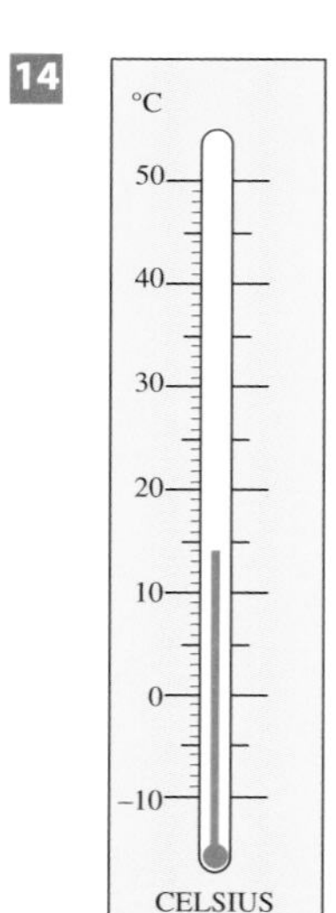

What temperature is shown on the thermometer?

15 When full, a tank holds 8 gallons. How many gallons are there in the tank now?

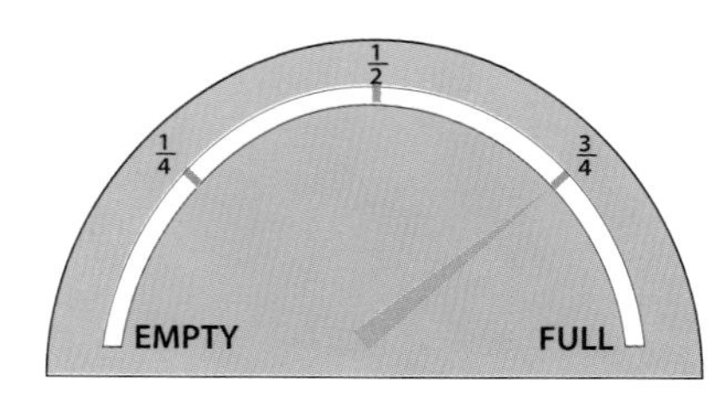

Symmetry

In this chapter you will learn:

- → about line symmetry
- → how to reflect shapes in a mirror line
- → about rotational symmetry

Starting points

In this picture all the creatures have something special about them.

Write down what you notice. Use a mirror to help you.
Discuss your findings with a partner.

Line symmetry

Exercise 1

Key fact
A picture is symmetrical if a line can be drawn on the shape so that one half is the mirror image of the other half.

1

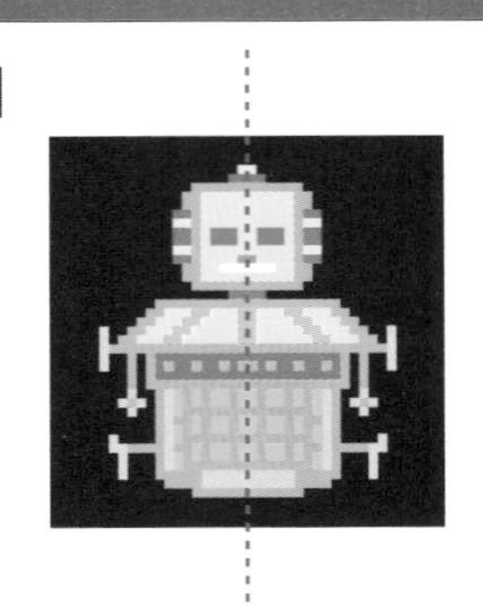

- Look at the face of this creature. Put a mirror along the dotted line. You should see that the face does not change.

The face has **reflection** or **line symmetry**.
The dotted line is the **line of symmetry**.

You can check for line symmetry using tracing paper.

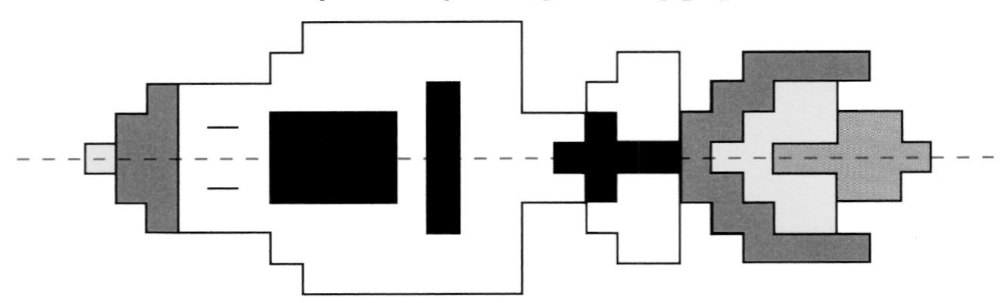

- Trace the picture and fold along the dotted line.
- Write down what you notice.
- Does the picture have reflection symmetry?

2 Which of these creatures have reflection symmetry?

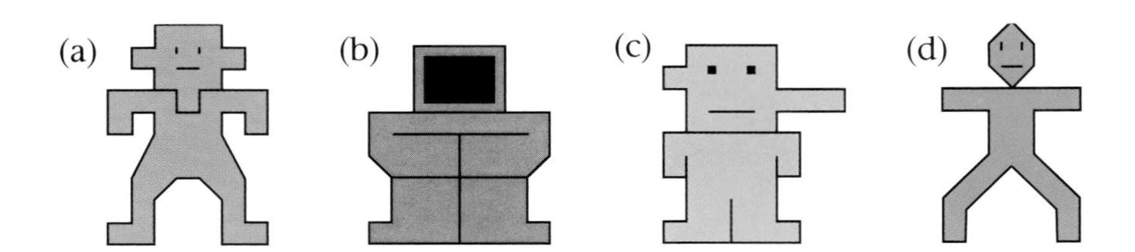

3 Here are some flags.
Which of these flags have a line of symmetry?

(a) Sweden

(b) France

(c) Morocco

(d) Barbados

4 Here are some pictures of everyday objects.
Which of these objects are symmetrical?

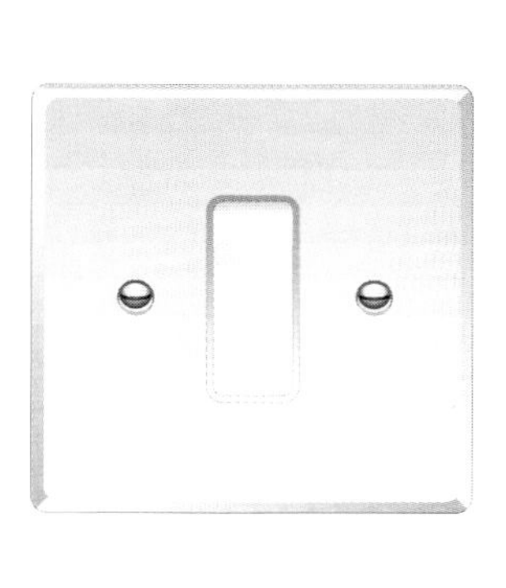

(a)

(b)

(c)

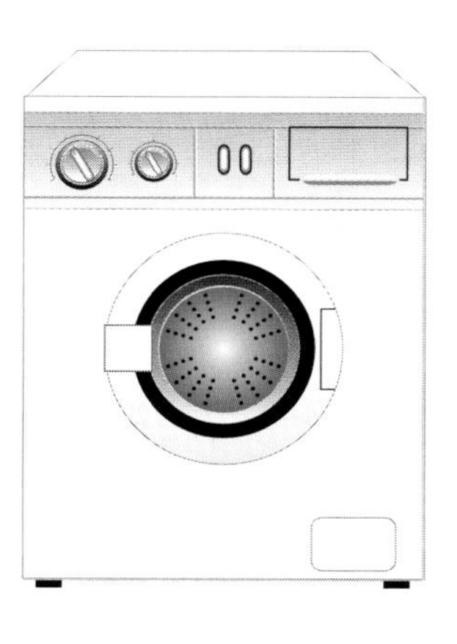

(d)

5 These symbols are used on maps.
Which of the symbols have reflection symmetry?

(a)

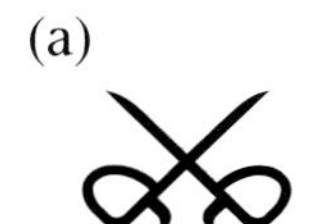

(b)

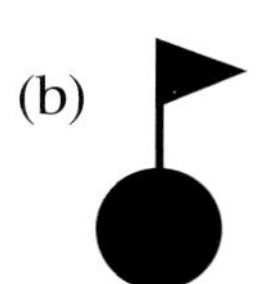

(c)

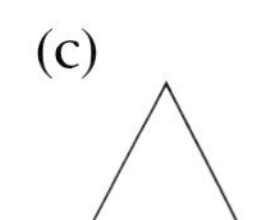

(d)

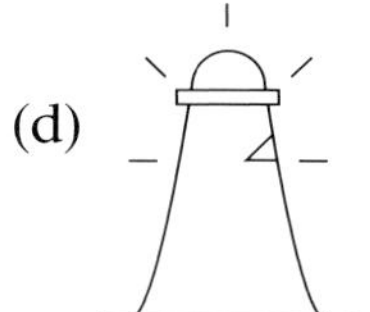

(e)

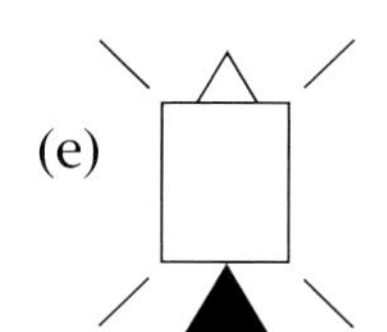

(f)

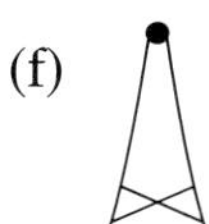

6 (a) Trace this pattern.
By folding, check that the dotted line is a line of symmetry.

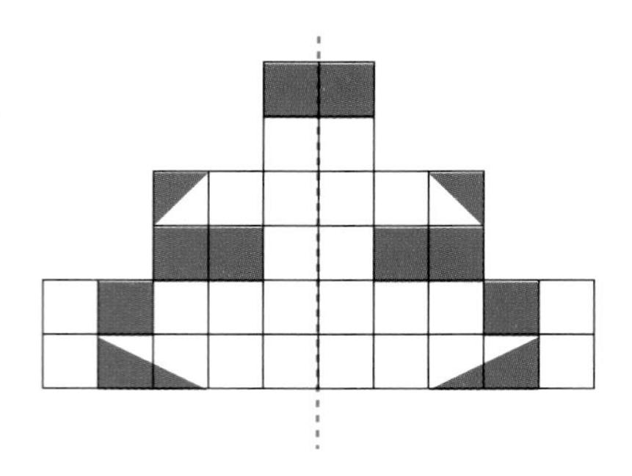

(b) Find which of these patterns have line symmetry.

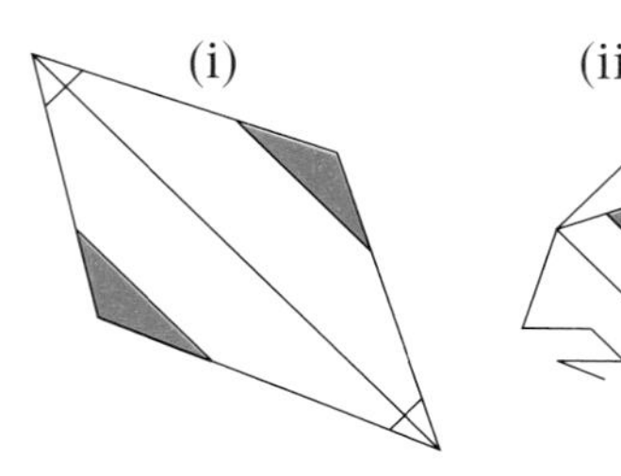

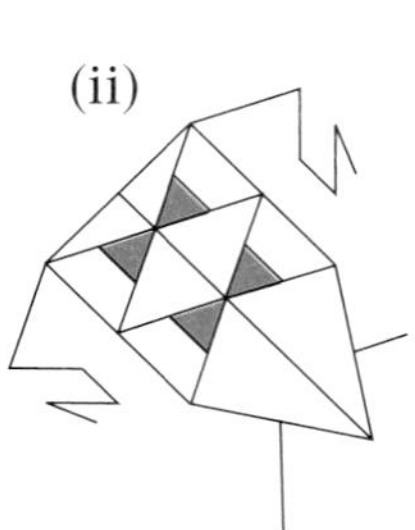

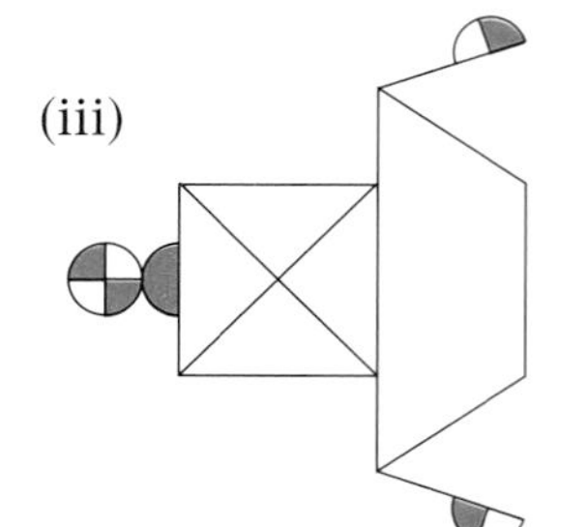

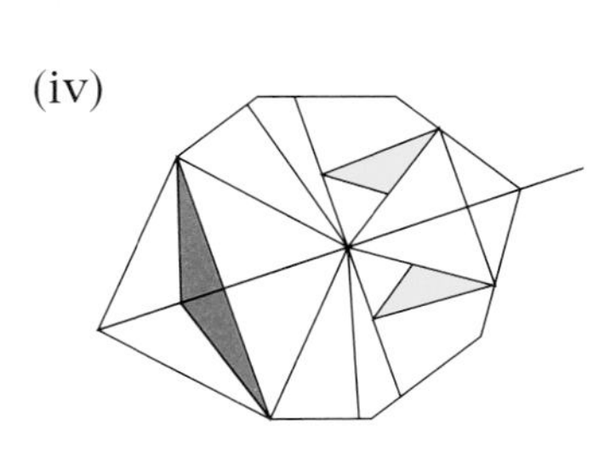

7 Copy these patterns onto squared paper and colour or shade some more squares so that the green line is a line of symmetry.

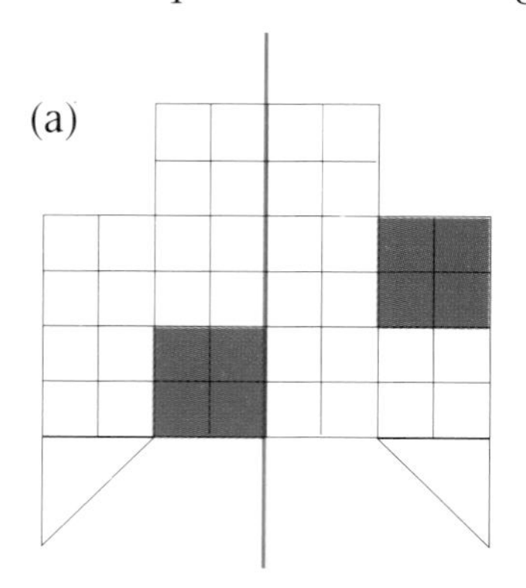

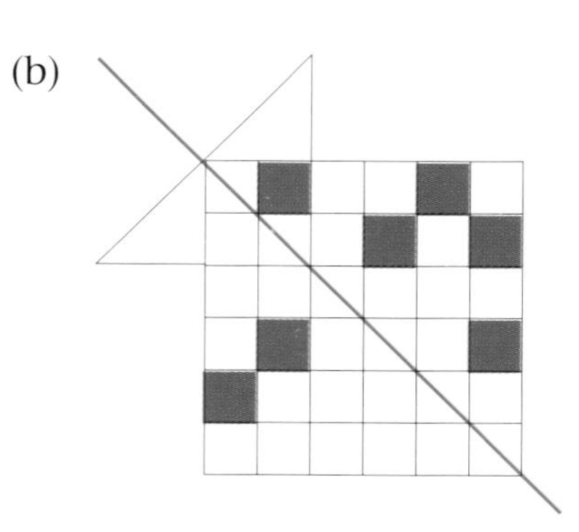

8 Using graph paper or a computer drawing program, make up your own symmetrical creature. If you use colours, make sure you colour the picture so that it stays symmetrical.

Draw the line of symmetry on your picture.

More than one line of symmetry

Exercise 2

1 Sometimes a picture may have more than one line of symmetry. Check that this pattern has two lines of symmetry.

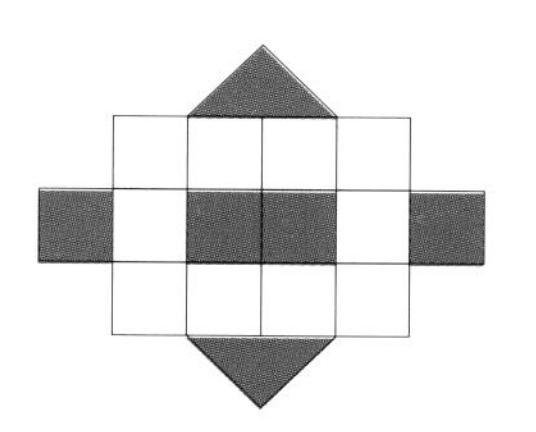

2 All the following shapes and patterns have two lines of symmetry. Trace or copy the shapes and draw in dotted lines to show the lines of symmetry.

(a)

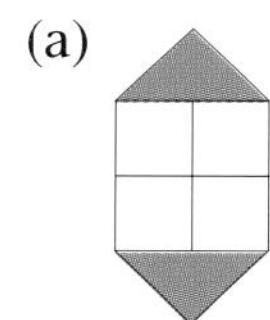

(b)

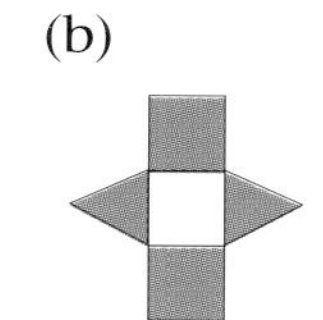

(c)

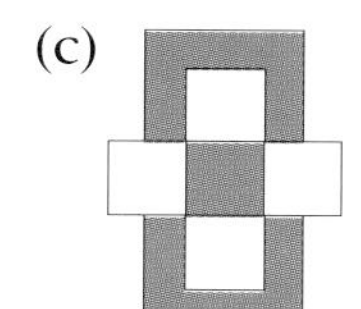

(d)

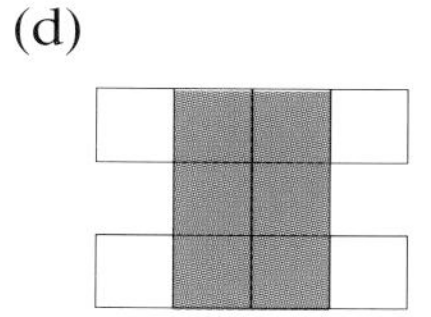

3 Here are some flags. Which flags have exactly two lines of symmetry? Look carefully at the colours.

(a) Japan

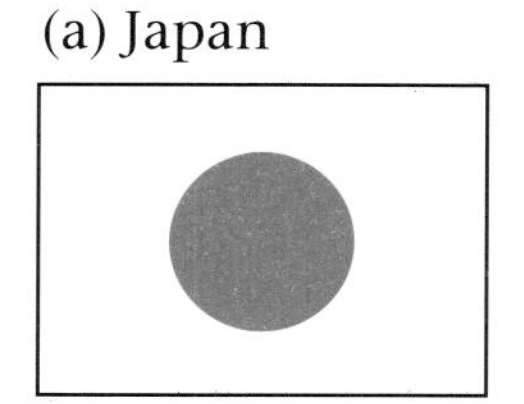

(b) Israel

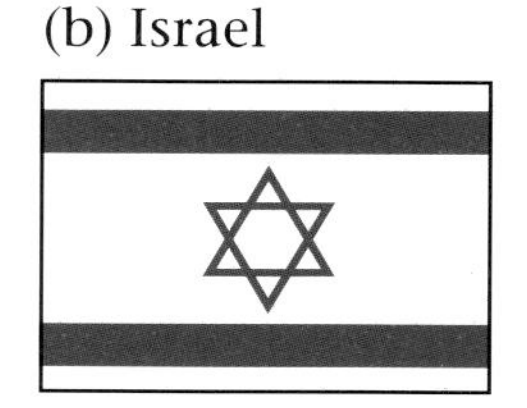

(c) Guinea

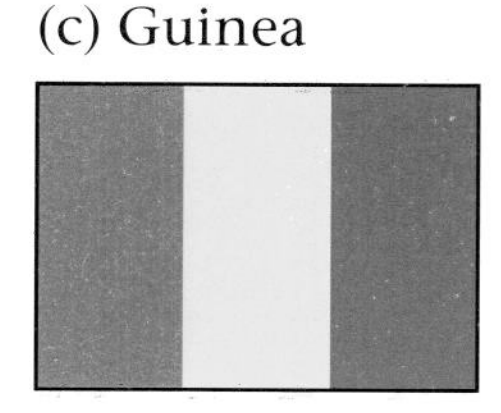

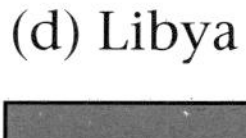

(d) Libya

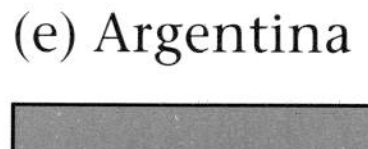

(e) Argentina

(f) Poland

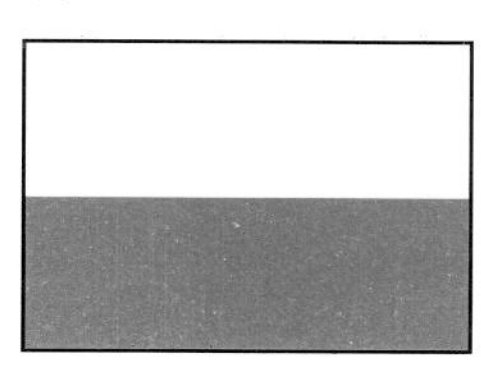

4 A square has four lines of symmetry. Draw a square with sides 4 cm and draw dotted lines to show all the lines of symmetry.

5 Which one of these flags has four lines of symmetry?

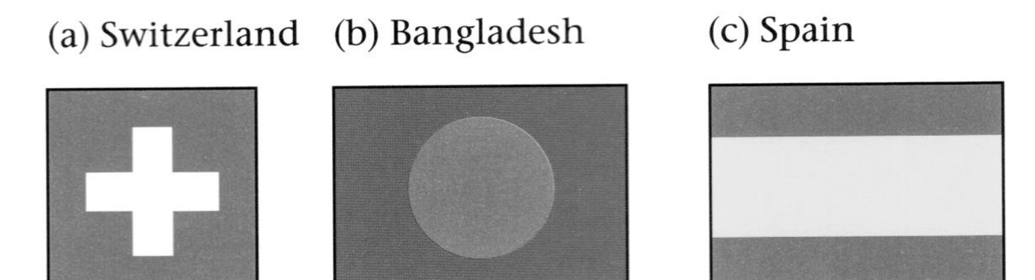

6 One of these patterns has four lines of symmetry.
Which one is it?

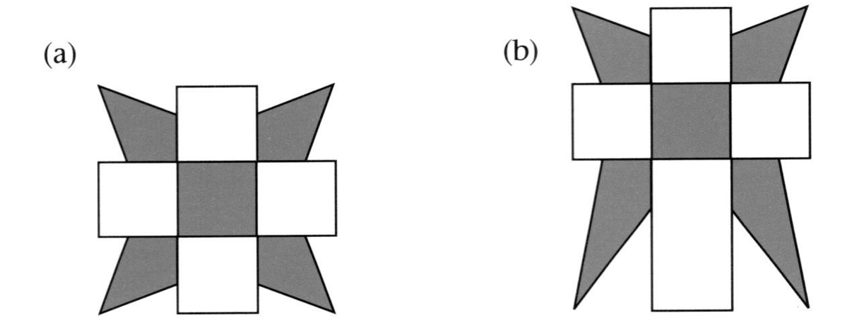

7 Copy these patterns on to squared paper and colour them with at least three colours so that they have exactly four lines of symmetry.

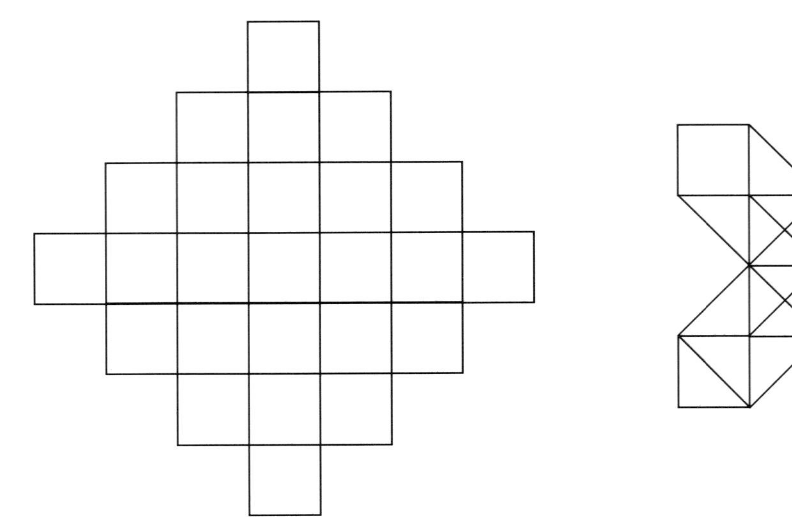

8 This pattern has four lines of symmetry. It is built from a square.

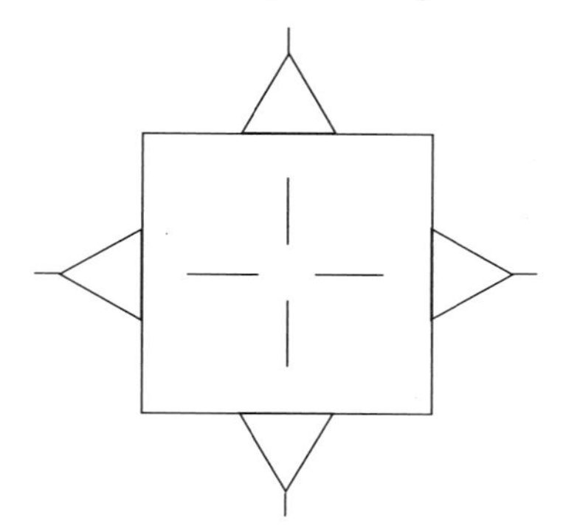

- Draw a square with sides 5 cm. Design your own pattern with four lines of symmetry based on the square.
- Draw dotted lines to show the lines of symmetry.
- Colour your pattern using as many colours as possible, making sure it still has four lines of symmetry.

Key fact

An equilateral triangle has three equal sides and three equal angles.

9 Here is an equilateral triangle.

(a) Without using a mirror or tracing paper, write down how many lines of symmetry you think the triangle has.

(b) Trace the triangle. Fold your tracing paper to show all the lines of symmetry.

(c) Write down what you notice.

10 How many lines of symmetry has each of the following shapes and patterns?

Use tracing paper to show the lines of symmetry.

(a)

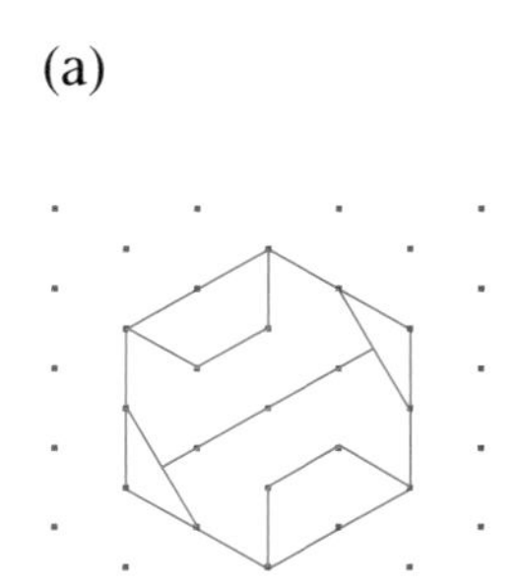

(b)

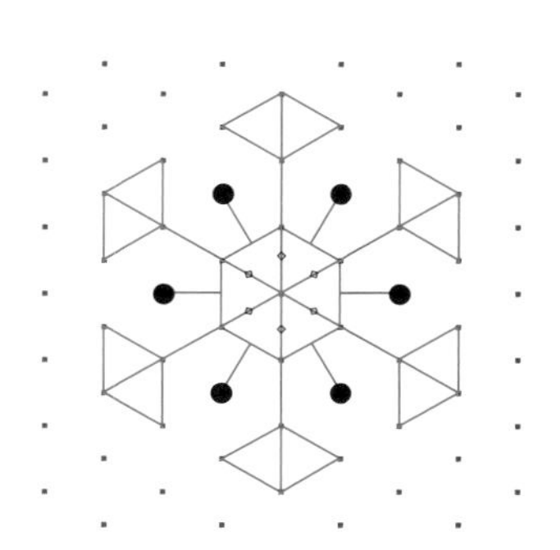

(c)

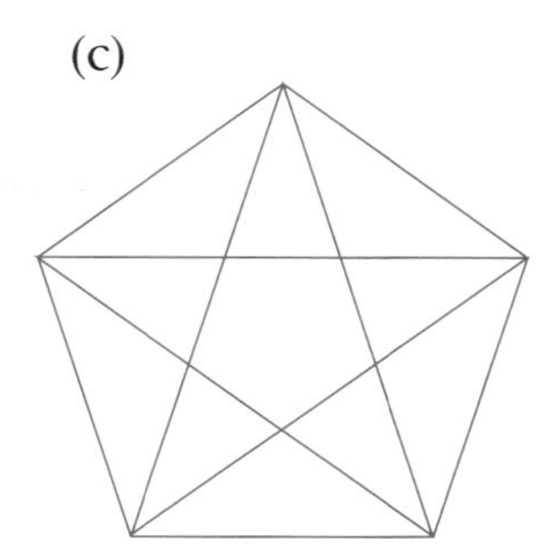

11 This pattern is based on a regular hexagon.

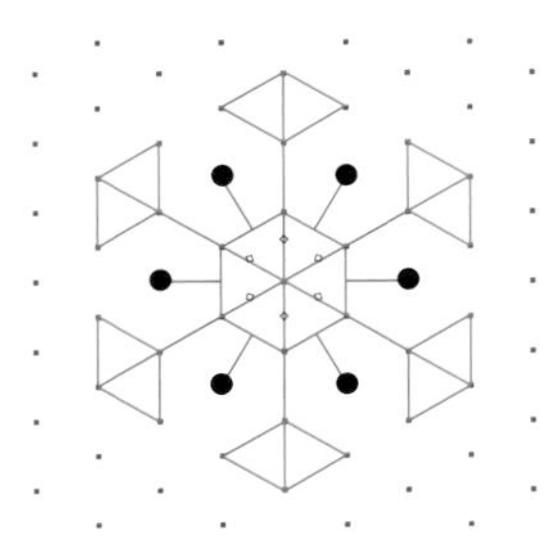

Remember:
A regular shape has equal sides and equal angles.

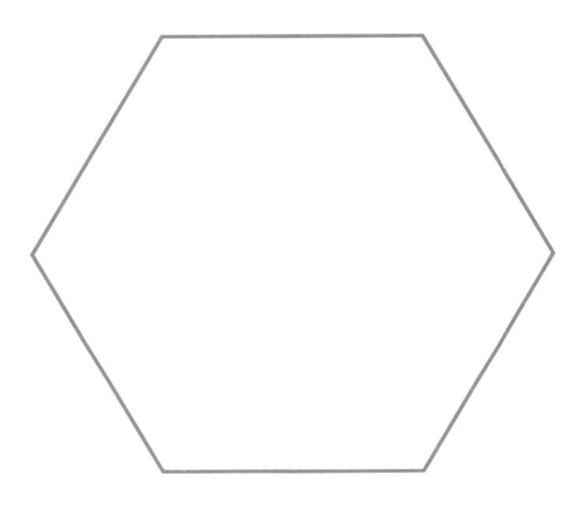

- Trace this hexagon.

  Draw on your tracing all the lines of symmetry.
- Add some more lines and shapes to create your own pattern with six lines of symmetry.

Drawing reflections

Exercise 3

1 In this diagram, the dotted line is a line of symmetry but the pattern is not complete.

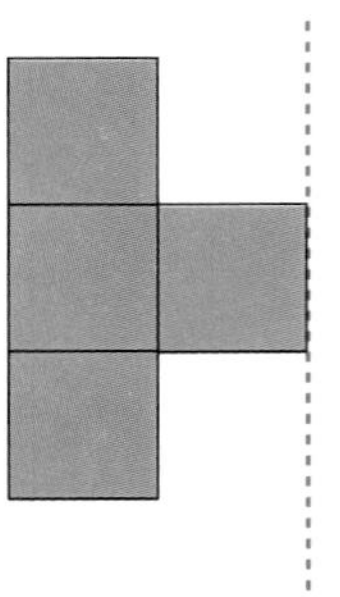

Here is how the diagram can be completed.

- Place a mirror on the dotted line and look at the reflection of the pattern in the mirror.
- Copy the diagram and its mirror image to complete the pattern.

You have drawn a reflection of the original diagram using the line of symmetry as a **mirror line.**

2 Copy each of these diagrams onto squared paper and complete them using the dotted line as a mirror line.

(a)

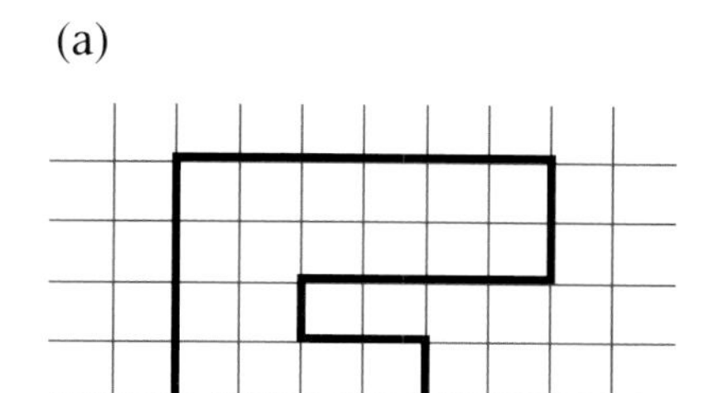

(b)

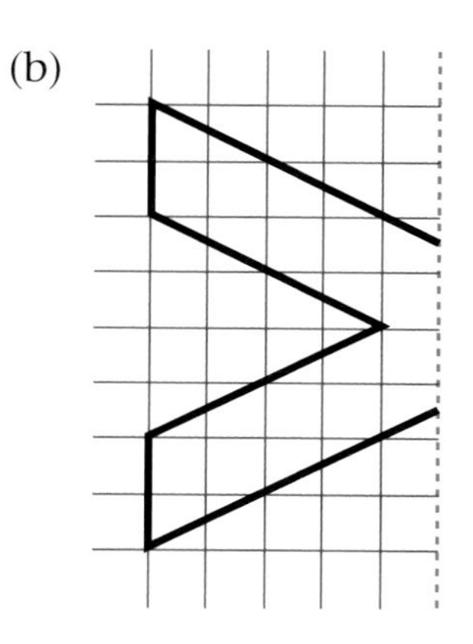

3 Symmetrical patterns can be made by reflecting points in a mirror line. When the dots are joined up, a symmetrical shape is created.

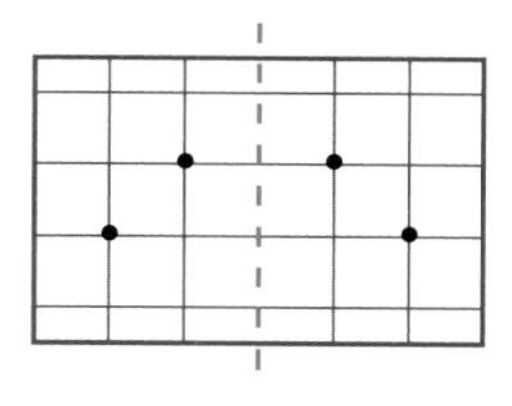

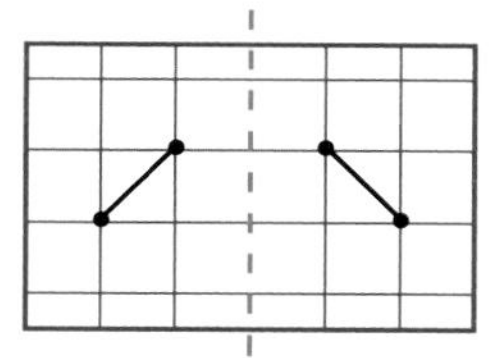

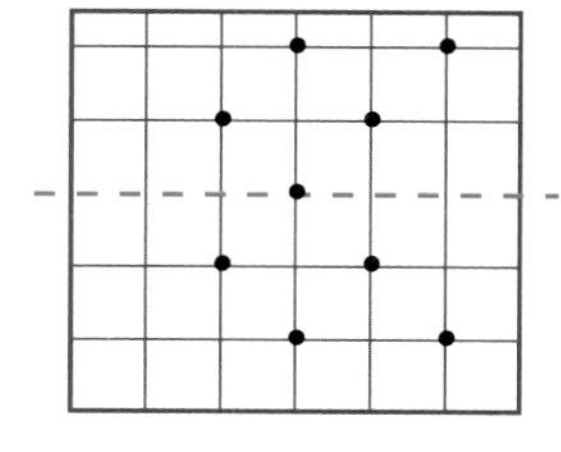

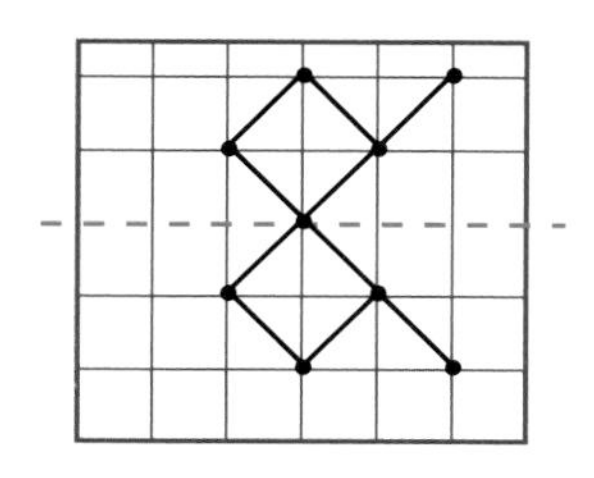

Copy each of these diagrams onto squared paper and reflect the points in the mirror line. Do *not* join up the points.

(a)

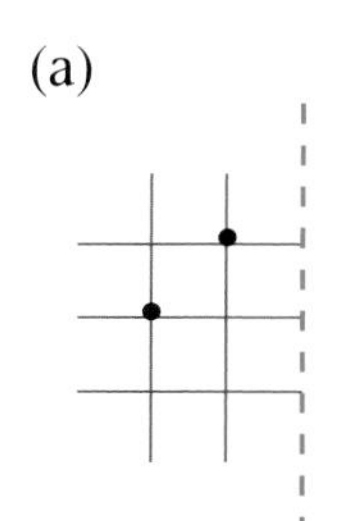

(b)

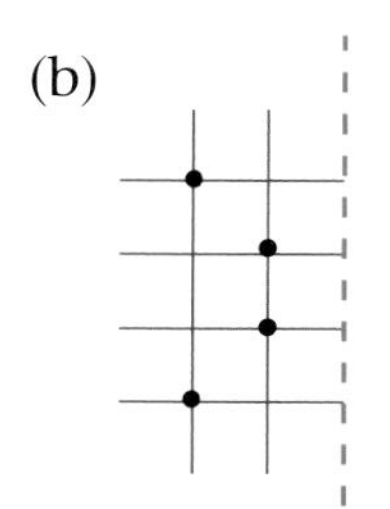

(c)

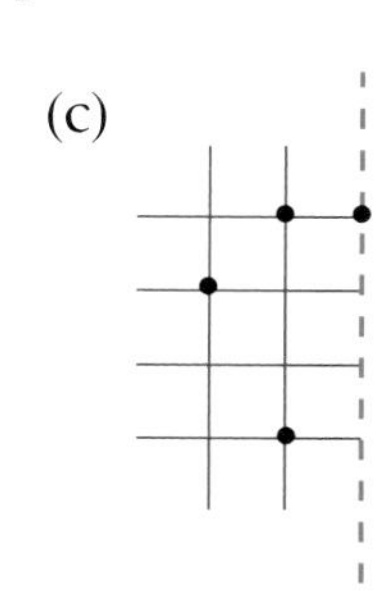

4 Copy each of these diagrams onto squared paper. Reflect the dots in the mirror line and join them up to make a symmetrical pattern.

(a)

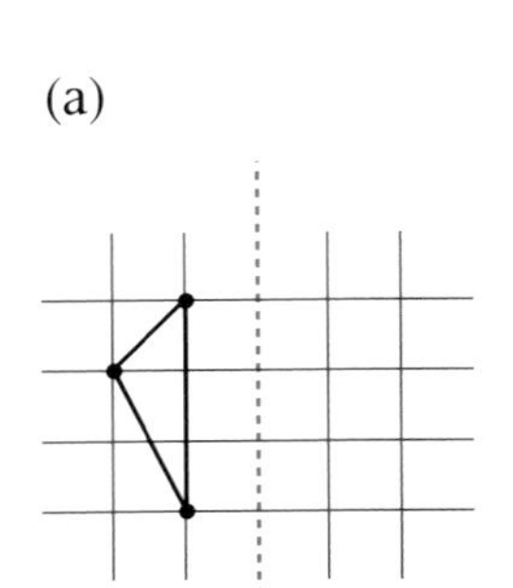

(b)

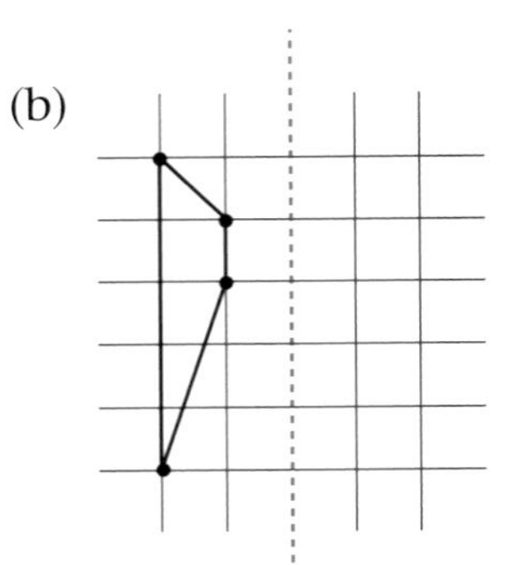

(c)

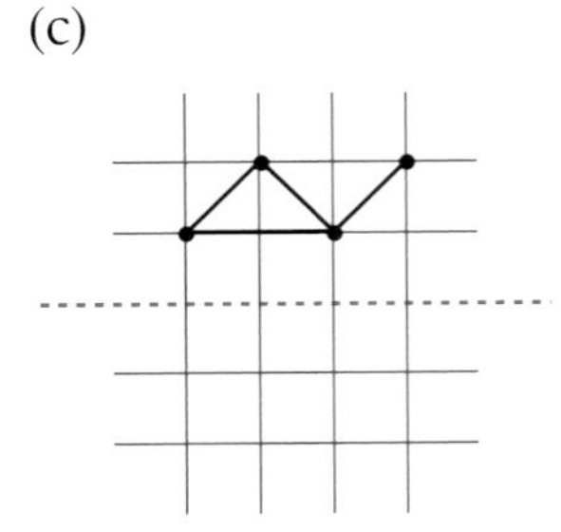

(d)

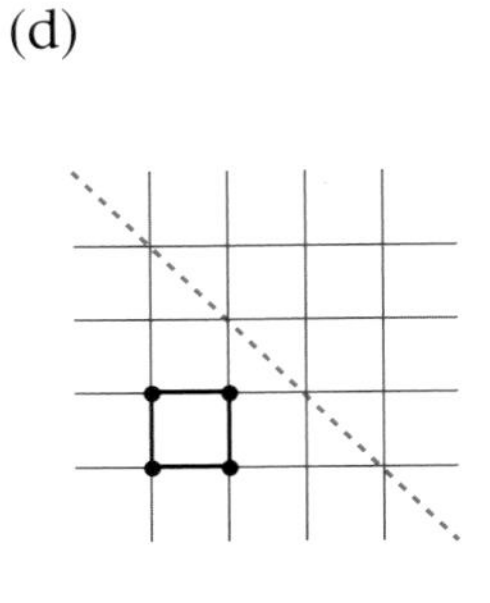

5 Copy these diagrams onto triangular dotted paper.
In each case reflect the dots in the mirror line and join them up to make a symmetrical pattern.

(a)

(b)

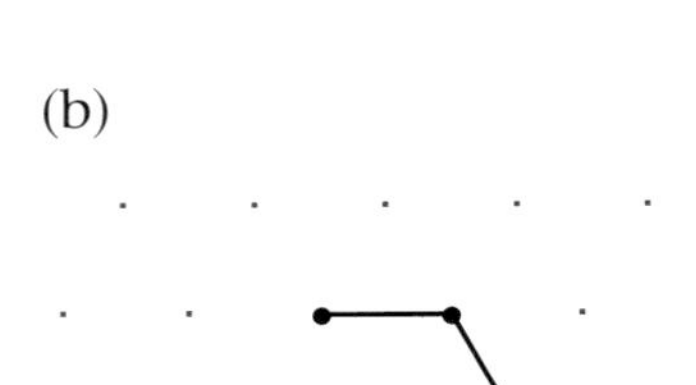

(c)

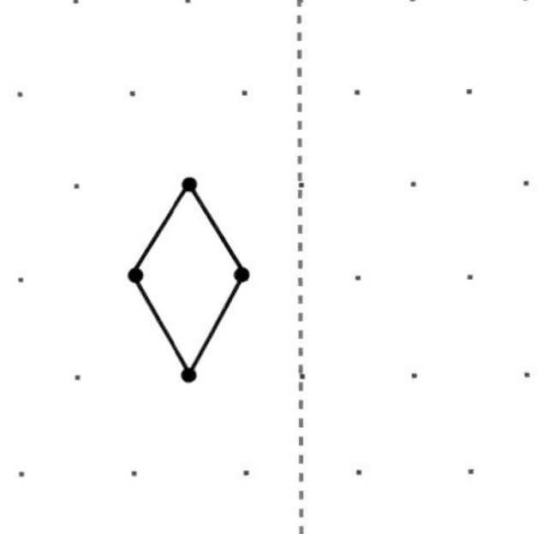

(d)

(e)

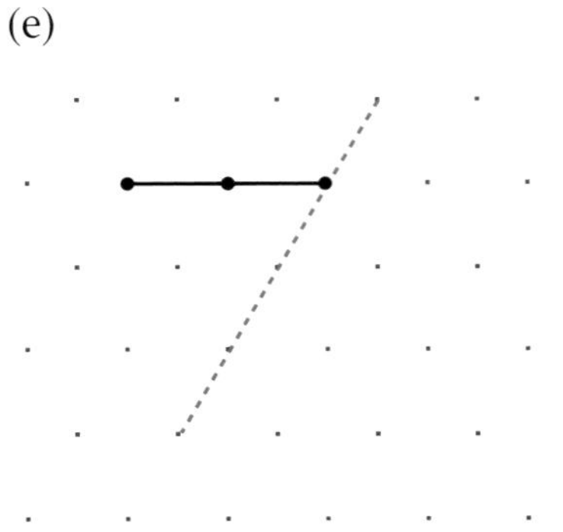

If it helps, put 'dots' on the shape first.

6 Copy these diagrams onto squared paper and draw the reflection of the given shape in the mirror line to complete the picture.

(a)

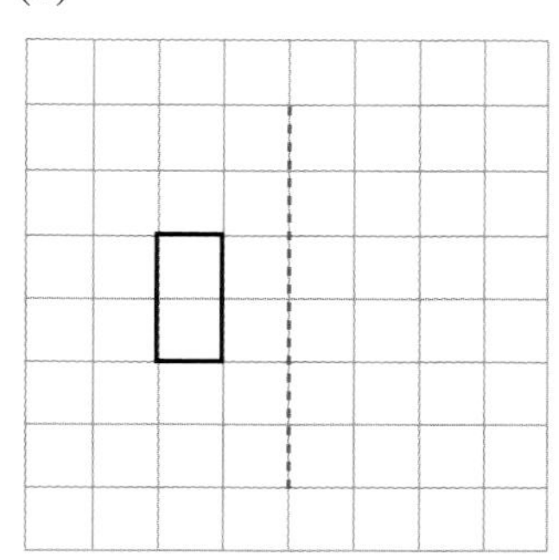

(b)

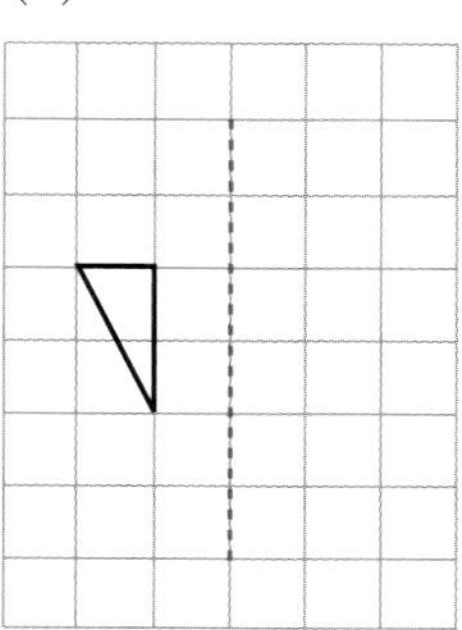

(c)

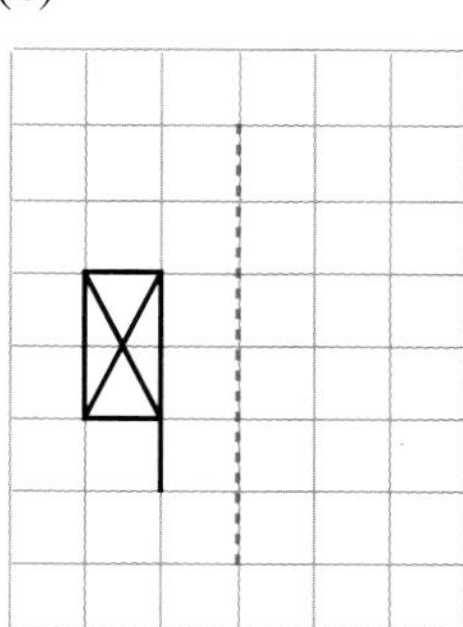

(d)

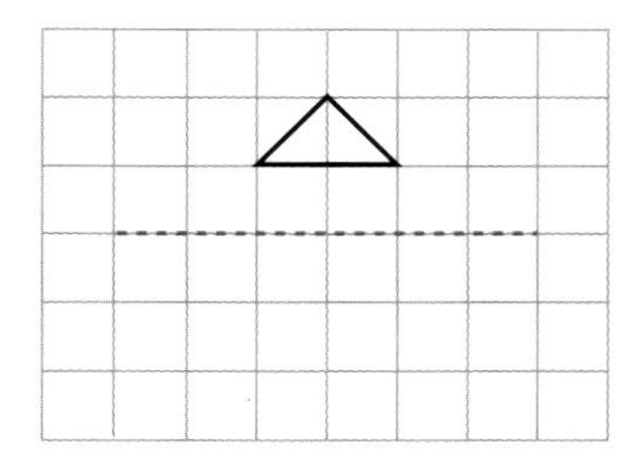

(e)

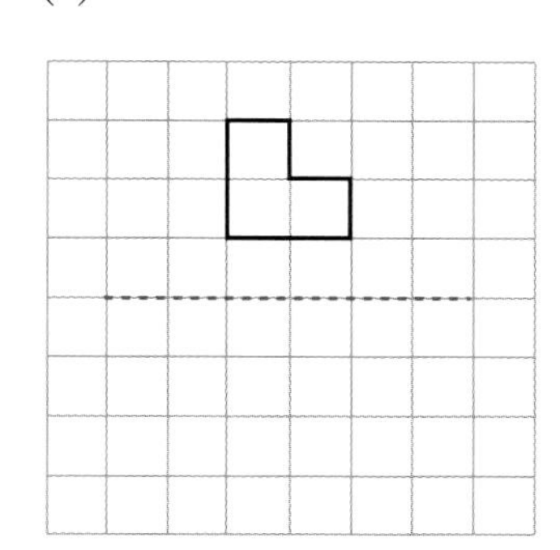

(f)

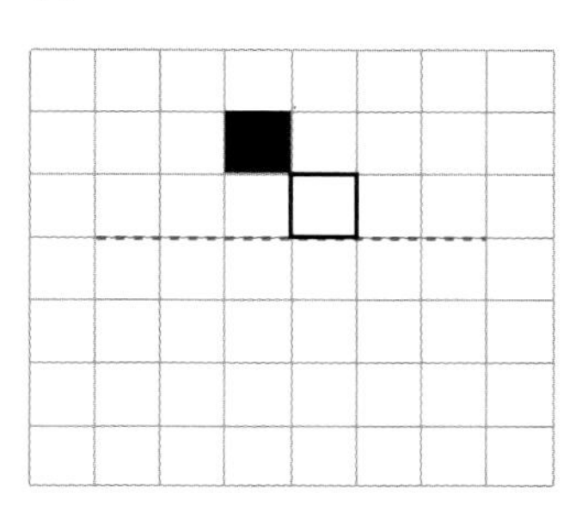

(g)

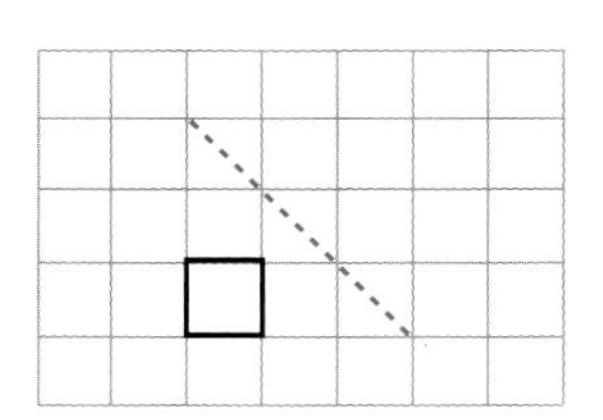

(h)

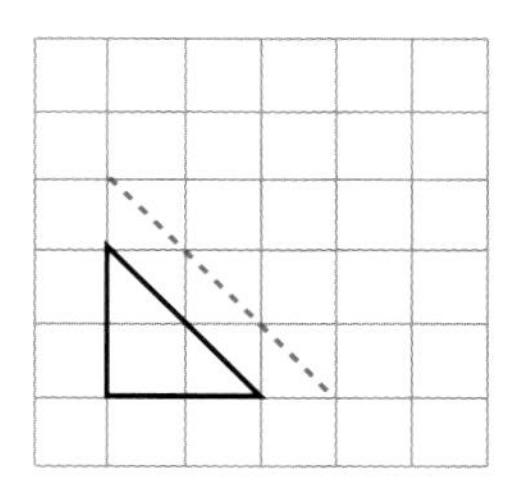

(i)

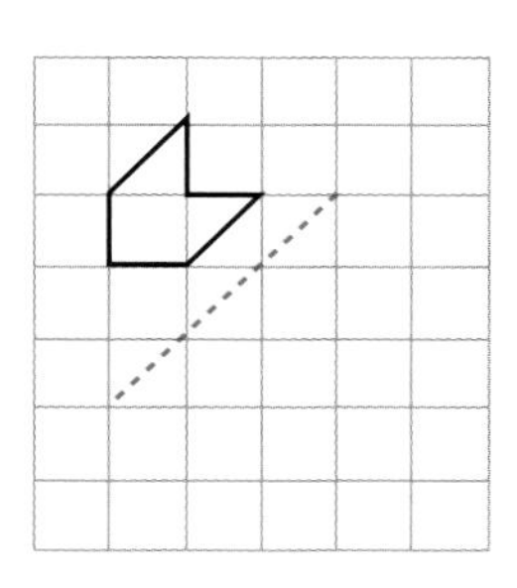

7 Copy this diagram onto squared paper. Colour it using at least three colours.

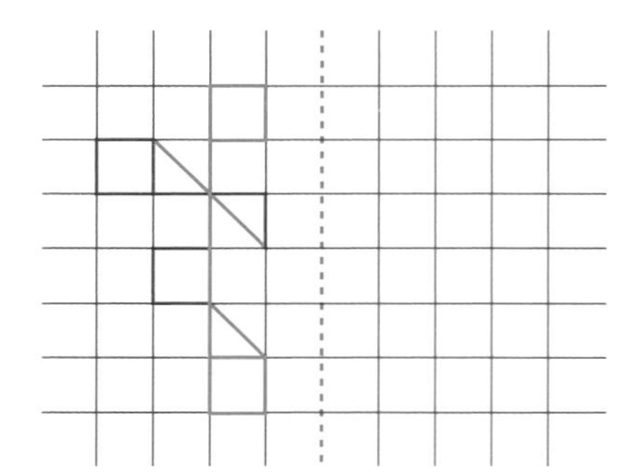

Reflect the pattern in the mirror line. Be careful to make sure your final picture is symmetrical.

8 Make up a pattern yourself on triangular dotted paper and colour it. Draw your own mirror line and reflect your pattern in the line.

Assignment 1 Pattern reflections

Make six copies of this diagram.

On each diagram, join up the five dots so that each dot is joined to at least one other.

Make six different patterns.

Reflect each pattern in the mirror line.

Colour your patterns, making sure the mirror line is still a line of symmetry in each case.

Example

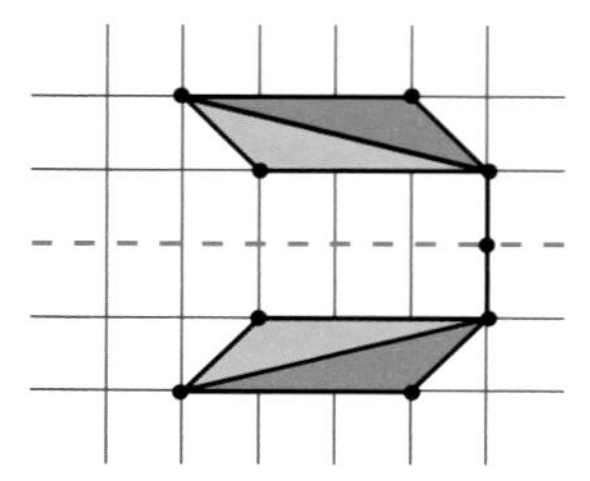

Assignment 2 Word reflections

You need a partner and a copy of the triangle sheet for this assignment.

- Complete the reflections of each letter in the word **TRIANGLE**.
- Cut along the dotted lines.

You can show the word **TRIANGLE** and its reflection in several ways, for example:

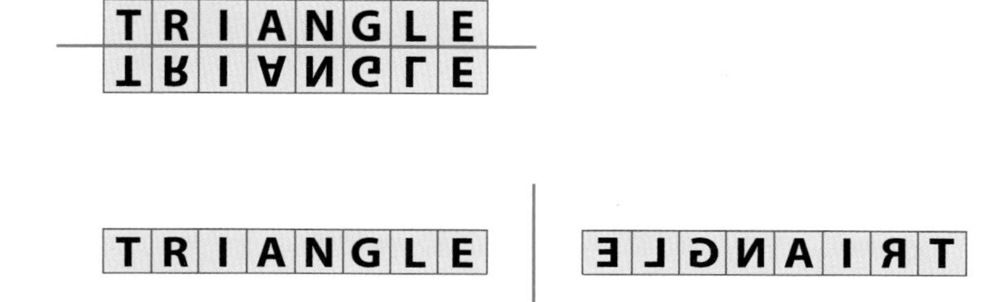

- Check that the dotted line is a line of symmetry in each case.

You can make other words from the word **TRIANGLE**,
for example: RAIN

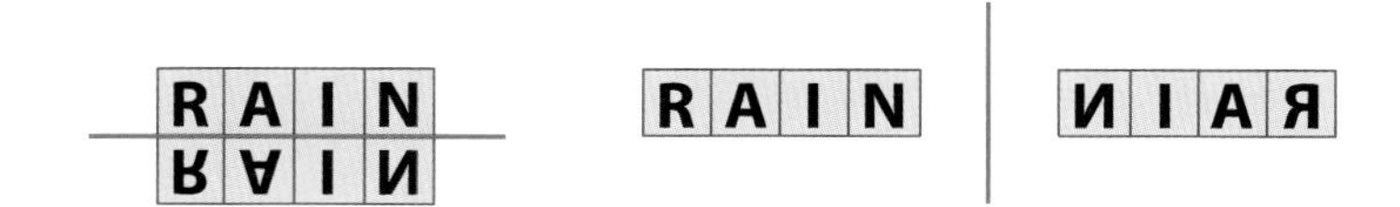

- Use the cards to give your partner a word made up from the letters in **TRIANGLE**, for example: L E G

  Your partner must use the reflected letters to give the reflection of the word in two ways.

Example

- Take it in turns to give each other a word until you have found at least six words.

Rotational symmetry

Exercise 4

Look at these two flags.
They look the same upside down.

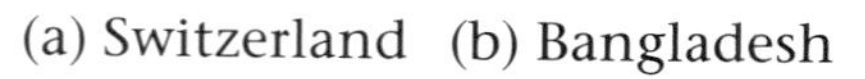

1 Here are some more flags.

(a) Thailand

(b) Peru

(c) Kenya

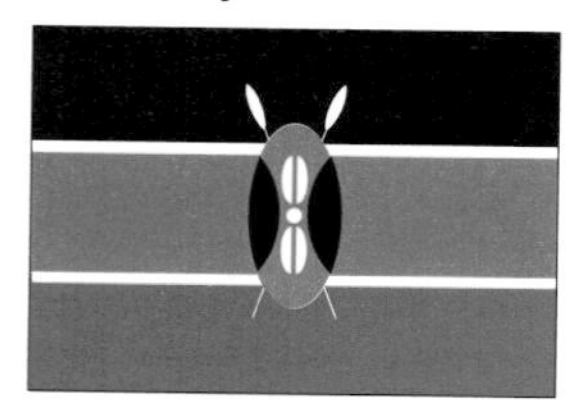

(d) Dominican Republic

(e) Honduras

(f) Austria

- Which of them look the same when they are turned upside down? Sometimes it is difficult to tell.

This is the flag of the State of Maryland in the United States of America.

This is the flag upside down.

Does this flag look exactly the same upside down? Look carefully.

Key fact
Rotational symmetry occurs when a pattern fits on top of itself as it is rotated about a point which is the centre of the rotation.

2 Here is how to check if a pattern is the same upside down.

- Trace this pattern.

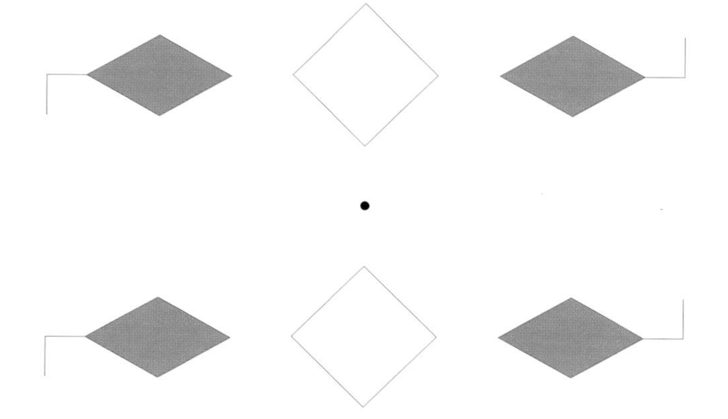

Key fact
The order of rotational symmetry is the number of times the pattern fits.

- Put a sharp pencil on the centre point.
- Turn the tracing round until it fits over the pattern again.

The tracing fits when it is upside down, so there are two positions where the tracing fits over the pattern.

The pattern has **rotational symmetry of order 2.**

The point where you put your pencil is called the **centre of rotation**.

Key fact

Every shape has rotational symmetry of at least order 1.

3 Look at this equilateral triangle.

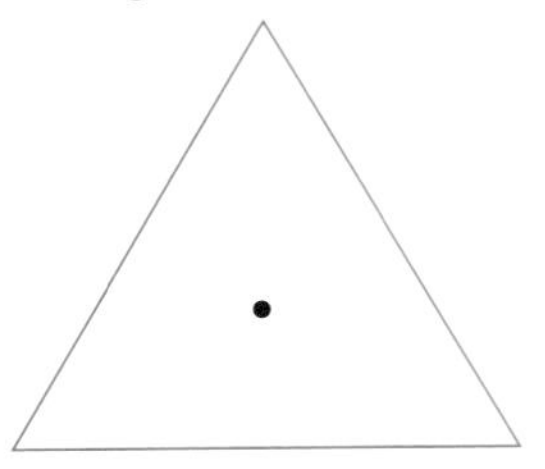

- Trace the triangle.
- Put a sharp pencil on the marked centre and rotate the tracing.
- How many positions can you find where the tracing fits exactly over the original shape?
- What order of symmetry does this triangle have?

4 Look carefully at this picture of the Union flag.

Check that the Union flag has rotational symmetry.

(a) What is the order of rotational symmetry?

(b) Does the Union flag have line symmetry?

(c) Now look at this picture of the Union flag.
What do you notice?

5 Which of these patterns have rotational symmetry of order 2?

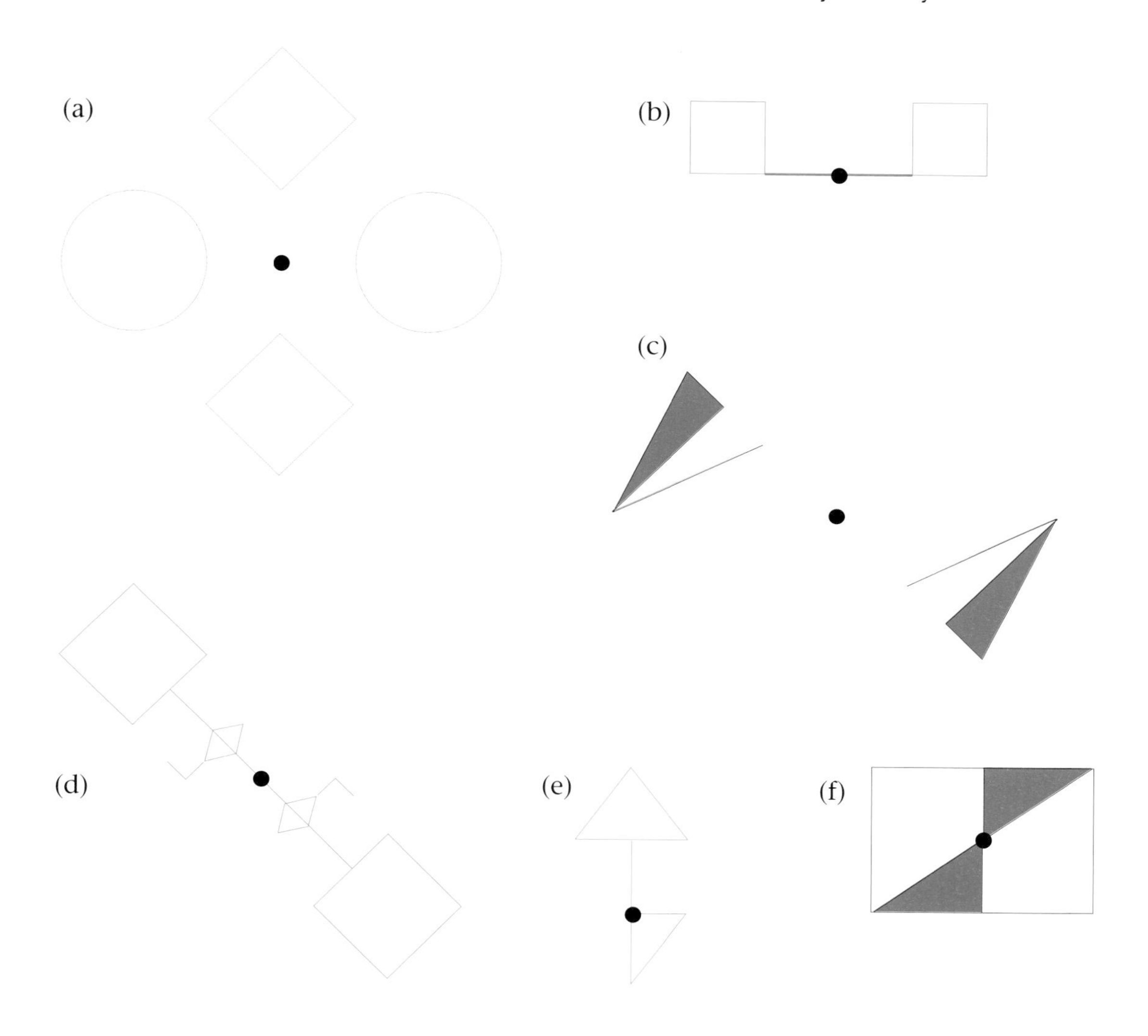

6 Copy these drawings onto squared paper. Add some shading so that the patterns have rotational symmetry of order 2.

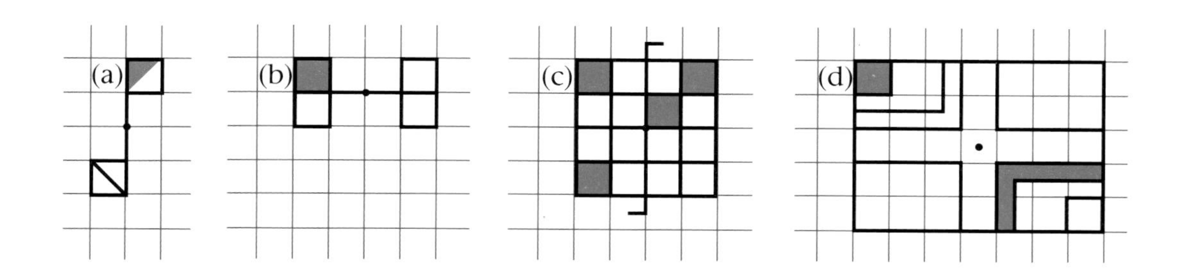

7 Copy each of these drawings onto squared paper.
Add another line to each drawing so that it has rotational symmetry of order 2.

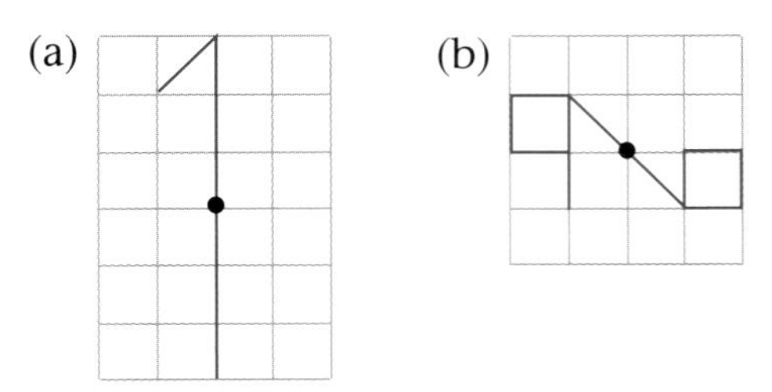

8 (a) Draw the letter Z and use tracing paper to show it has rotational symmetry of order 2.
(b) Draw some other capital letters with rotational symmetry of order 2.

9 Look at the letters in the word **TRIANGLE**.

(a) Which letters have rotational symmetry of order greater than 1?

(b) What is the order of rotational symmetry in each case?

10 Four of these patterns have rotational symmetry of order 3.
Use tracing paper to find them, and then draw them.

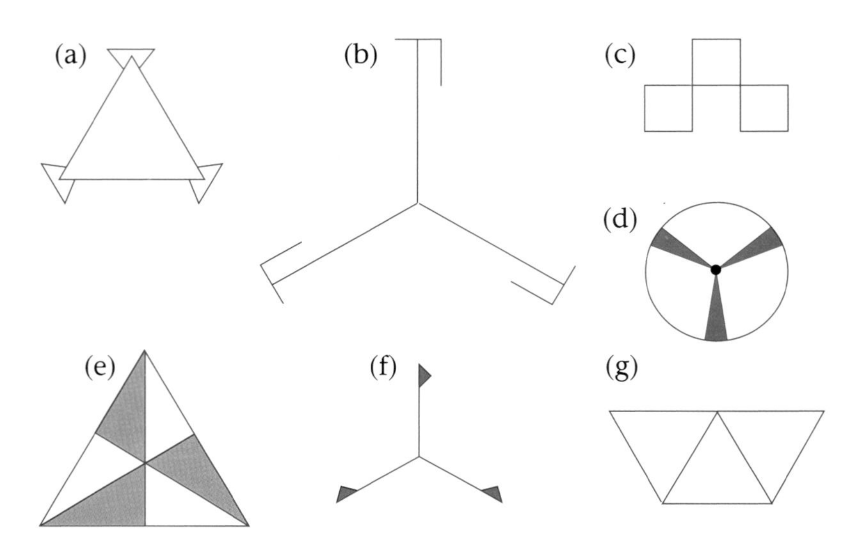

11 Write down the order of rotational symmetry for each of the following patterns.

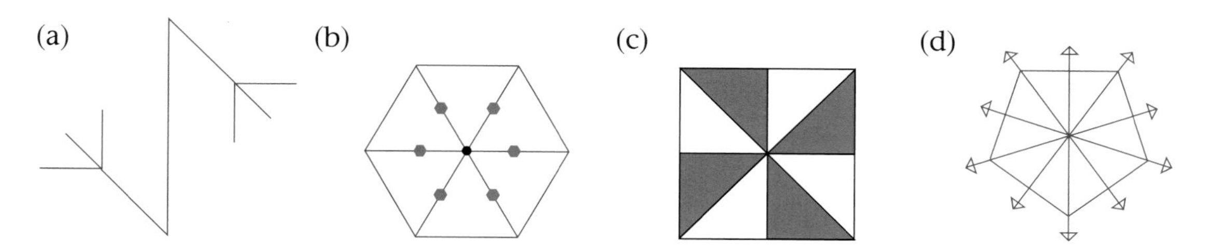

12 Copy and complete these drawings so that they have rotational symmetry of order 4.

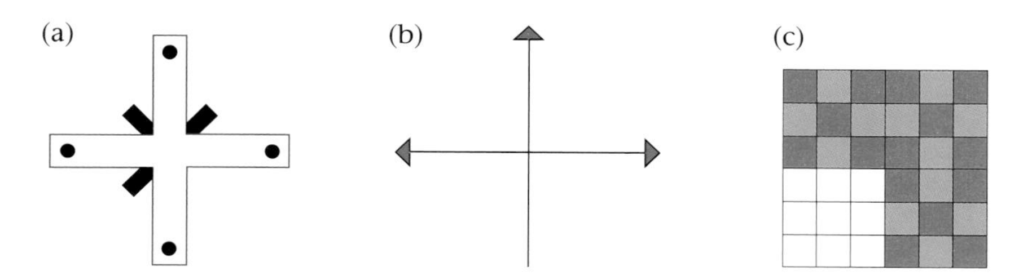

Assignment 3 Rotating patterns

This pattern has been made by rotating an arrow around a point and drawing it at regular intervals.

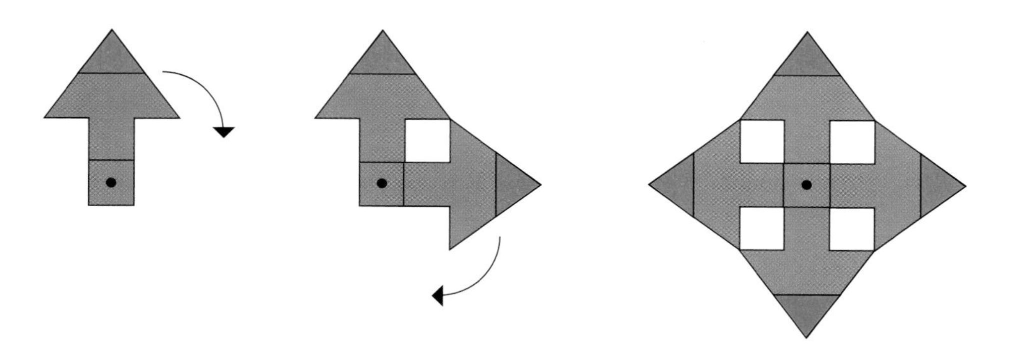

- What is the order of rotational symmetry of this pattern?

- Make a pattern similar to this. Copy or trace one of these shapes onto some coloured paper.

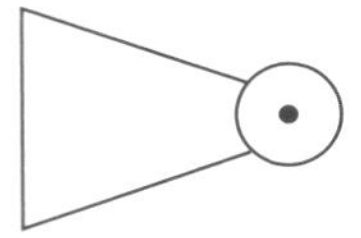
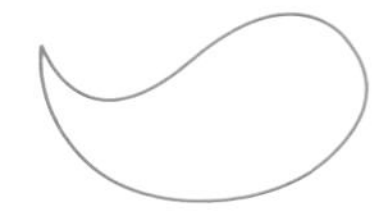

By folding the paper, cut out four copies of the same shape.

Glue your cut-out shapes to make a pattern with rotational symmetry of order 4. Remember to space the shapes evenly.

- Cut out some more shapes and make some more patterns with rotational symmetry of different orders. You could put lettering in the shape to make it more interesting.

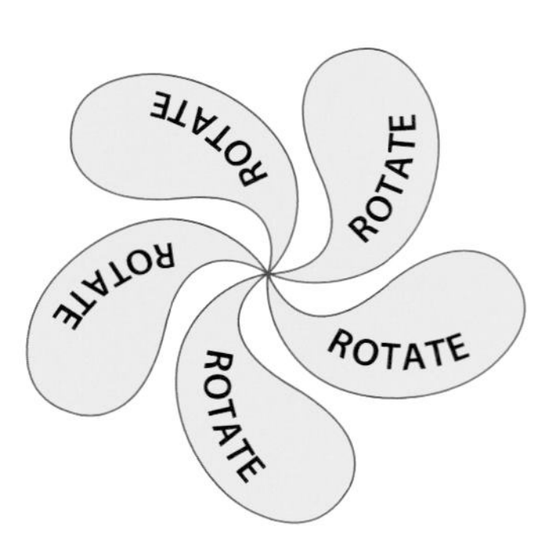

Review Exercise

1 Make three copies of this diagram.

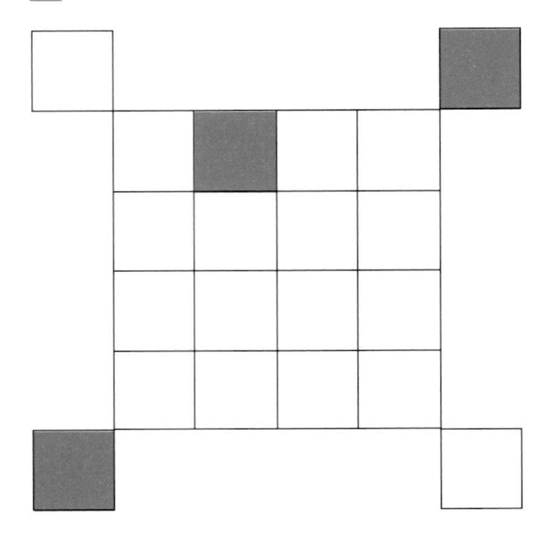

Shade some more squares so that:

(a) the first diagram has one line of symmetry

(b) the second diagram has two lines of symmetry

(c) the third diagram has four lines of symmetry.

2 Copy each of these diagrams onto squared paper. Draw the reflections of each shape in its mirror line, to make a symmetrical pattern.

(a)

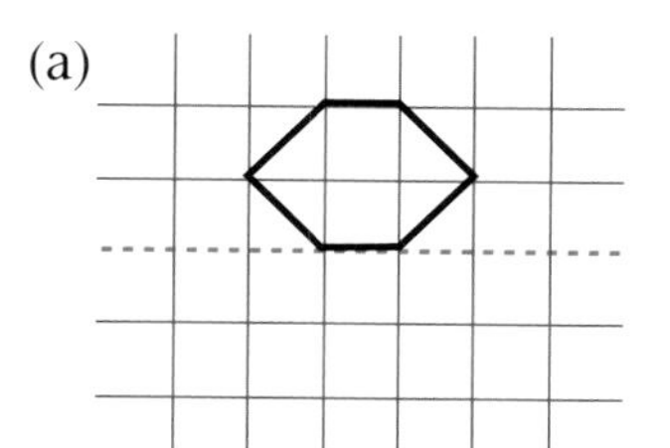

(b)

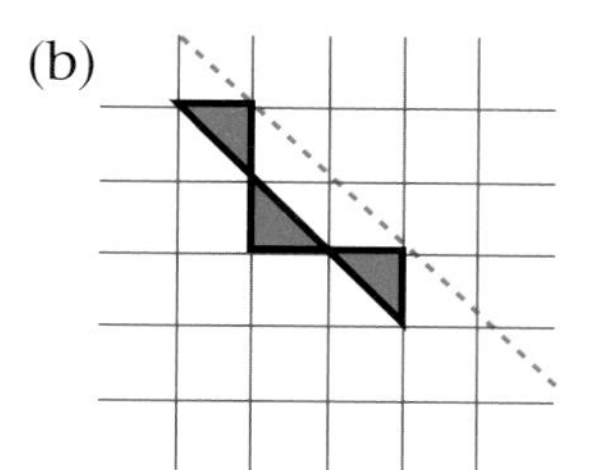

3 Copy each of these shapes onto triangular dotted paper. For each of them, reflect the dots in the mirror line and join them up to make a symmetrical pattern.

(a)

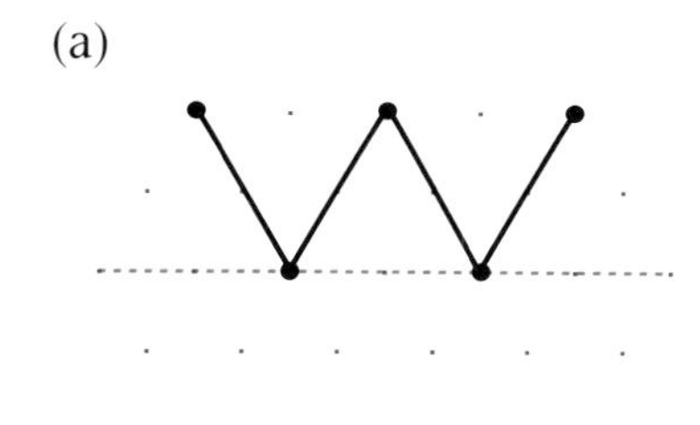

(b)

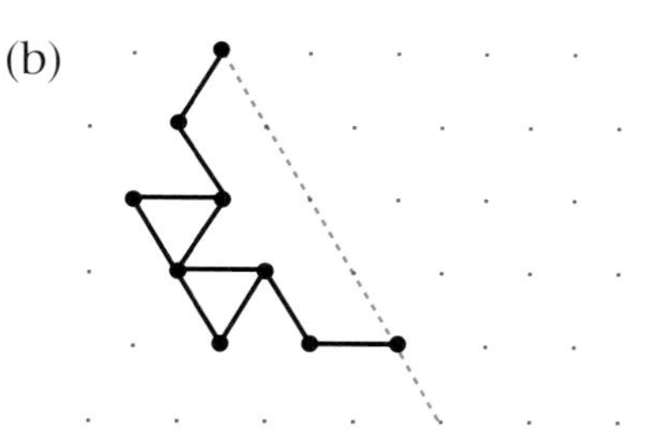

4 State the order of rotational symmetry for each of these patterns and whether or not they have line symmetry.

(a)

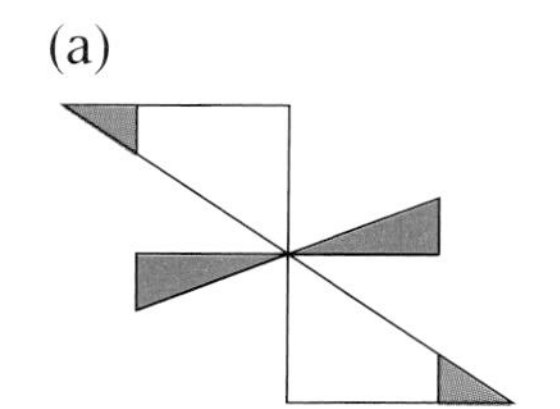

(b)

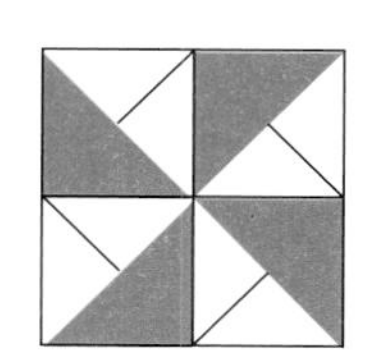

(c)

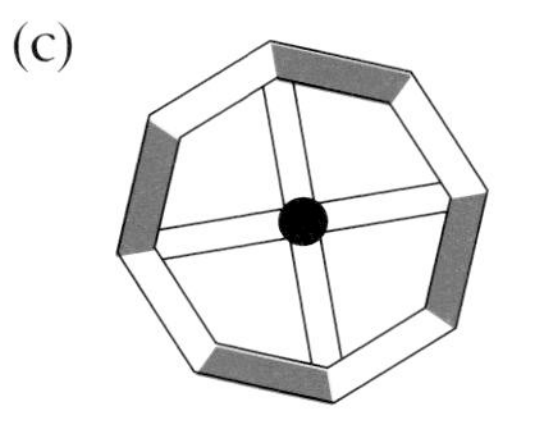

(d)

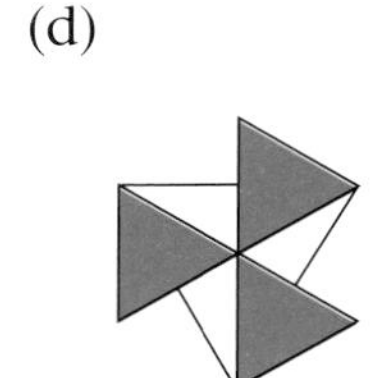

(e)

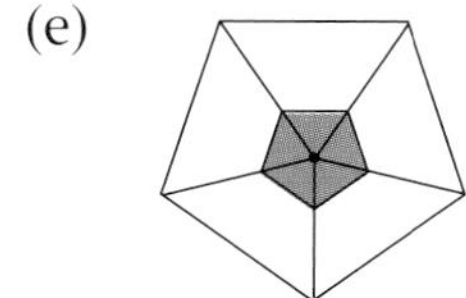

(f)

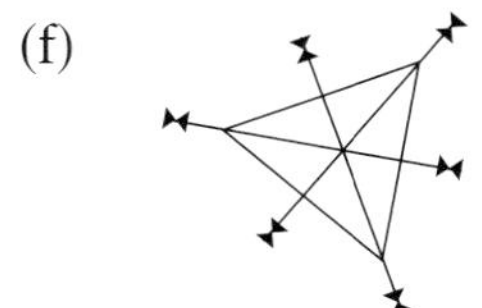

(g)

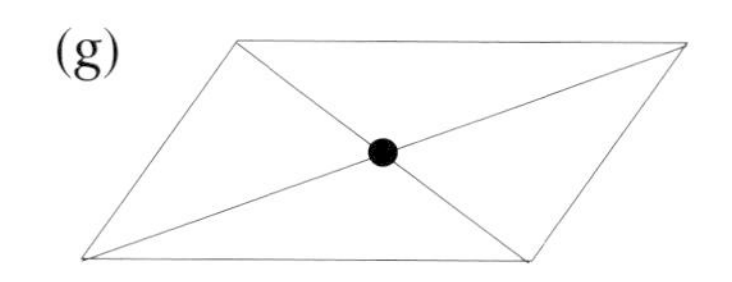

5 Design and colour your own flag. Give it two lines of symmetry and rotational symmetry of order 2.

Assignment 4 Symmetry pairs

Work in a group of three or four for this task. You need some blank cards and a polygon stencil might be useful.

This pair of cards 'match'.

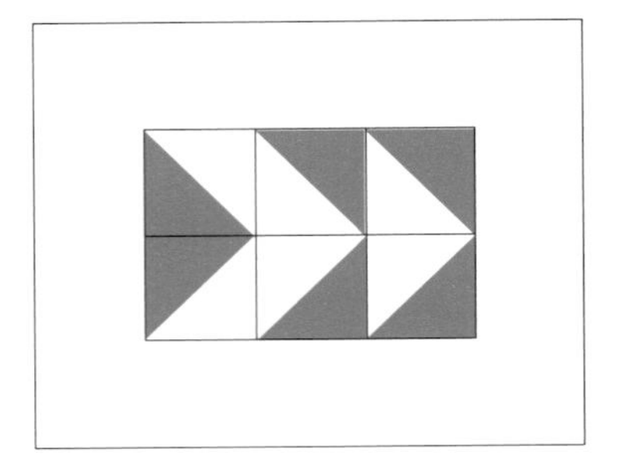

This pattern has one line of symmetry

- Make some more matching pairs of cards for shapes or patterns which have line or rotational symmetry, or both. Colour your diagrams, being careful to keep your symmetry statement correct.

In your group, make at least ten pairs of matching cards.

- Swap your set of cards with those made by another group. Sort the cards into matching pairs.

- When you have finished, make a display with your cards.

Fractions, percentages and decimals

In this chapter you will learn:

- → **about equivalent fractions**
- → **how to simplify and compare fractions**
- → **how to simplify and compare percentages**
- → **how to write decimals**
- → **the importance of the position of a digit in a number**

Starting points

Key fact
A fraction is a part of something.

Before starting this chapter you will need to be able to:

- recognise a simple fraction
- use decimals in money problems.

Fractions

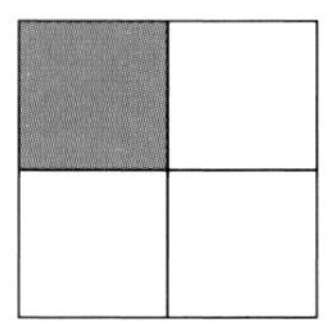
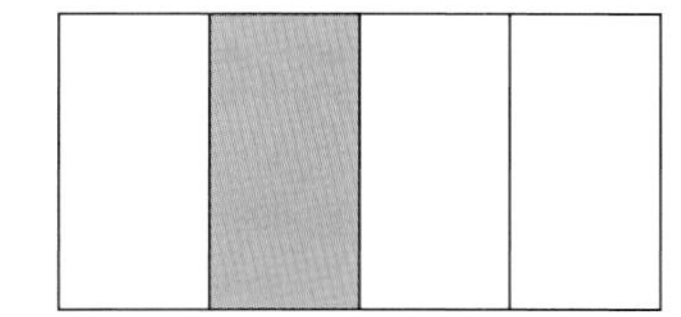
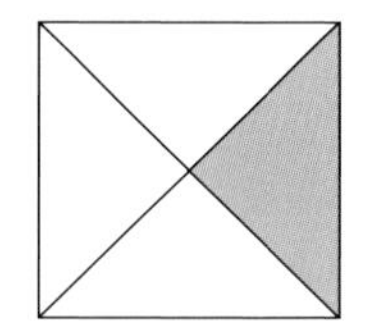
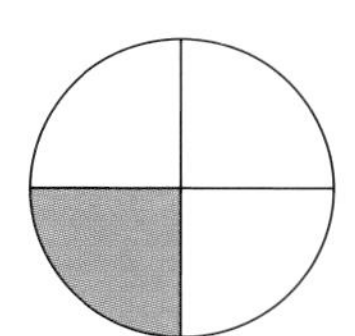

Each of these shapes is divided into four equal parts.

In each shape one part **out of** the four equal parts is shaded.

This is written as: $\frac{1}{4}$

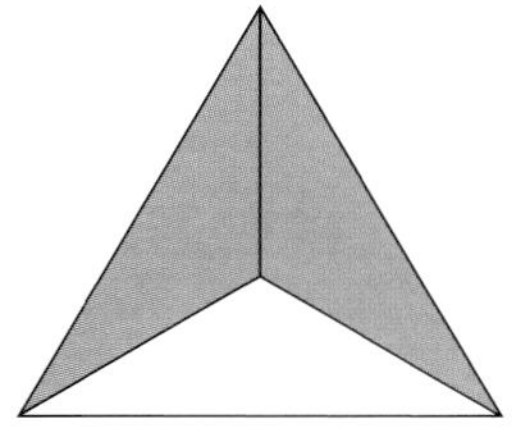

In this shape two parts **out of** the three equal parts are shaded.

This is written as: $\frac{2}{3}$

The expressions $\frac{1}{4}$ and $\frac{2}{3}$ are called **fractions**.

Exercise 1

1 Write down the fraction of each shape that is shaded.

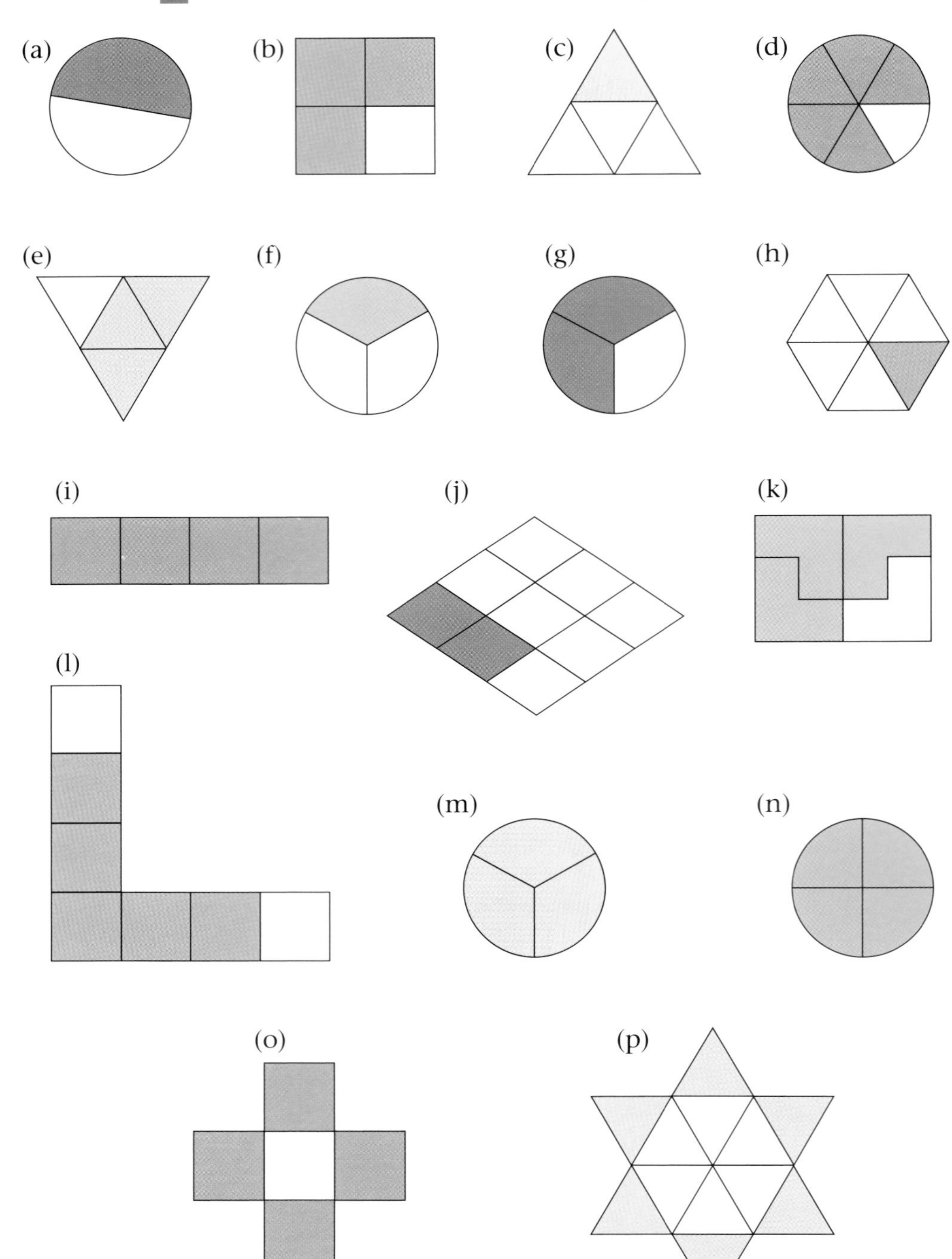

2 Which shapes have the following fractions shaded?

(a) $\frac{1}{2}$ (b) $\frac{1}{4}$ (c) $\frac{1}{3}$ (d) $\frac{1}{5}$ (e) none of these

A
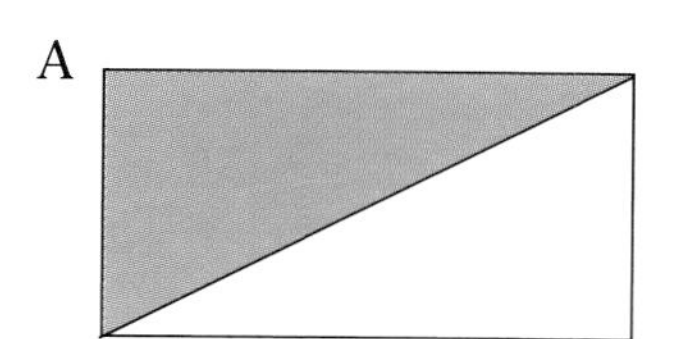

B
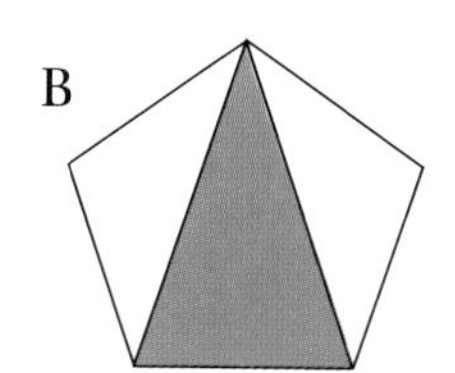

C
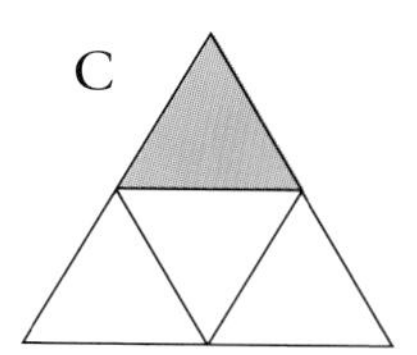

D
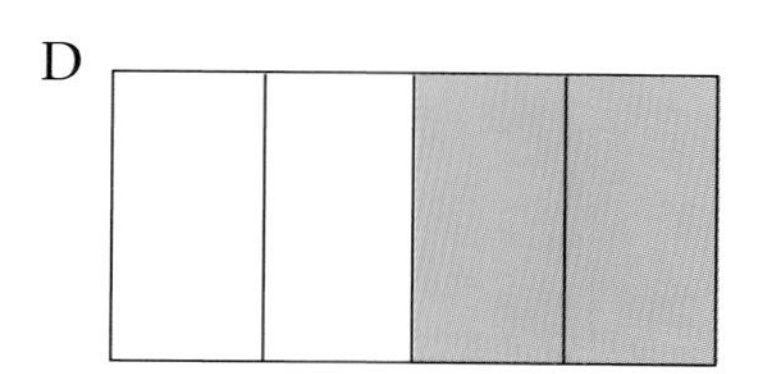

E
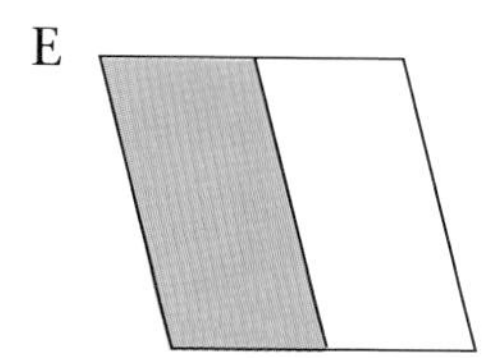

F
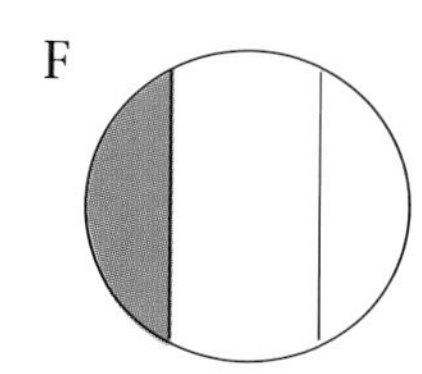

G
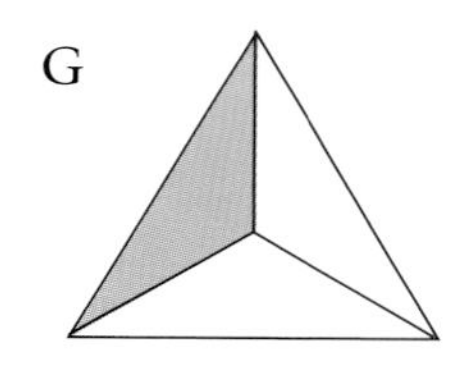

H
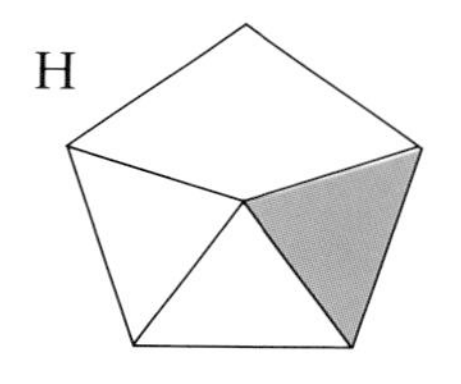

I
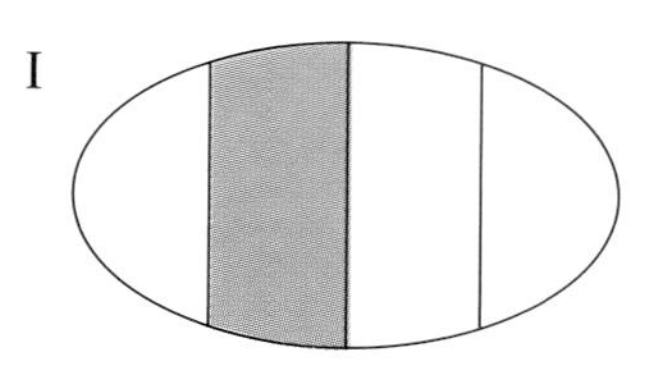

J
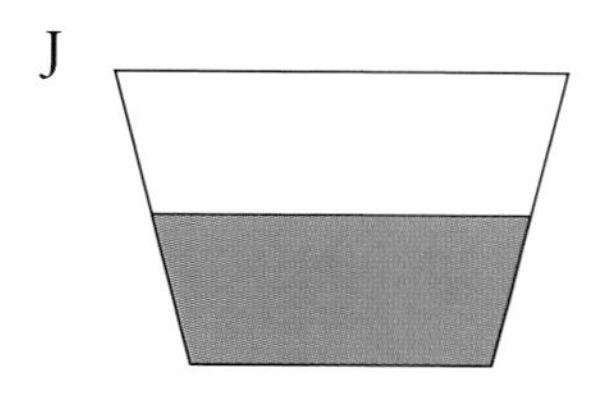

K
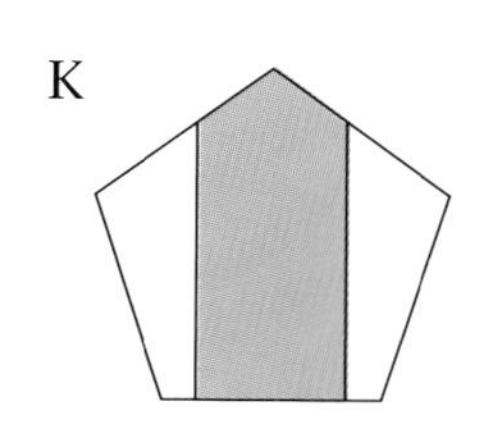

L
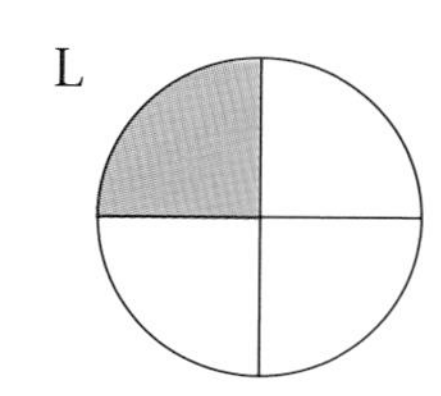

M

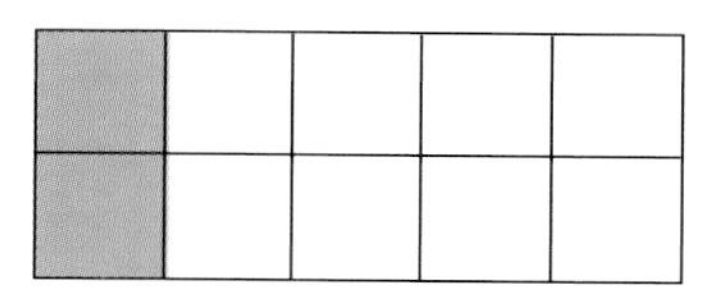

N
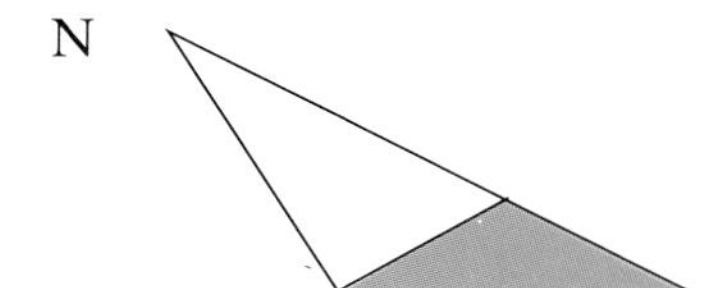

O
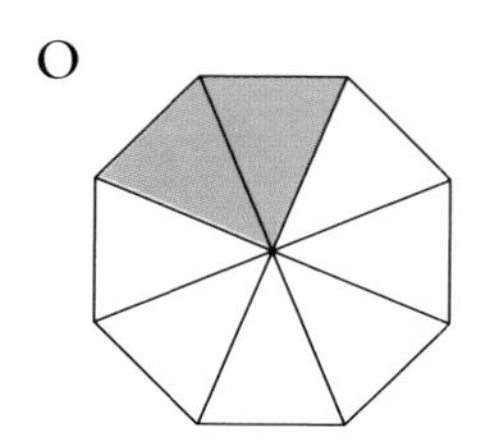

P
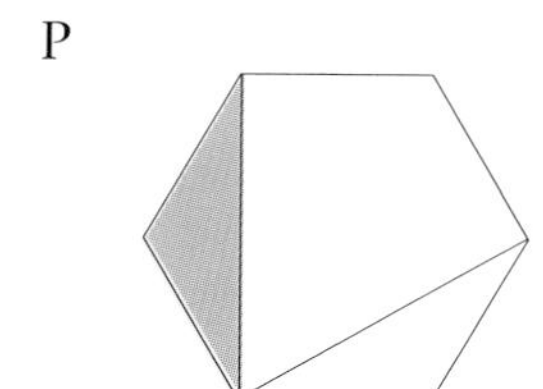

Shading fractions

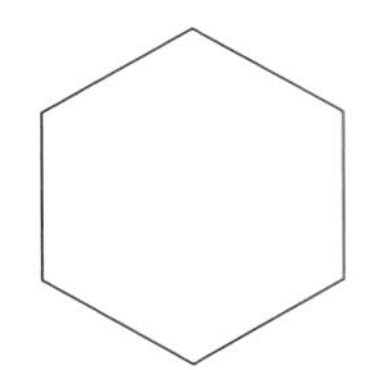

$\frac{5}{6}$ of this shape is shaded.

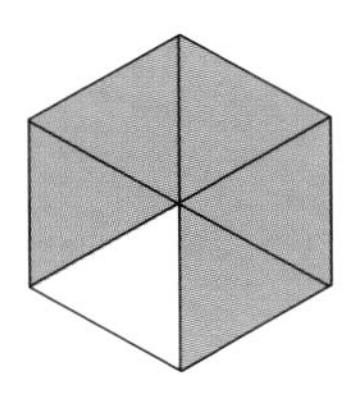

Each shape has been divided into equal parts.

$\frac{3}{4}$ of this shape is shaded.

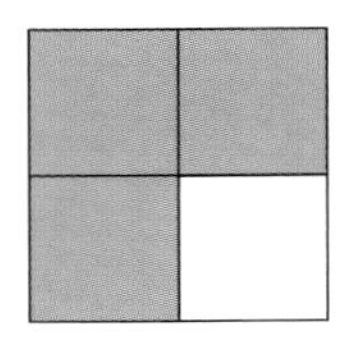

Exercise 2

1 Copy each of the following shapes. On your copy, shade in the fractional part given.

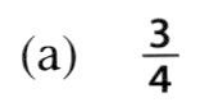

(a) $\frac{3}{4}$

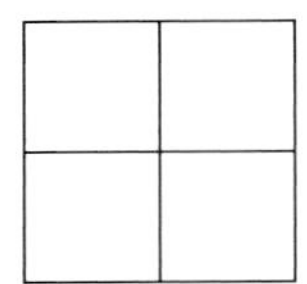

(b) $\frac{5}{8}$

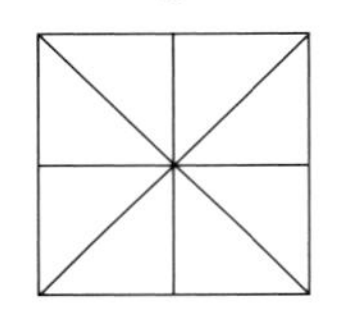

(c) $\frac{3}{8}$

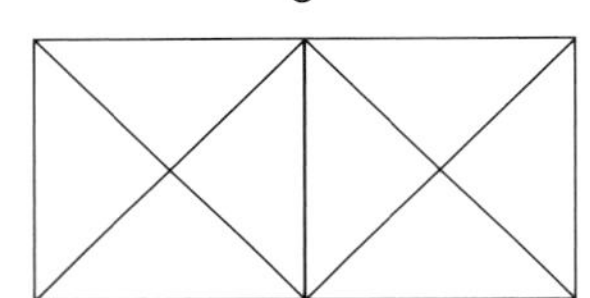

(d) $\frac{1}{4}$

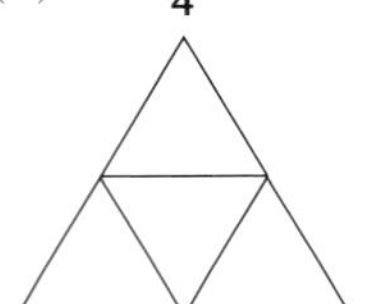

(e) $\frac{1}{2}$

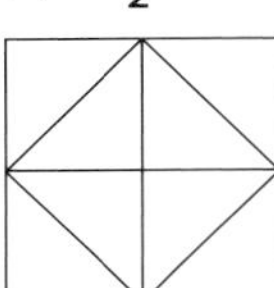

(f) $\frac{5}{6}$

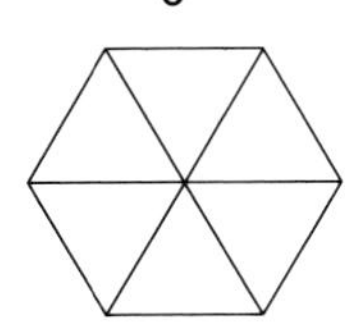

(g) $\frac{1}{2}$

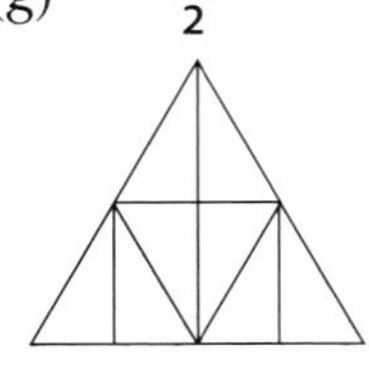

(h) $\frac{1}{3}$

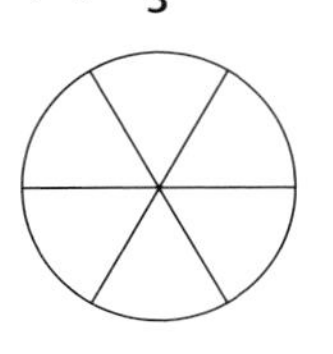

(i) $\frac{3}{4}$

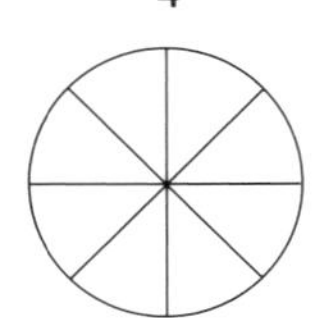

(j) $\frac{5}{6}$

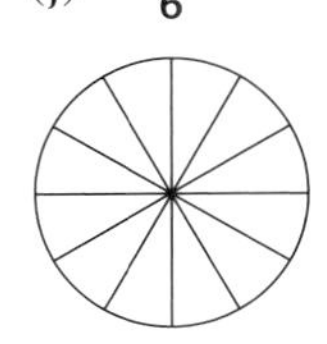

(k) $\frac{2}{3}$

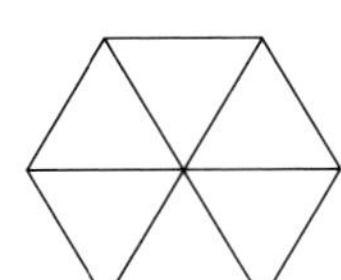

(l) $\frac{4}{5}$

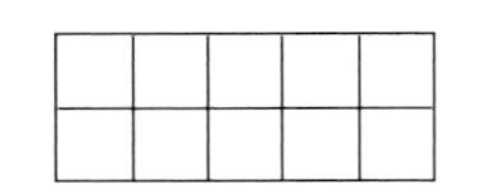

(m) $\frac{5}{8}$

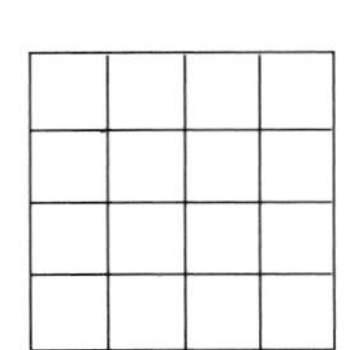

(n) $\frac{3}{5}$

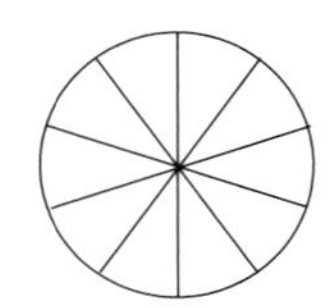

(o) $\frac{2}{9}$

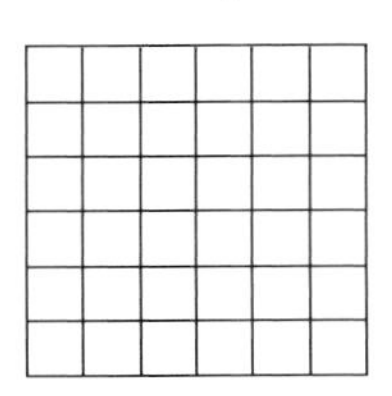

2 Copy these squares and rectangles. On your copy, shade the fractions given.

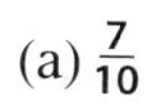

(a) $\frac{7}{10}$

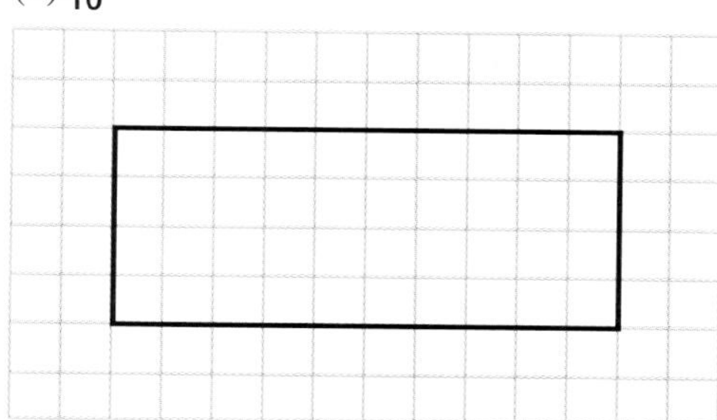

(b) $\frac{2}{3}$

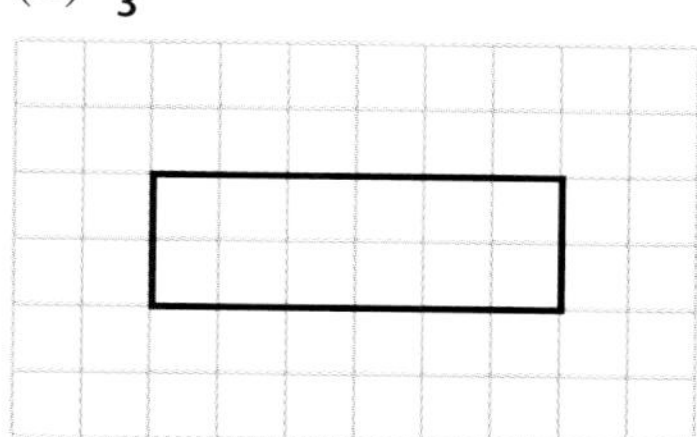

(c) $\frac{3}{4}$

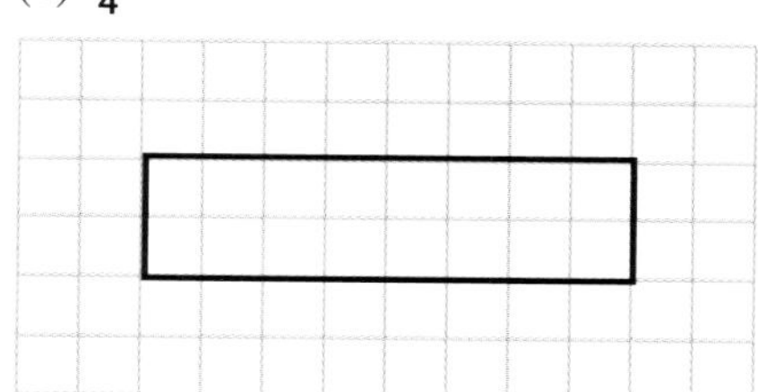

(d) $\frac{4}{5}$

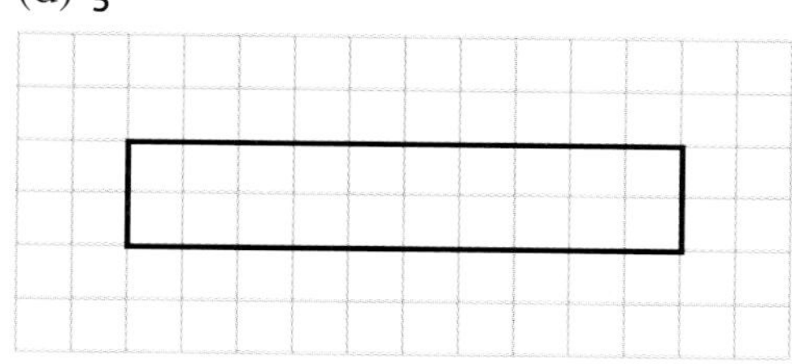

(e) $\frac{5}{6}$

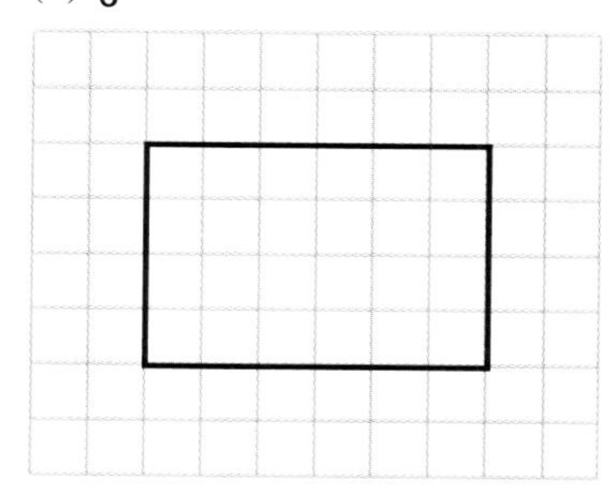

(f) $\frac{3}{8}$

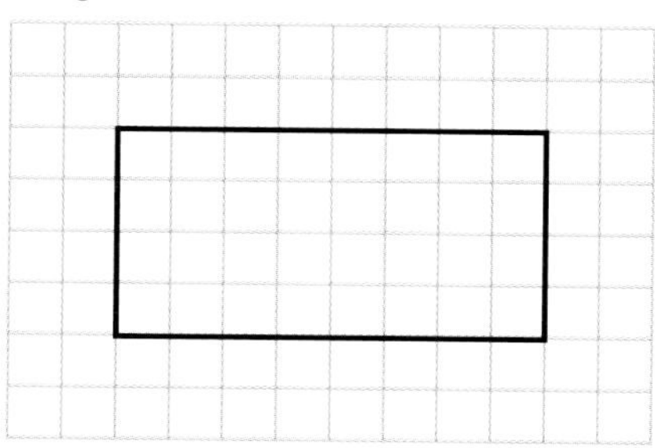

(g) $\frac{5}{9}$

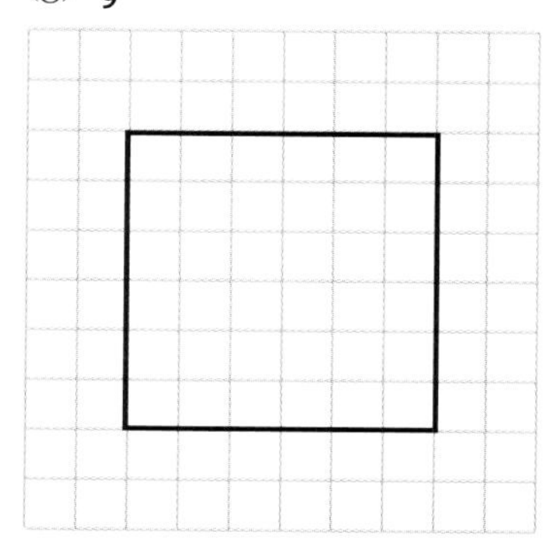

(h) $\frac{4}{7}$

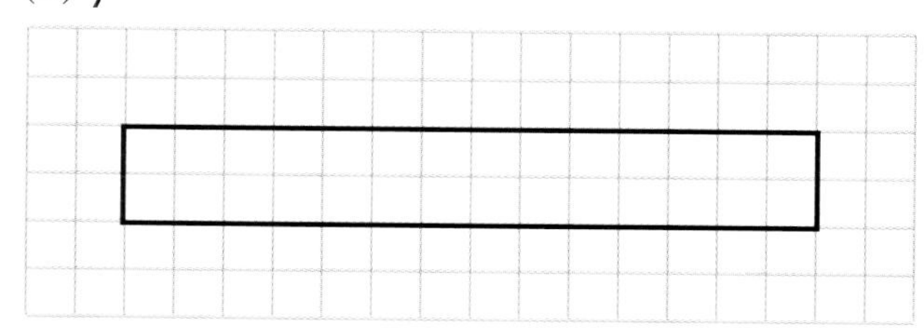

The fraction wall

1 whole	1
2 halves	$\frac{1}{2}$ $\frac{1}{2}$
3 thirds	$\frac{1}{3}$ $\frac{1}{3}$ $\frac{1}{3}$
4 fourths	$\frac{1}{4}$ $\frac{1}{4}$ $\frac{1}{4}$ $\frac{1}{4}$
5 fifths	$\frac{1}{5}$ $\frac{1}{5}$ $\frac{1}{5}$ $\frac{1}{5}$ $\frac{1}{5}$
6 sixths	$\frac{1}{6}$ $\frac{1}{6}$ $\frac{1}{6}$ $\frac{1}{6}$ $\frac{1}{6}$ $\frac{1}{6}$
7 sevenths	$\frac{1}{7}$ $\frac{1}{7}$ $\frac{1}{7}$ $\frac{1}{7}$ $\frac{1}{7}$ $\frac{1}{7}$ $\frac{1}{7}$
8 eighths	$\frac{1}{8}$ $\frac{1}{8}$ $\frac{1}{8}$ $\frac{1}{8}$ $\frac{1}{8}$ $\frac{1}{8}$ $\frac{1}{8}$ $\frac{1}{8}$
9 ninths	$\frac{1}{9}$ $\frac{1}{9}$ $\frac{1}{9}$ $\frac{1}{9}$ $\frac{1}{9}$ $\frac{1}{9}$ $\frac{1}{9}$ $\frac{1}{9}$ $\frac{1}{9}$
10 tenths	$\frac{1}{10}$ $\frac{1}{10}$ $\frac{1}{10}$ $\frac{1}{10}$ $\frac{1}{10}$ $\frac{1}{10}$ $\frac{1}{10}$ $\frac{1}{10}$ $\frac{1}{10}$ $\frac{1}{10}$

Exercise 3

Copy the following fractions and use the wall to complete the answers. The first one is done for you.

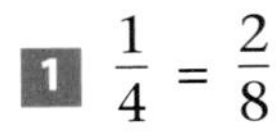

1 $\frac{1}{4} = \frac{2}{8}$

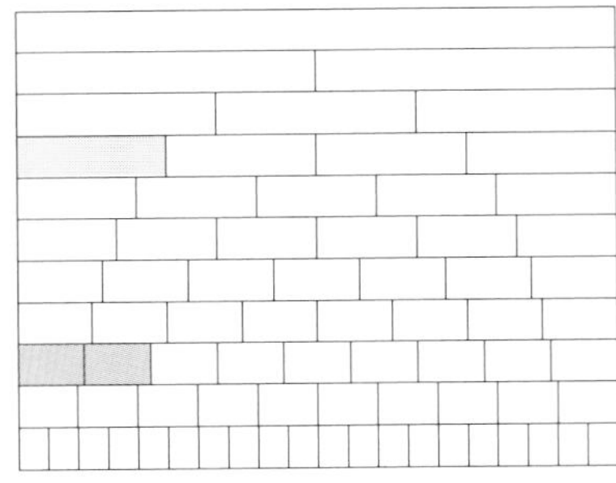

2 $\frac{1}{5} =$

3 $\frac{3}{4} =$

4 $\frac{3}{5} =$

5 $\frac{4}{10} =$

6 $\frac{4}{5} =$

7 $\frac{1}{3} =$

8 $\frac{2}{3} =$

9 $\frac{1}{2} =$

10 Questions 7, 8 and 9 have more than one answer. Write down all the other answers you can find.

Assignment 1 Shading grids

On squared paper draw some 4 by 4 grids.
Shade half of the squares for each grid.

Here is an example:

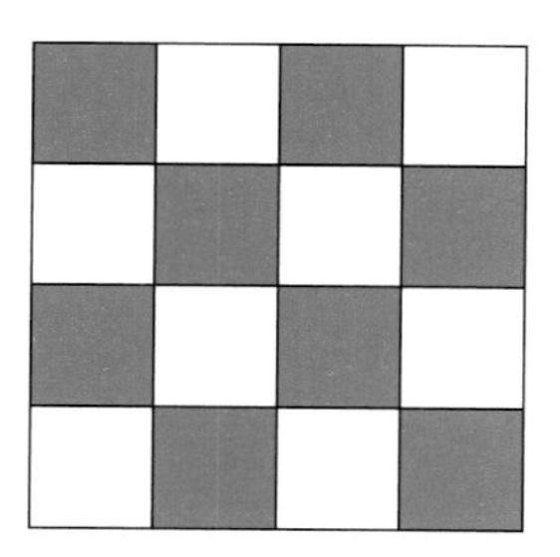

A reflected or rotated shape is *not* different.

- How many different ways can you find to shade half of the grid?
- On a new set of 4 by 4 grids show in how many different ways you can shade $\frac{3}{4}$ of the shape.

Equivalent fractions

Key fact
Equivalent fractions are fractions which are the same size.

Exercise 4

Copy the fractions below, and fill in the boxes. You may find the fraction wall useful.

1 $\frac{1}{3} = \frac{\square}{6}$ **2** $\frac{3}{6} = \frac{\square}{2}$ **3** $\frac{6}{\square} = \frac{3}{4}$ **4** $\frac{5}{10} = \frac{1}{\square}$

5 $\frac{3}{4} = \frac{6}{\square}$ **6** $\frac{3}{\square} = \frac{6}{10}$ **7** $\frac{6}{9} = \frac{2}{\square}$ **8** $\frac{4}{10} = \frac{\square}{5}$

9 $\frac{8}{8} = \frac{\square}{1}$ **10** $\frac{1}{2} = \frac{\square}{4} = \frac{3}{\square} = \frac{\square}{8} = \frac{5}{\square}$

11 $1 = \frac{1}{1} = \frac{\square}{2} = \frac{\square}{3} = \frac{\square}{4} = \frac{5}{\square} = \frac{6}{\square} = \frac{\square}{7} = \frac{\square}{8}$

One whole is equivalent to $\frac{1}{1}$.

Exercise 5

Copy and complete the following by writing down an equivalent fraction for each fraction given.

1 $\frac{8}{10} = \frac{\square}{\square}$ **2** $\frac{3}{9} = \frac{\square}{\square}$ **3** $\frac{2}{6} = \frac{\square}{\square}$ **4** $\frac{2}{10} = \frac{\square}{\square}$

5 $\frac{4}{5} = \frac{\square}{\square}$ **6** $\frac{2}{3} = \frac{\square}{\square}$ **7** $\frac{6}{8} = \frac{\square}{\square}$ **8** $\frac{6}{9} = \frac{\square}{\square}$

Finding equivalent fractions

Consider:

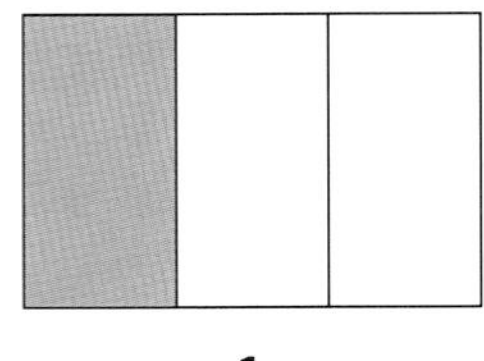

$\frac{1}{3}$

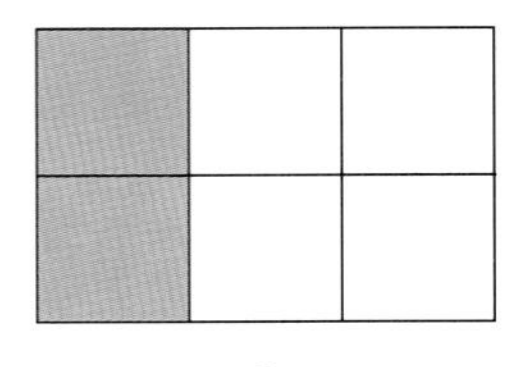

$\frac{2}{6}$

From the diagram you can see that:

$$\frac{1}{3} = \frac{2}{6}$$

You can write this:

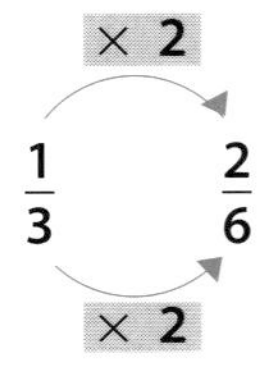

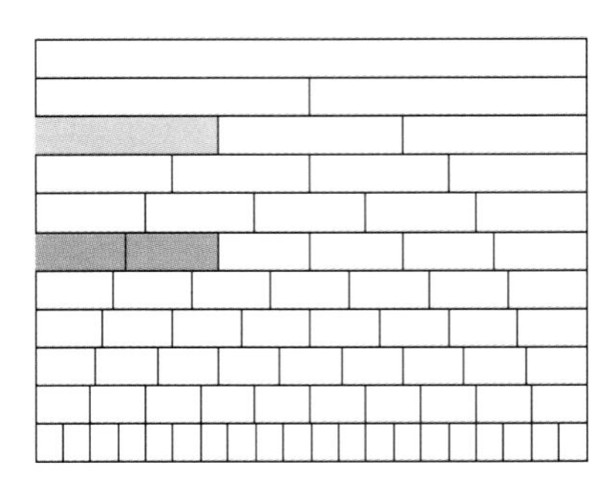

Similarly:

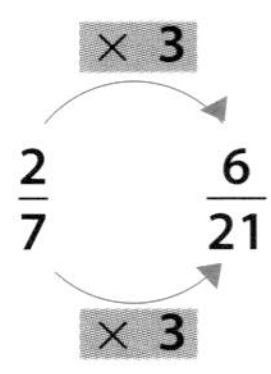

$$\frac{2}{7} = \frac{6}{21}$$

Exercise 6

Copy and complete the following fractions.

1

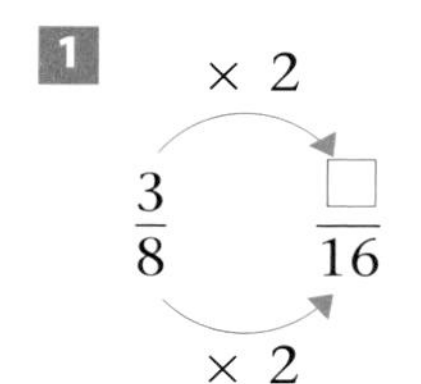

2 × 3 $\quad \frac{5}{6} \quad \frac{\square}{18} \quad$ × 3

3 × 4 $\quad \frac{2}{3} \quad \frac{\square}{12} \quad$ × 4

4 × 4 $\quad \frac{7}{8} \quad \frac{\square}{\square} \quad$ × 4

5 × 5 $\quad \frac{3}{4} \quad \frac{\square}{\square} \quad$ × 5

6 × 2 $\quad \frac{8}{9} \quad \frac{\square}{\square} \quad$ × 2

7 × $\square$ $\quad \frac{3}{7} \quad \frac{\square}{14} \quad$ × $\square$

8

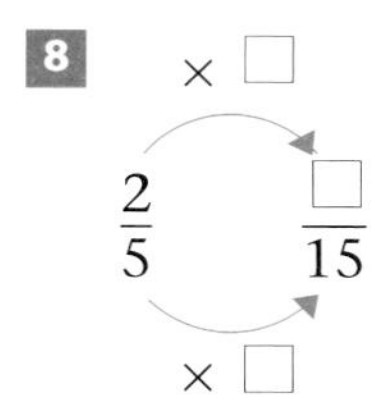

9 × $\square$ $\quad \frac{4}{7} \quad \frac{\square}{21} \quad$ × $\square$

10 × 7 $\quad \frac{\square}{5} \quad \frac{21}{\square} \quad$ × 7

11

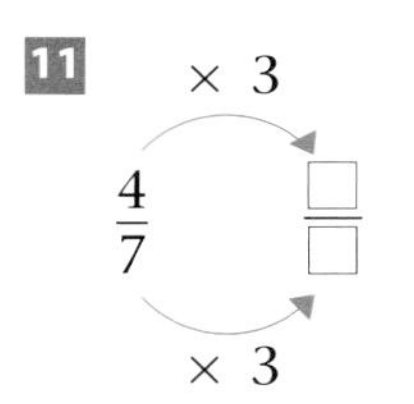

12 × 4 $\quad \frac{8}{\square} \quad \frac{\square}{44} \quad$ × 4

13 × 2 $\quad \frac{6}{7} \quad \frac{\square}{\square} \quad$ × $\square$

14 × 5 $\quad \frac{4}{9} \quad \frac{20}{\square} \quad$ × $\square$

15 × $\square$ $\quad \frac{13}{20} \quad \frac{\square}{40} \quad$ × 2

16

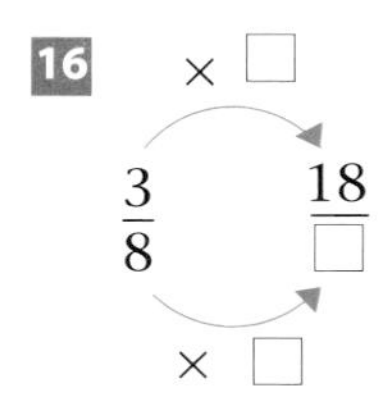

17

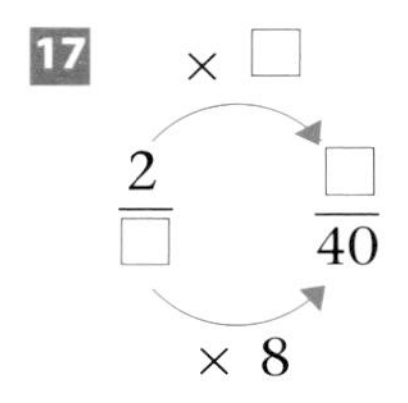

18

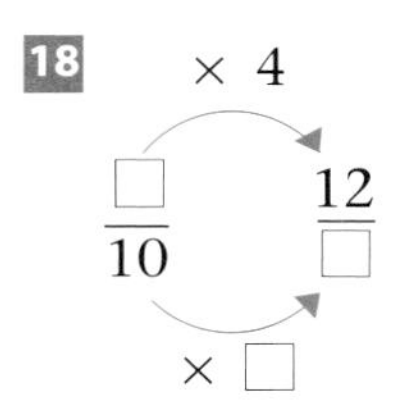

19

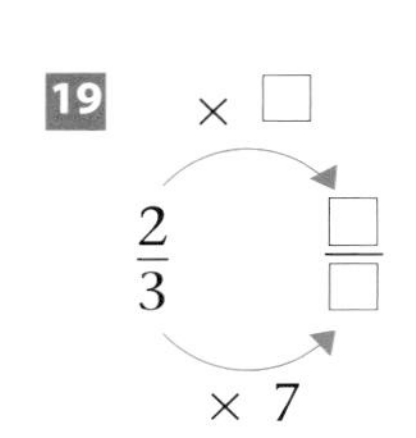

20

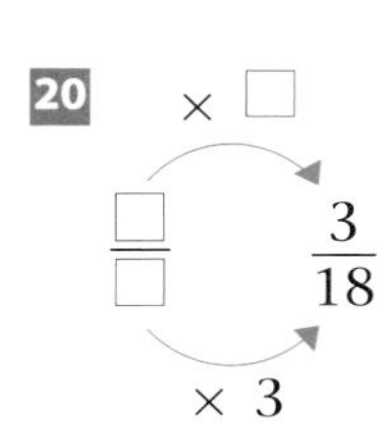

Simplifying fractions

The fractions

$$\frac{5}{12} \quad \frac{3}{6} \quad \frac{4}{8} \quad \frac{6}{12} \quad \frac{1}{2} \quad \frac{2}{4}$$

are equivalent ways of saying a half.

Fractions, however, are usually written in their **simplest form,** for example:

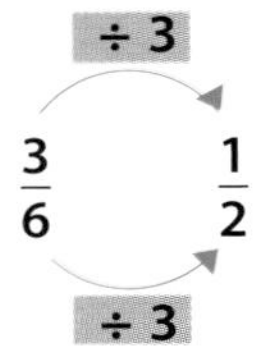

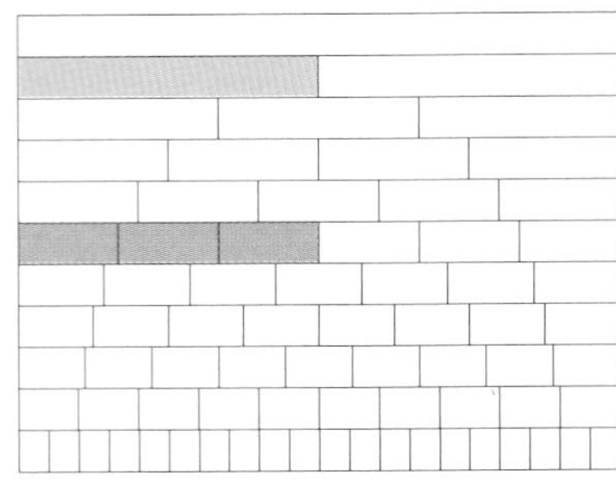

This is the simplest form as no other number will divide into both the top and bottom numbers.

The process of dividing both the top and bottom number by the same number is called simplifying.

Similarly:

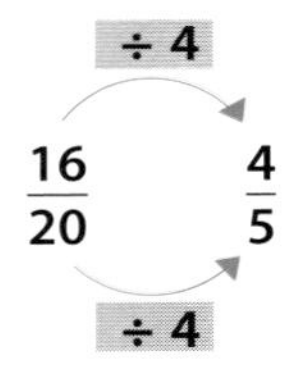

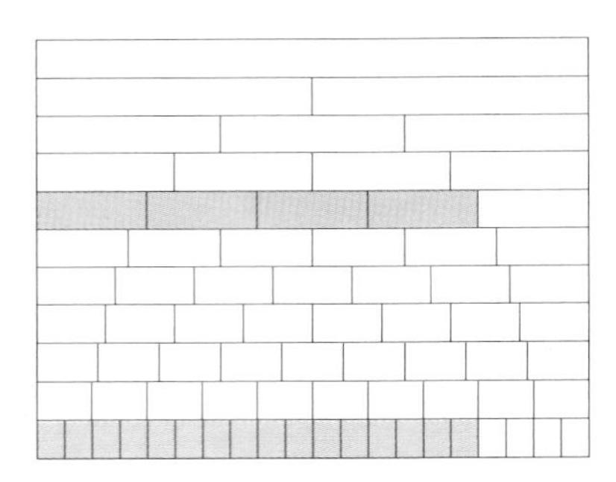

The fraction $\frac{16}{20}$ will simplify to $\frac{4}{5}$.

Exercise 7

Simplify each of the following fractions.

1. $\frac{3}{9}$
2. $\frac{9}{12}$
3. $\frac{16}{32}$
4. $\frac{16}{24}$
5. $\frac{12}{36}$
6. $\frac{18}{24}$
7. $\frac{21}{35}$
8. $\frac{45}{50}$
9. $\frac{15}{75}$
10. $\frac{56}{72}$
11. $\frac{22}{33}$
12. $\frac{21}{28}$

Assignment 2 Equal tiling patterns

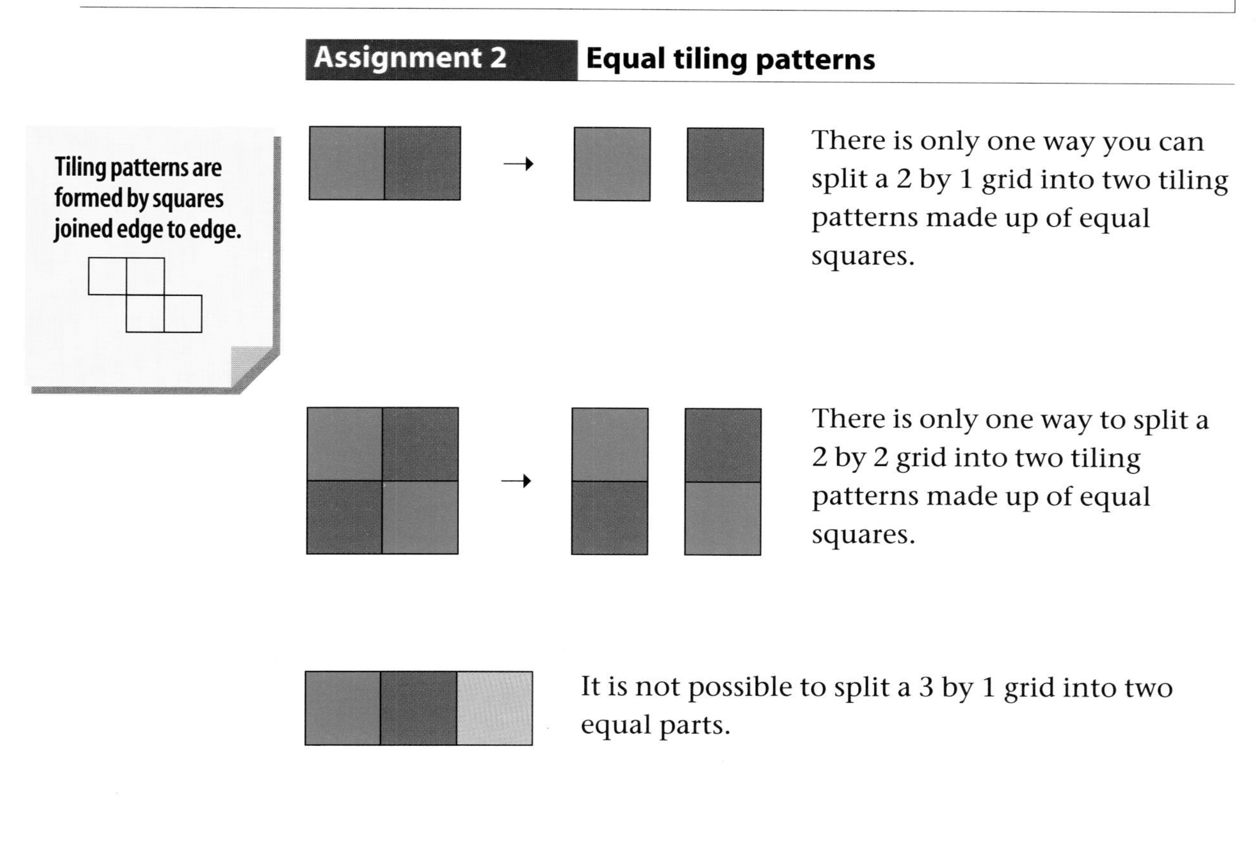

Tiling patterns are formed by squares joined edge to edge.

There is only one way you can split a 2 by 1 grid into two tiling patterns made up of equal squares.

There is only one way to split a 2 by 2 grid into two tiling patterns made up of equal squares.

It is not possible to split a 3 by 1 grid into two equal parts.

'Equal parts' means that both parts have the same number of squares but are not necessarily the same shapes.

This diagram shows one way of splitting a 3 by 2 grid into two equal parts.

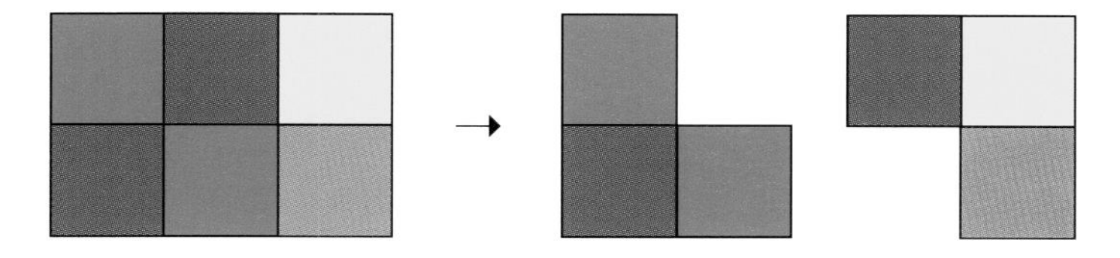

- How many other ways can you find of splitting a 3 by 2 grid? Record your answers on squared paper.
- In how many ways can each of the following grids be split into two equal parts?

  Record all your answers on squared paper.

  (a) 4 by 2 (b) 3 by 3 (c) 4 by 3 (d) 4 by 4

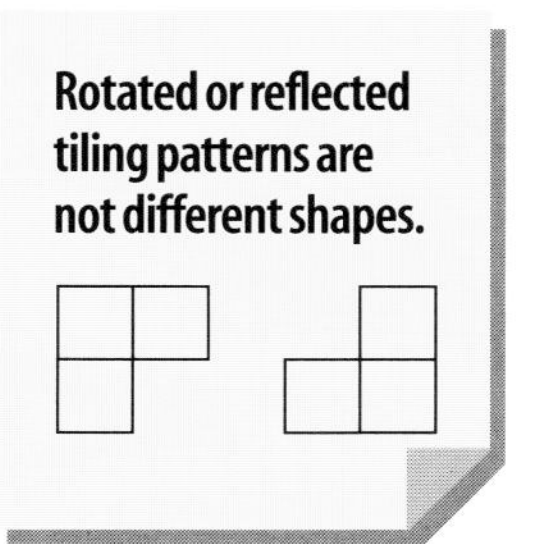

- Write down anything you notice.

Assignment 3 Equivalent dominoes

Most dominoes can be put down to show two different fractions.

Example

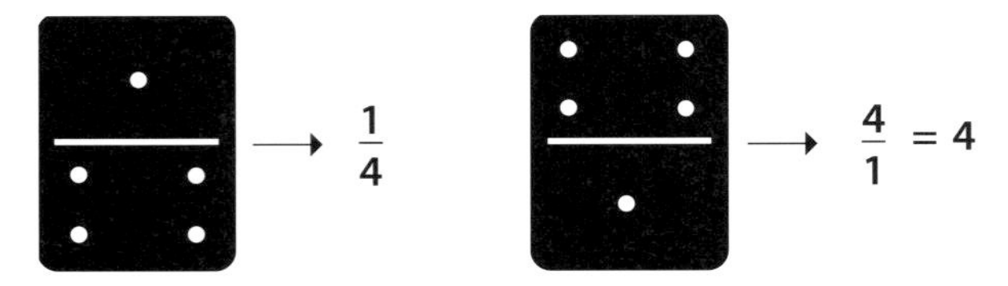

The 21 tiles can be used to show 36 fractions.

- Which tiles cannot show two different fractions?
- How many equivalent fractions can you find from the domino tiles?
- Write all the fractions in order, starting from the largest.

You might find it helpful to use the fraction wall.

Decimals

Tenths and hundredths

There are 10 millimetres in a centimetre.

10 mm = 1 cm

1 millimetre is $\frac{1}{10}$ of a centimetre.

In decimals, 1 millimetre is written as:

units		tenths	
0	.	1	centimetres

1 mm = 0.1 cm

There are 100 centimetres in a metre.

1 m = 100 cm

1 centimetre is $\frac{1}{100}$ of a metre.

In decimals, this is written as:

units		tenths	hundredths	
0	.	0	1	metres

1 cm = 0.01 m

12 cm = $\frac{12}{100}$ m

This is the same as $\frac{10}{100}$ plus $\frac{2}{100}$ of a metre.

In decimals, this is written as:

units		tenths	hundredths	
0	.	1	2	metres

12 cm = 0.12 m

Remember

$\frac{1}{10} = \frac{10}{100}$

Exercise 8

Write the following as decimals of a metre.

1 70 cm **2** 90 cm **3** 32 cm **4** 64 cm

5 42 cm **6** 24 cm **7** 8 cm **8** 2 cm

9 1 m 70 cm **10** 1 m 12 cm **11** 3 m 2 cm **12** 4 m 1 cm

Exercise 9

1 There are three numbers in each of the following lists.
Write these in order of size, starting with the largest.

(a) 12.09 m 21.90 m 22.19 m

(b) 1001.10 m 998 m 909.99 m

(c) 35.34 metres 34.35 metres 34.33 metres

(d) 2.02 1.99 2.2

2 Write the following numbers in order of size, starting with the largest.

(a) 1007.9 109.7 1070.9 1009.7

(b) 9071 7910.0 791 191.70

(c) 1.09 9.01 7.9 1.01

Exercise 10

Give the value of the 7 in each of the following numbers.
For example, the 7 in the number 73.1 has the value 7 tens or 70.

1 7.81 **2** 71.0 **3** 11.71 **4** 20.71

5 701.8 **6** 123.07 **7** 72.8 **8** 0.07

Assignment 4 Making numbers

On an old calculator, only a few keys work.

They are:  [1] [0] [•] [+] [–] [=]

Using only these keys, you can make up other numbers.

Example

Make 3.9 by pressing: [1] [+] [1] [+] [1] [+] [1] [–] [•] [1]

- Make up the following numbers, using only the keys that work on the calculator. Use as few key presses as possible.

(a) 2.2 (b) 7.5 (c) 6.8 (d) 3.01 (e) 5.22

(f) 10.9 (g) 0.89 (h) 8.89 (i) 21.29 (j) 10.97

Remember

$\frac{5}{10} = \frac{1}{2}$

and

$\frac{5}{10} = 0.5$

Changing fractions to decimals

There are two halves in one whole.

$\frac{1}{2}$ means $1 \div 2$

If this is worked out on a calculator, the answer is 0.5.

Similarly, on a calculator: $\frac{3}{4}$ gives 0.75

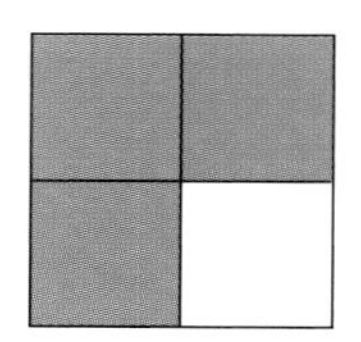

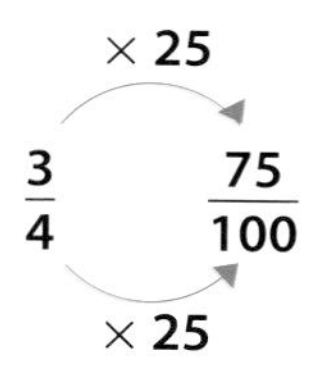

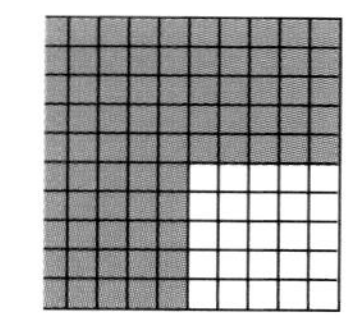

$\frac{1}{4} = \frac{75}{100} =$

units		tenths	hundredths
0	•	7	5

Exercise 11

Use your calculator to convert the following fractions to decimals.

1 (a) $\frac{6}{10}$ (b) $\frac{3}{5}$ (c) $\frac{7}{20}$ (d) $\frac{17}{25}$

(e) $\frac{1}{4}$ (f) $\frac{3}{8}$ (g) $\frac{4}{5}$ (h) $\frac{9}{12}$

(i) $\frac{9}{15}$ (j) $\frac{18}{24}$ (k) $\frac{14}{20}$ (l) $\frac{6}{25}$

2 (a) $\frac{64}{200}$ (b) $\frac{45}{60}$ (c) $\frac{45}{75}$ (d) $\frac{160}{250}$

(e) $\frac{68}{80}$ (f) $\frac{39}{60}$ (g) $\frac{48}{72}$ (h) $\frac{108}{144}$

(i) $\frac{64}{256}$ (j) $\frac{125}{625}$ (k) $\frac{128}{1024}$ (l) $\frac{975}{1250}$

Percentages

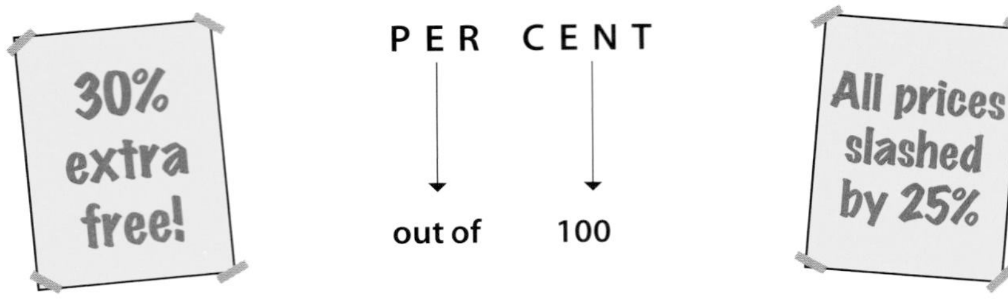

So 30 per cent means $\frac{30}{100}$ and is written 30%.

Exercise 12

1 Below are six words that all start with the word 'cent' and six sentences. Copy each sentence and fill in the missing word with one from the list.

centipede **centenarian** **century**
centimetres **centimes** **centurion**

(a) A .... is a period of 100 years.

(b) There are 100 .... in a metre.

(c) A creature with 100 legs is a ....

(d) A .... is the leader of 100 Roman soldiers.

(e) There are 100 .... in one French franc.

(f) A person who is 100 years old is a ....

2 In the end-of-year exams each of six subjects was marked out of 100. Give each of the results as a percentage. The first one is done for you.

(a) *Maths:* *50 marks* $\longrightarrow$ $\frac{50}{100} = 50\%$

(b) *English:* *71 marks*

(c) *Science:* *35 marks*

(d) *French:* *58 marks*

(e) *Geography:* *98 marks*

(f) *History:* *24 marks*

Changing decimals to percentages

Marks are not always out of 100.

Example

The result of the Art exam was $\frac{15}{20}$.

This can be changed into a percentage.

$\frac{15}{20} = 15 \div 20 = 0.75$

units		tenths	hundredths
0	•	7	5

Next change the decimal to a percentage.

$0.75 = \frac{75}{100}$

$\frac{75}{100} = 75\%$

Similarly $\frac{13}{25} = 0.52 = 52\%$

Exercise 13

1 Using your calculator, change the following fractions to percentages.

(a) $\frac{7}{10}$ (b) $\frac{2}{5}$ (c) $\frac{3}{20}$ (d) $\frac{19}{25}$ (e) $\frac{1}{4}$ (f) $\frac{28}{200}$

2 Six pupils took a maths exam that was marked out of 50.
Using your calculator, change each mark to a percentage.

(a) $\frac{40}{50}$ (b) $\frac{23}{50}$ (c) $\frac{41}{50}$ (d) $\frac{46}{50}$ (e) $\frac{18}{50}$ (f) $\frac{9}{50}$

3 The same pupils also took a French exam that was marked out of 20.
Change each mark to a percentage.

(a) $\frac{12}{20}$ (b) $\frac{10}{20}$ (c) $\frac{6}{20}$ (d) $\frac{19}{20}$ (e) $\frac{17}{20}$ (f) $\frac{7}{20}$

4 The science exam was marked out of 200.
Change each mark to a percentage.

(a) $\frac{160}{200}$ (b) $\frac{48}{200}$ (c) $\frac{190}{200}$ (d) $\frac{176}{200}$ (e) $\frac{85}{200}$ (f) $\frac{71}{200}$

5 Copy these tables of exam results and complete them.
Some of them are done for you.

(a)

English	
out of 100	%
52	
92	
43	
38	
87	
$63\frac{1}{2}$	

(b)

Maths	
out of 50	%
30	
37	
41	82
49	
$28\frac{1}{2}$	
11	

(c)

French	
out of 20	%
11	
8	
2	
14	
$12\frac{1}{2}$	
20	

(d)

Science	
out of 200	%
180	
154	
165	$82\frac{1}{2}$
96	
101	
$88\frac{1}{2}$	

Comparing results

Exercise 14

These tables show a pupil's half-yearly exam results.

Winter term		
Subject	Mark	%
English	$\frac{33}{50}$	
Maths	$\frac{7}{10}$	70
Science	$\frac{164}{200}$	
French	$\frac{9}{20}$	
Art	$\frac{20}{25}$	
Music	$\frac{8}{10}$	

Summer term		
Subject	Mark	%
English	$\frac{13}{20}$	
Maths	$\frac{16}{25}$	64
Science	$\frac{81}{200}$	
French	$\frac{31}{50}$	
Art	$\frac{78}{100}$	
Music	$\frac{192}{200}$	

1 Copy the tables and complete them.

2 In which subject was the highest mark gained:

(a) in the winter term?
(b) in the summer term?

3 In which term was the mark for the English exam better?

4 In which subject do the marks improve the most between the winter and summer exams?

Review Exercise

1 What fraction of each of the following shapes is shaded?

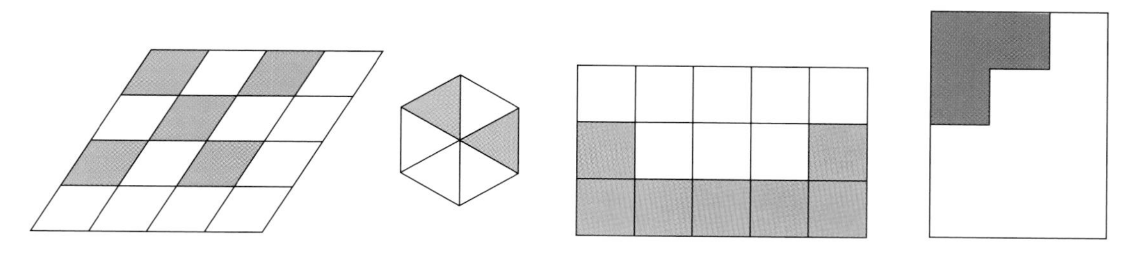

2 Copy each of the shapes below. Divide each shape into the number of parts stated.

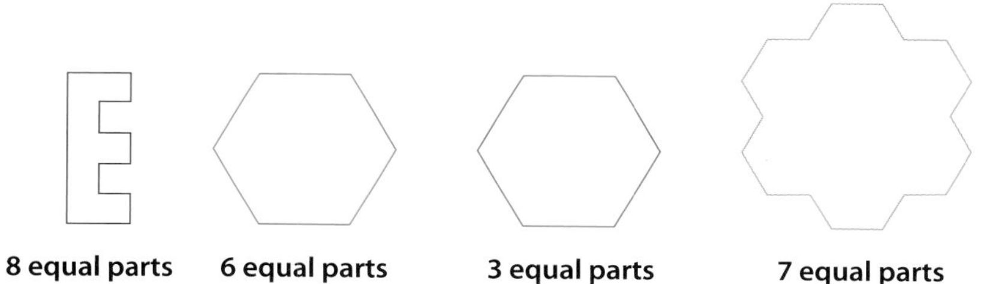

8 equal parts 6 equal parts 3 equal parts 7 equal parts

3 Copy and complete the following equivalent fractions.

(a) $\frac{4}{5} = \frac{\square}{10}$ (b) $\frac{2}{3} = \frac{\square}{6} = \frac{\square}{9}$ (c) $\frac{8}{8} = \frac{\square}{10} = \frac{37}{\square}$

(d) $\frac{4}{10} = \frac{\square}{20} = \frac{\square}{100}$ (e) $\frac{1}{4} = \frac{\square}{100}$ (f) $\frac{1}{2} = \frac{\square}{4} = \frac{\square}{10} = \frac{\square}{20}$

4 Which fraction is bigger in each of the following pairs of fractions? Change them to decimals to find out.

(a) $\frac{2}{5}$ $\frac{1}{2}$ (b) $\frac{7}{8}$ $\frac{3}{4}$ (c) $\frac{2}{3}$ $\frac{5}{7}$ (d) $\frac{1}{5}$ $\frac{1}{4}$

5 A notepad has 135 pages, and 90 of these have been used. What fraction of the pad remains?

6 Using your calculator, change the following fractions to percentages.

(a) $\frac{15}{50}$ (b) $\frac{7}{25}$ (c) $\frac{3}{5}$ (d) $\frac{1}{20}$ (e) $\frac{7}{10}$ (f) $\frac{38}{200}$

7 What is the value of the digit 5 in each of the following numbers?

(a) 520.03 (b) 102.05 (c) 253.4

(d) 435.34 (e) 36.57 (f) 5000.67

Assignment 5 Halving a pinboard

It is only possible to halve a 2×2 pinboard one way.

A 3×3 pinboard can be halved two different ways.

- How many ways can you halve a 4×4 pinboard so that each part is the same?

Nets and 3-D shapes

In this chapter you will learn:

- ➜ **the names of some 3-D shapes**
- ➜ **how to make some 3-D shapes**
- ➜ **some facts about them**
- ➜ **how to represent them on paper**

A polygon is a 2-D (flat) shape.
It encloses a single space.
It is made from straight lines.

Starting points

Before starting this chapter you will need to know:

- the names of polygons
- some facts about polygons

Try this exercise to check yourself.

Exercise 1

1 Which of these are polygons?
If a shape is not a polygon, write down why not.

(a)
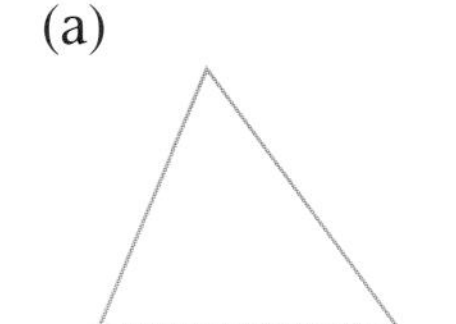
(b)
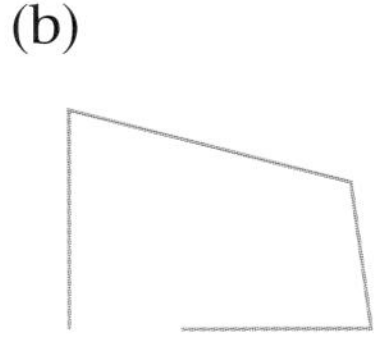
(c)
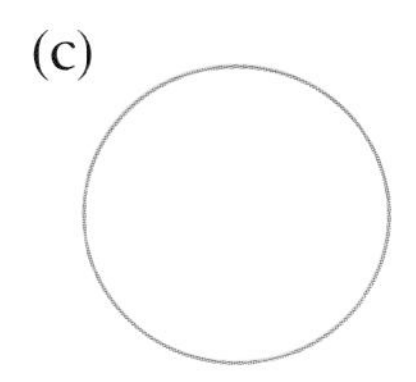
(d)
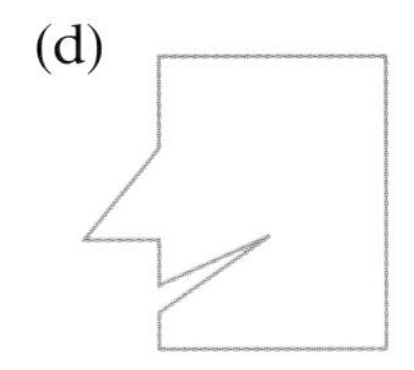
(e)
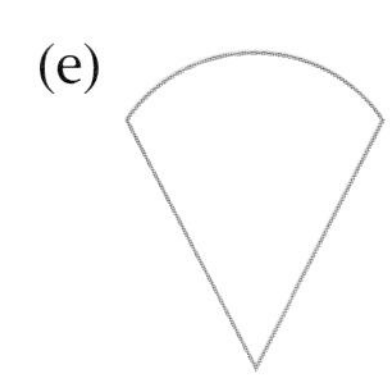

(f)

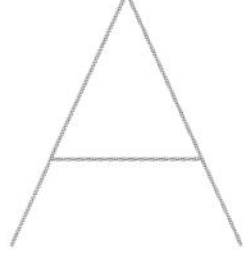
(g)
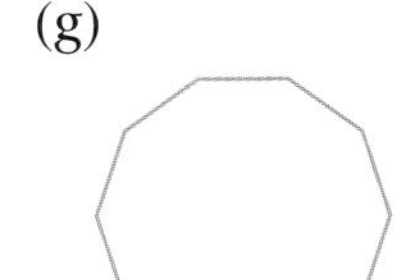
(h)
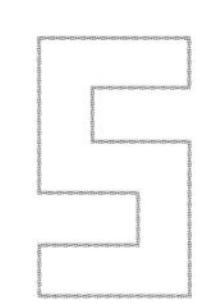
(i)
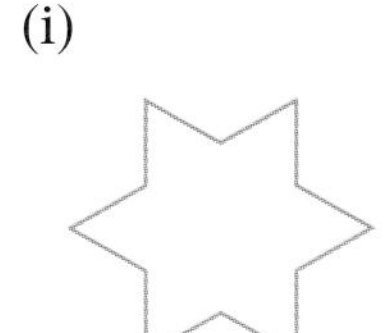
(j)
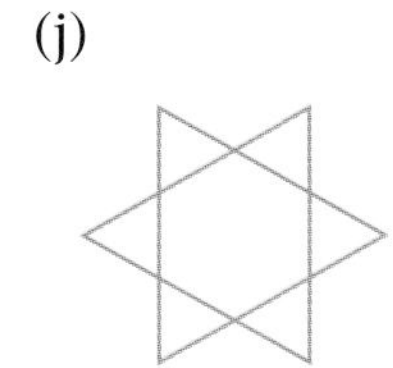

A pentagon has five sides

2 Name each of these polygons.

A heptagon has seven sides

(a) (b) (c) (d)

(e) (f) (g) (h)

A nonagon has nine sides

A decagon has ten sides

3 Draw and name five of your own polygons.

4 Write whether or not the following are regular polygons.

(a)

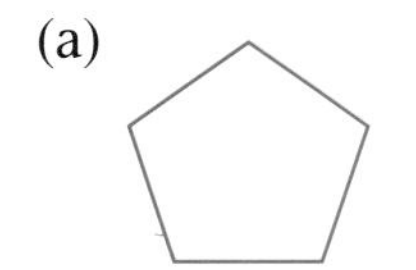

(b)

(c)

(d)

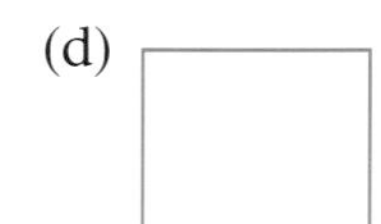

(e)

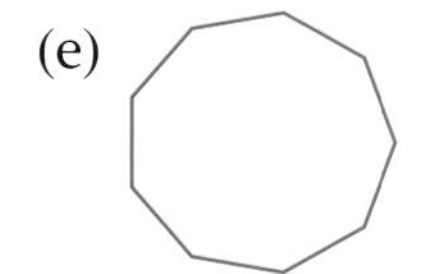

(f)

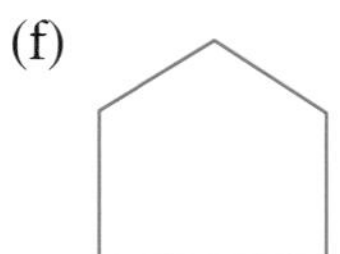

(g)

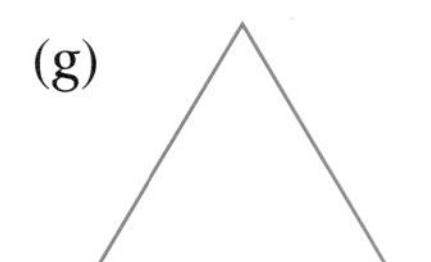

(h)

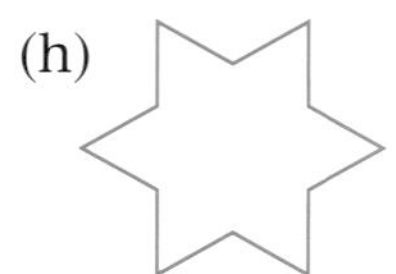

5 How many lines of symmetry has each of these shapes?

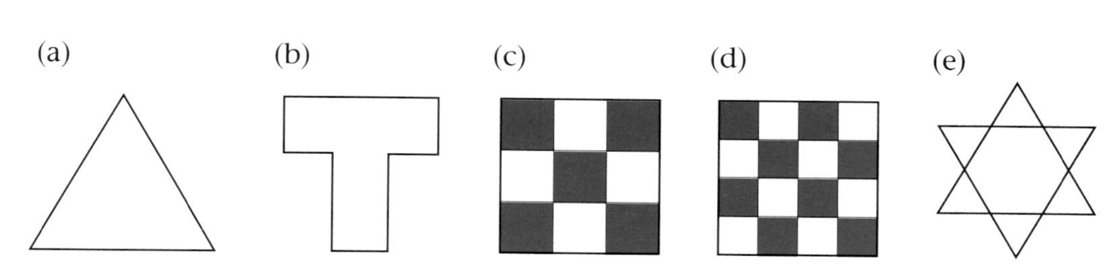

6 What is the order of rotational symmetry for each of these shapes?

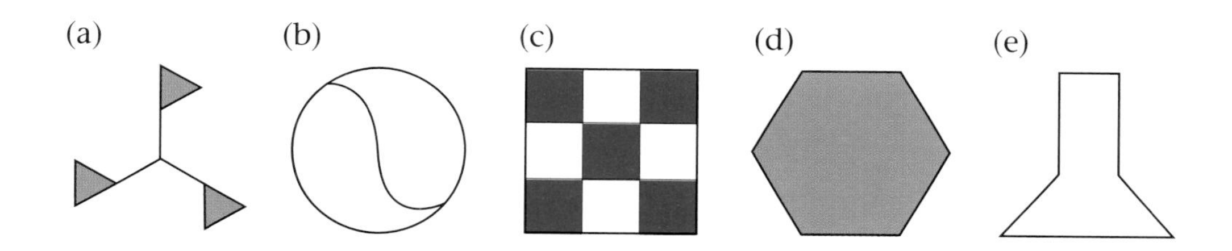

7 How many lines of symmetry has each of these shapes?

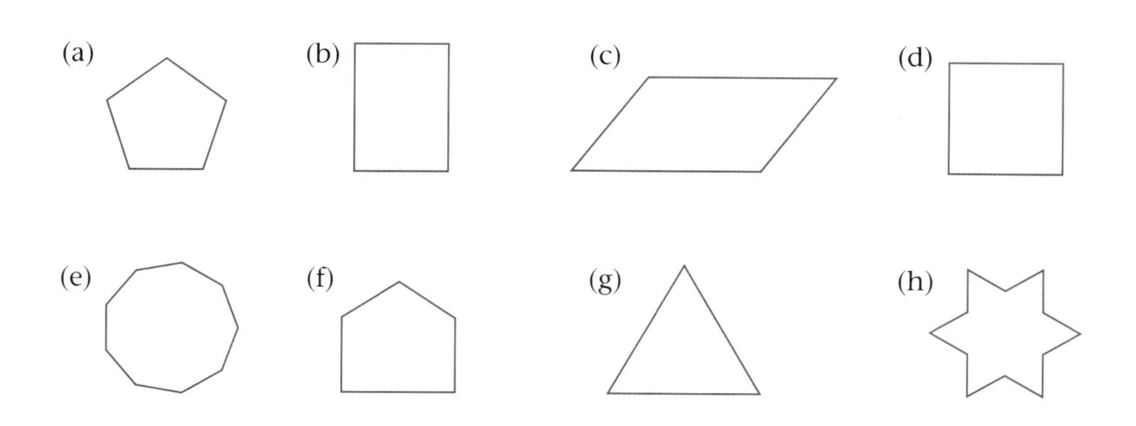

8 What is the order of rotational symmetry for each of the shapes in question 7?

Polyhedra

Key fact
A polyhedron is a shape which has many faces.

A shape which has many flat surfaces is called a **polyhedron**.

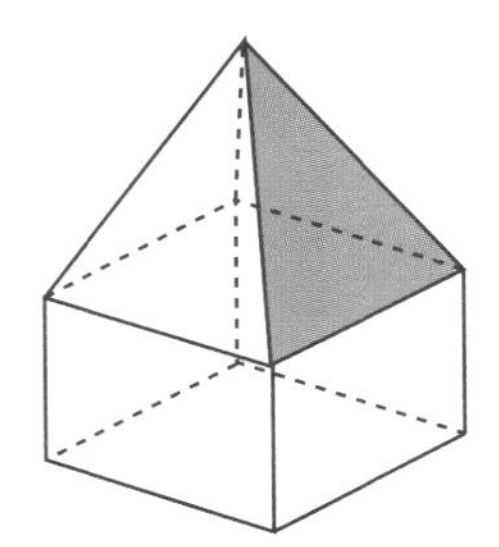

Flat surfaces are often called faces.

The name comes from the Greek words *poly* (many) and *hedra* (seats). 'Polyhedra' is the plural of polyhedron.

Polyhedra are called **3-dimensional** or **3-D** shapes, or **solids**.

Exercise 2

The name of a polyhedron can be given by the number of faces it has:

tetra – 4	penta – 5	hexa – 6	hepta – 7
octa – 8	nona – 9	deca – 10	dodeca – 12
icosa – 20	poly – many		

This shape has 5 faces.
It is a pentahedron.

Some polyhedra have their own names, for example **cube** and **pyramid**.

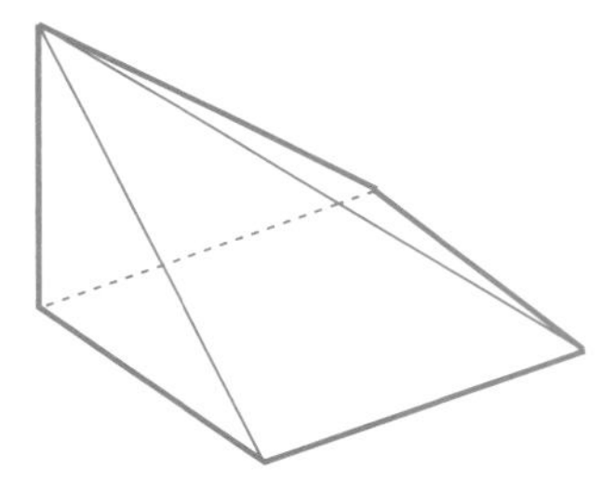

In drawings of polyhedra, the dotted lines show edges that are hidden from view.

1 Name the polyhedra below.

(a)

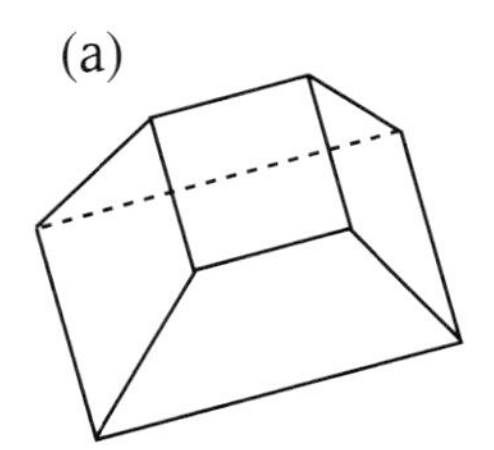

(b)

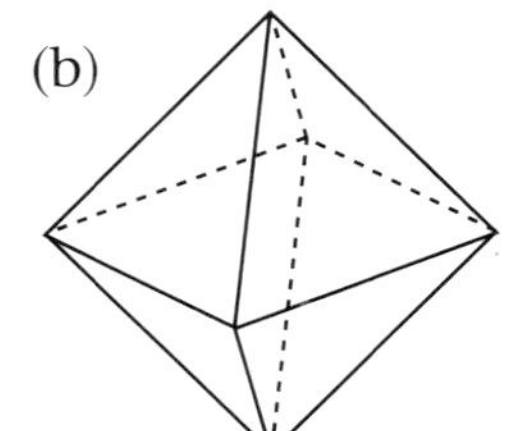

(c)

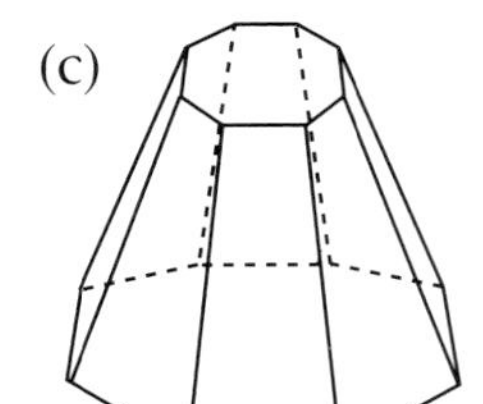

(d)

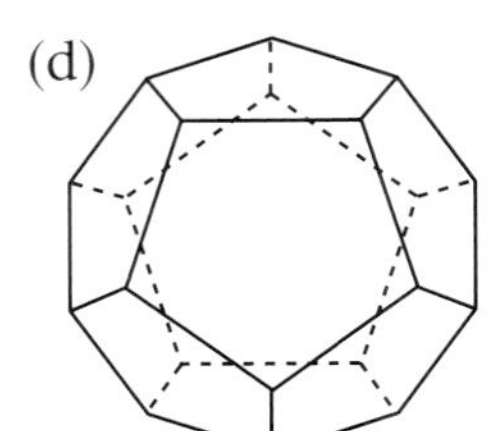

Key fact

A regular polyhedron is one in which all the faces are identical.

2 A polygon where all the faces are identical is called **regular**.

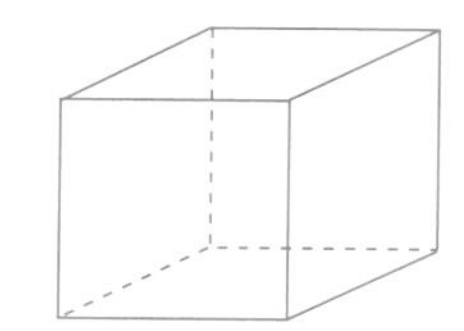

For example, this shape has six identical faces so it is called a **regular polyhedron**.

Because it is such a special shape, it has another name: a **cube**.

Which of these shapes are regular polyhedra?

(a)

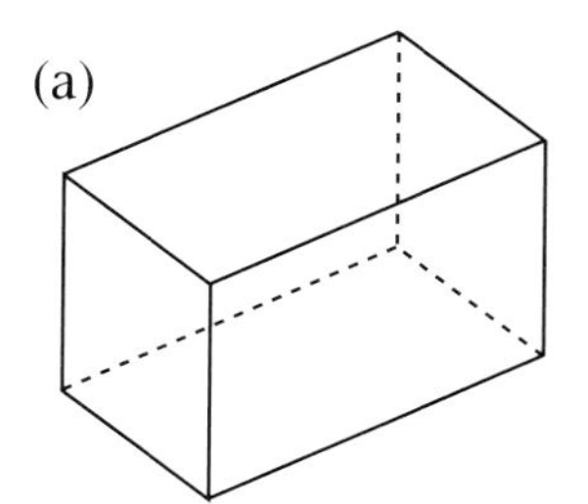

(b)

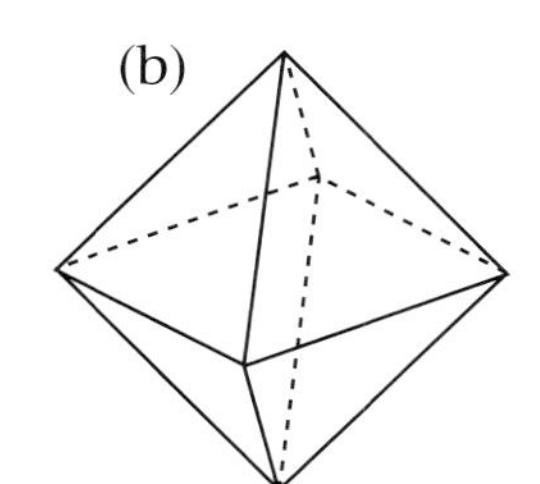

(c)

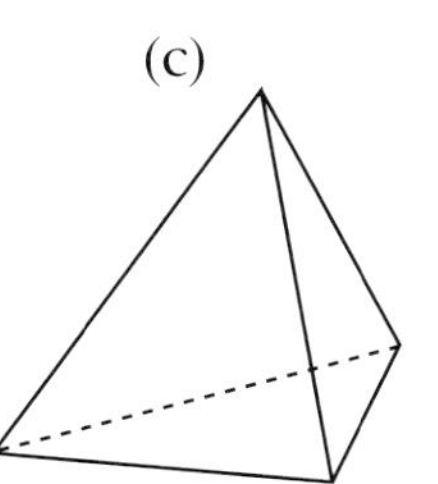

(d)

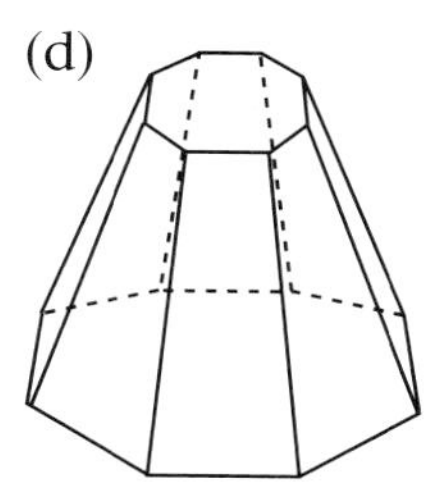

(e)

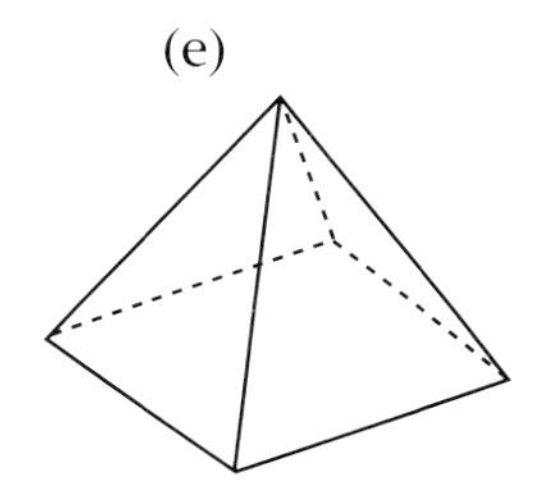

(f)

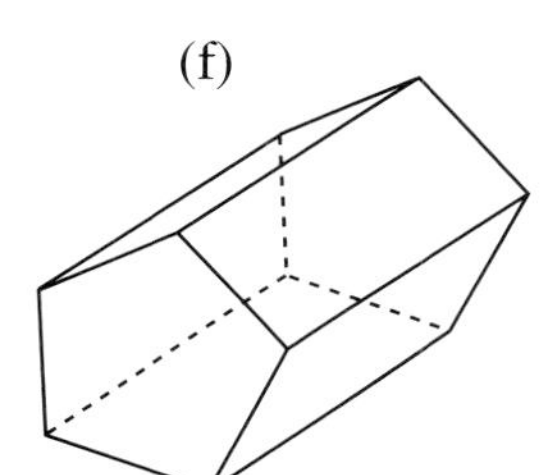

(g)

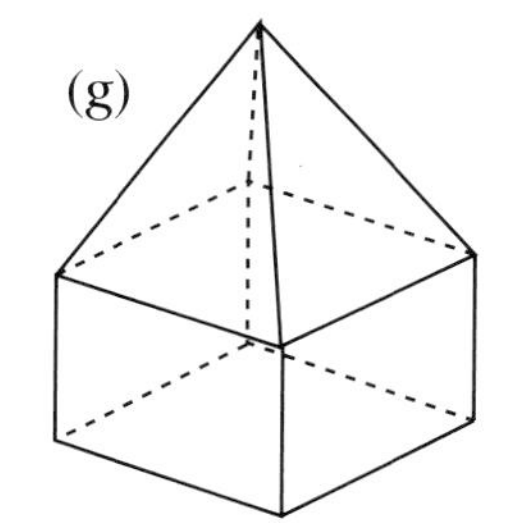

(h)

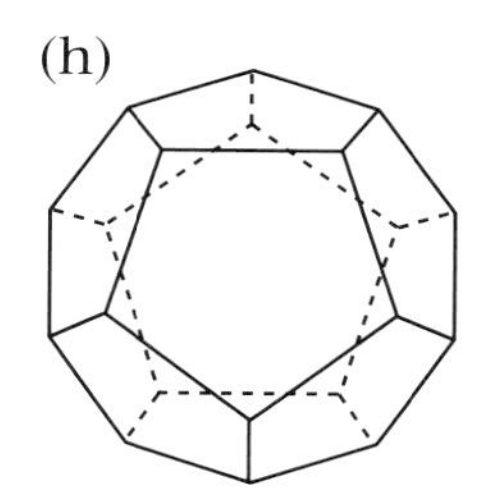

3 These three shapes are examples of **pyramids**.

(a)

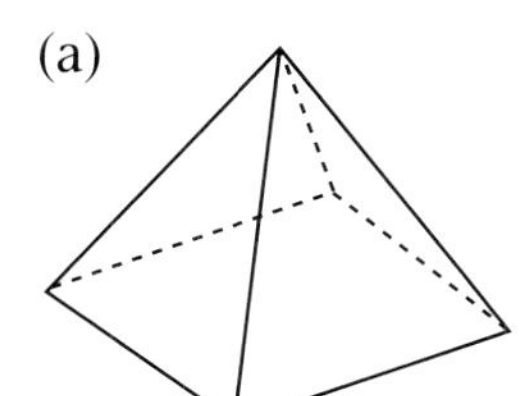

(b)

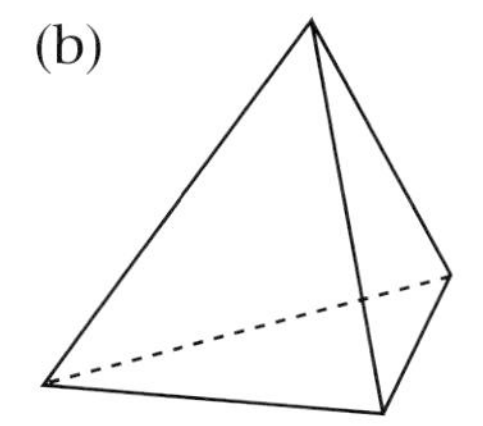

(c)

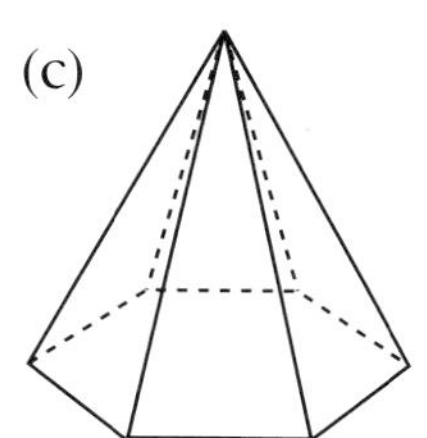

(i) What is special about the shape of pyramids?

(ii) Give a special name to each one. The name should give a clue as to what type of pyramid it is. (A pyramid is usually named after the shape of its base.)

Prisms

Key fact

A prism is a solid with an identical shape throughout its length.

A solid with an identical shape throughout its length is called a **prism**.
For example, this shape is called a **triangular prism**.

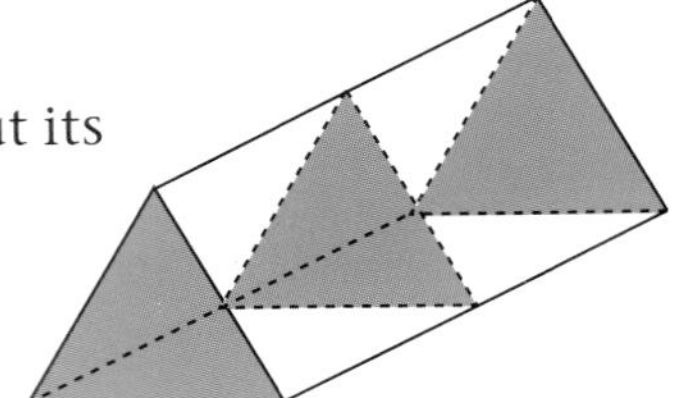

Exercise 3

1 Name these three shapes.

(a)

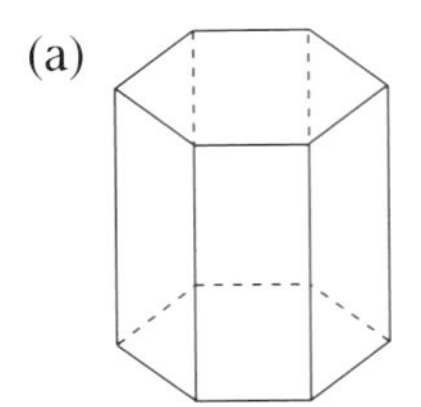

(b)

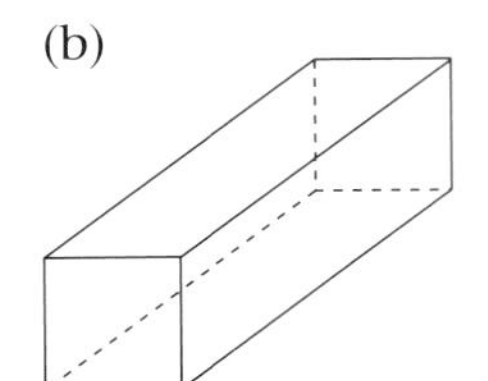

(c)

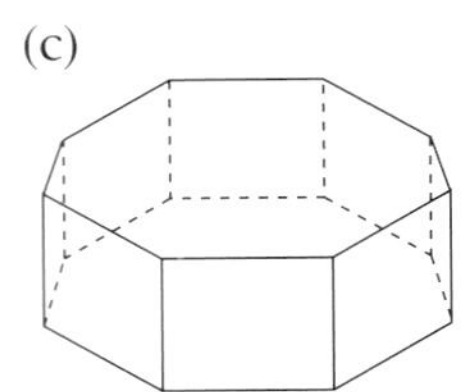

2 A prism with circular ends is called a **cylinder**.
Write down three everyday examples of cylinders.

Nets of polyhedra

The net of a cube

A paper cube is opened out so that its faces lie on a flat surface.
The result is a **net**.

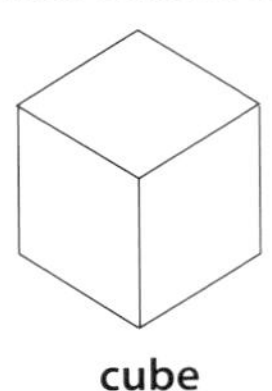

cube

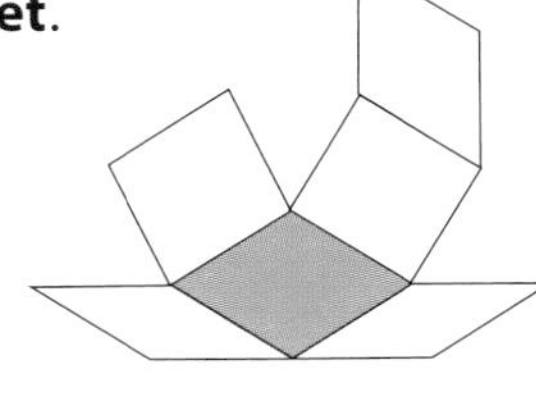

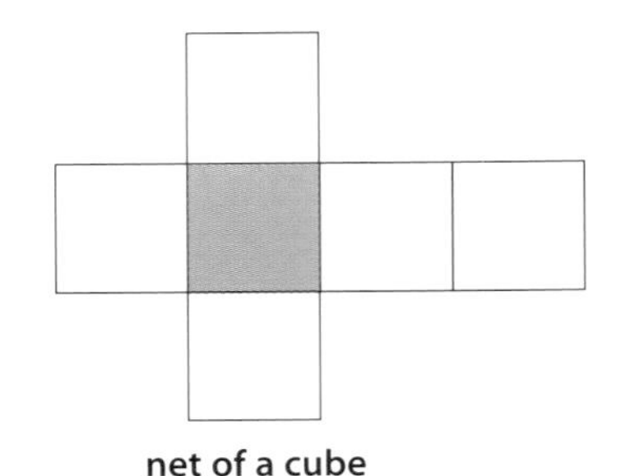

net of a cube

The net is made from six squares.
There are different ways to arrange six squares.
Here are some examples.

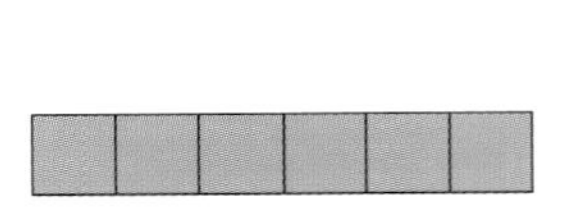

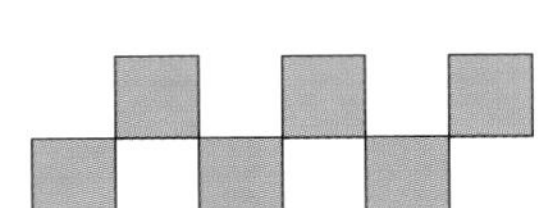

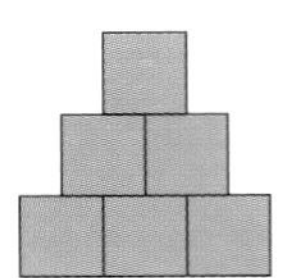

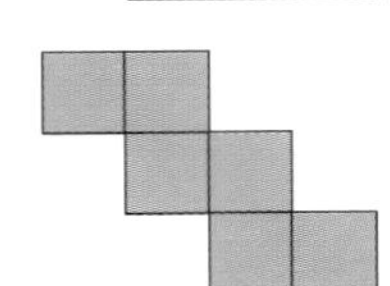

Not all of these can be folded to make a cube.

Assignment 1 Nets for a cube

- How many different nets for a cube can you find?
- Are these nets all the same? You will need to consider your work on symmetry.

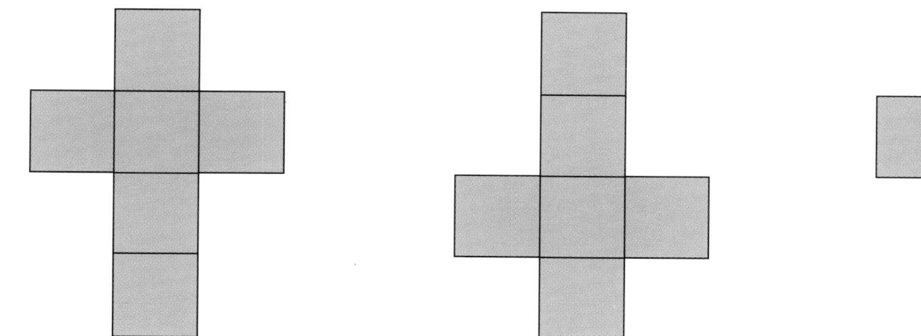

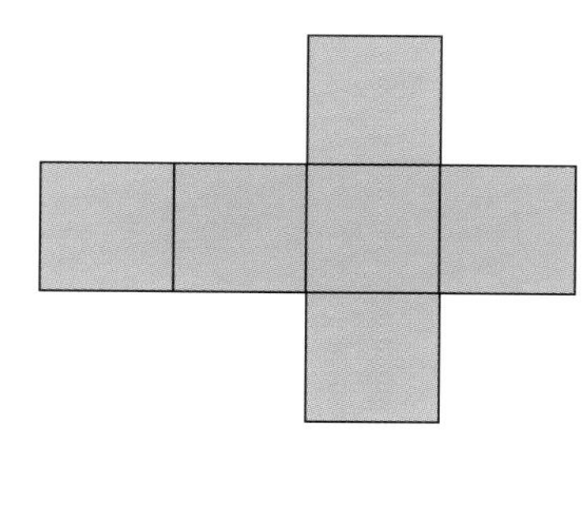

Nets of other shapes

Exercise 4

1 Match each name with the correct net.

pentagonal-based pyramid
tetrahedron
cuboid
square-based pyramid
triangular prism
octahedron

(a)

(b)

(c)

(d)

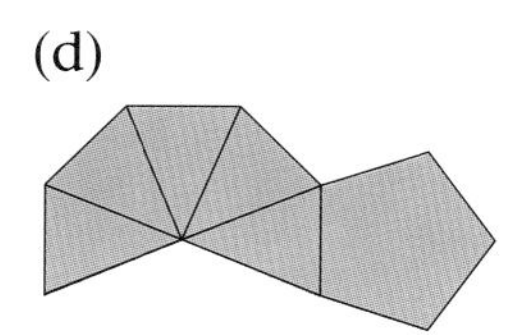

(e)

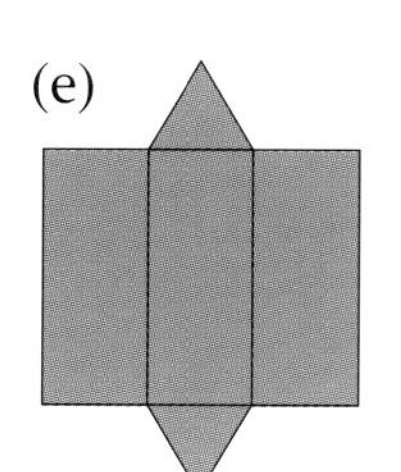

(f)

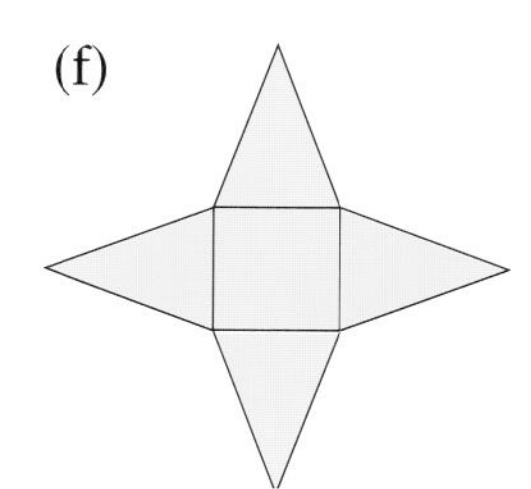

The edges of the shape won't meet if the net isn't accurately drawn.

2 Choose a simple shape, for example a cuboid or pyramid.

Design a suitable net on thin card or dotted paper.
Remember to add tabs. Fold and glue your net to make the shape.

3 Design a net for a more complicated shape. You could try to make a shape with more faces. Experiment.

Assignment 2 Seagulls

Take a **square** piece of paper and fold it as shown below.

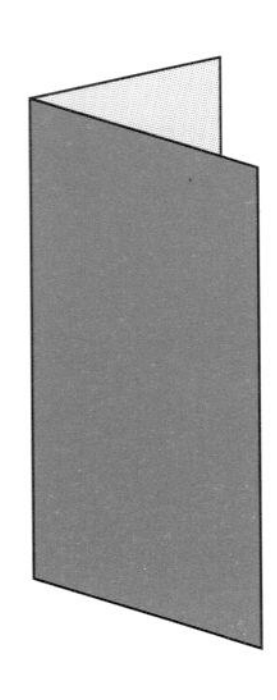

1
Fold it in half.

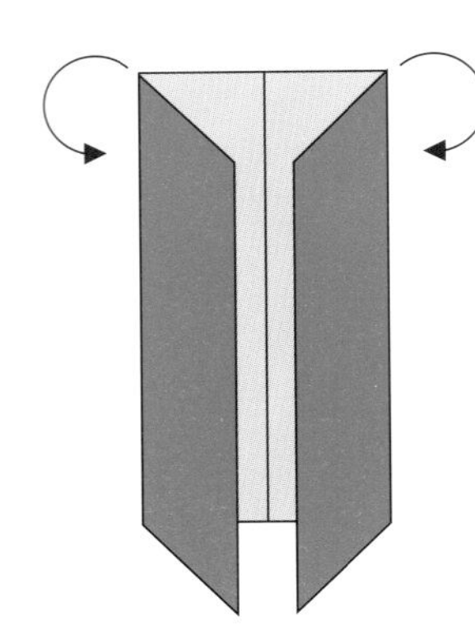

2
Fold the sides to the middle.

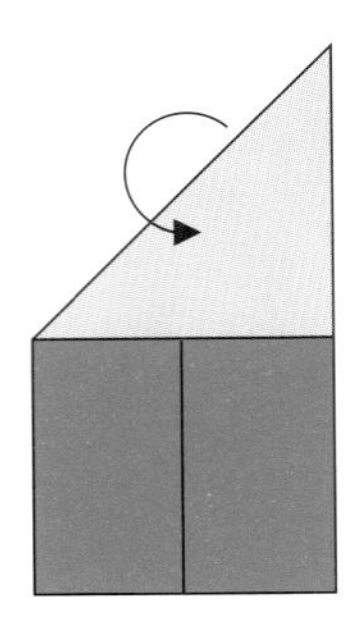

3
Fold down the top left corner.

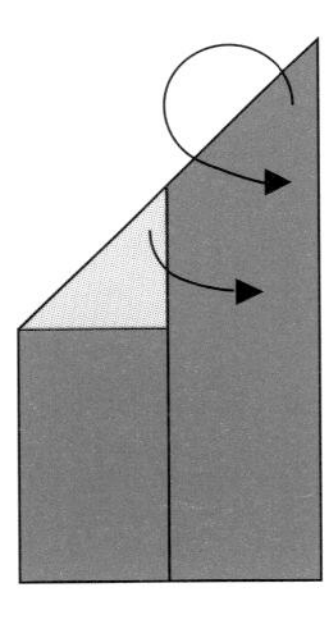

4
Tuck the big corner under and the little flap in.

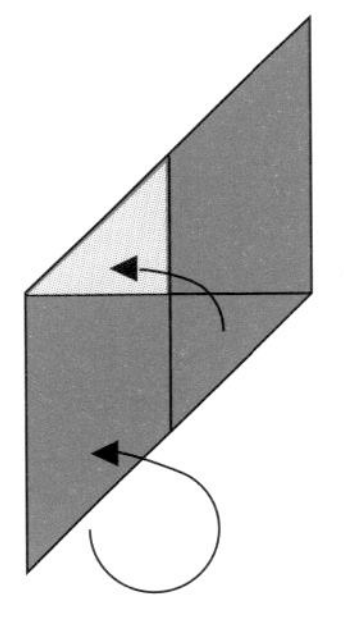

5
Fold up the bottom right corner and tuck under.

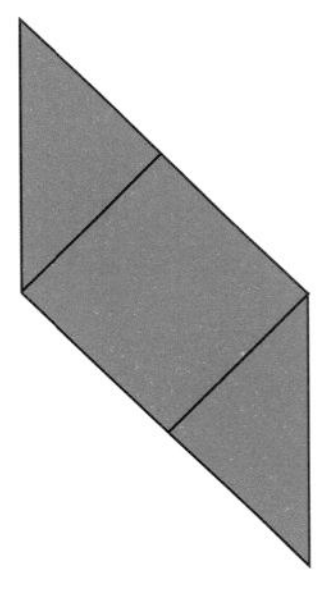

6
Turn over and fold the points forward along the dotted lines to form a square.

The completed shape is called a **seagull**.

When you have six seagulls, you can link them together to make a cube.

Try joining two cubes together.

If you have more seagulls, try making some other shapes.

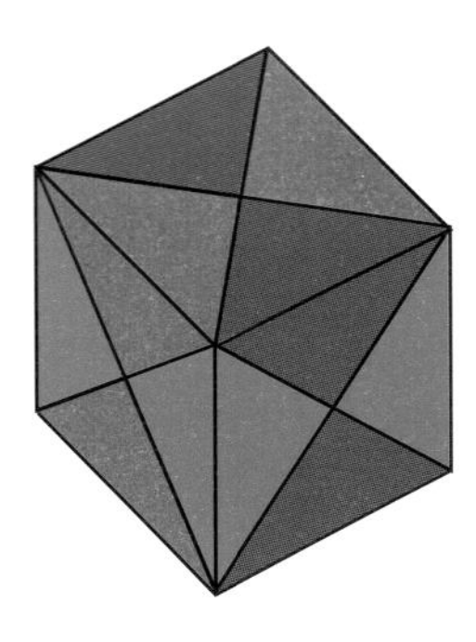

Drawing 3-D shapes in 2-D

The diagrams show how a cuboid, which is 3-dimensional, can be drawn on paper. This is in 2 dimensions.

You could use squared paper (dotted or lined).

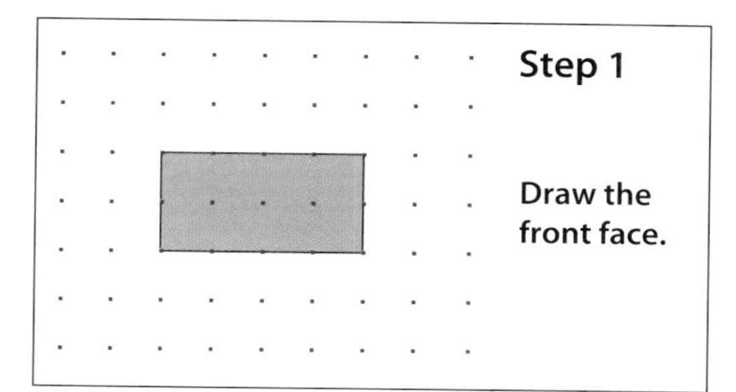

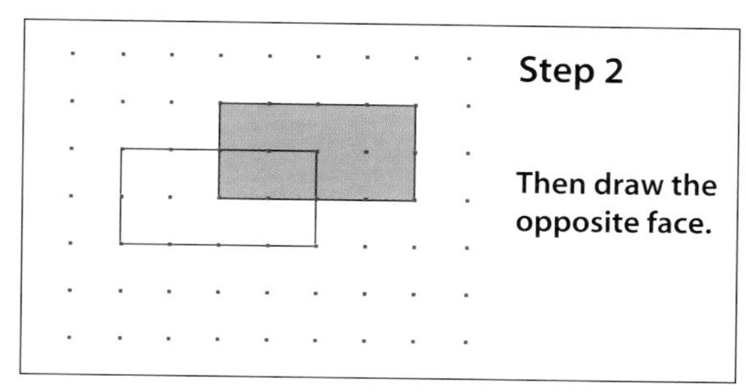

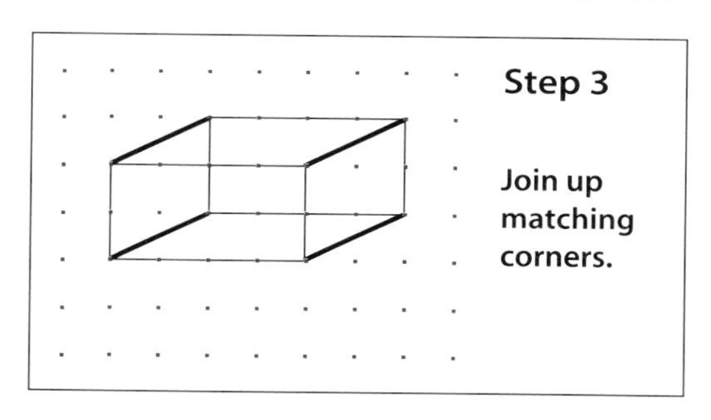

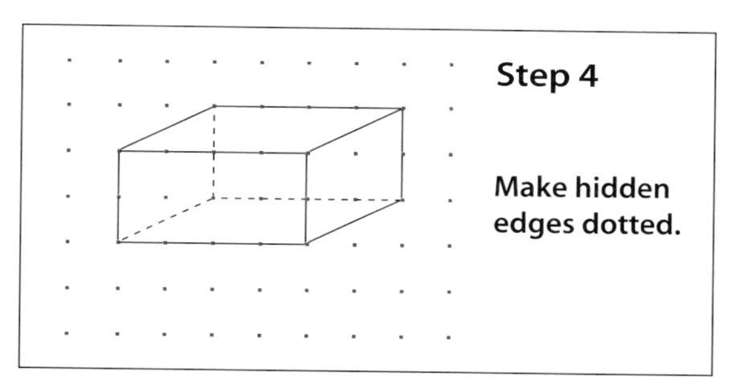

Or you could use isometric paper (triangular).

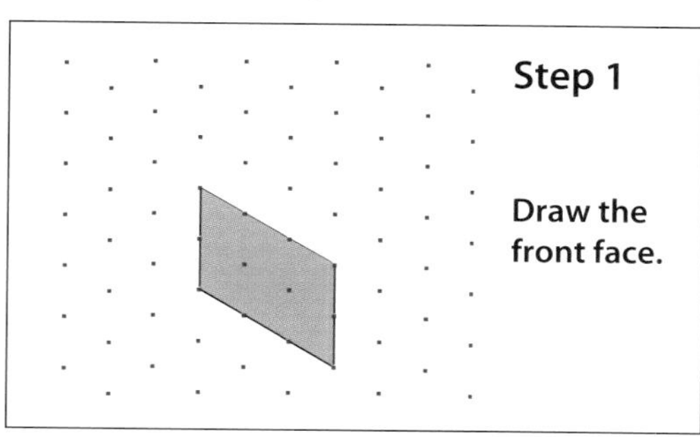

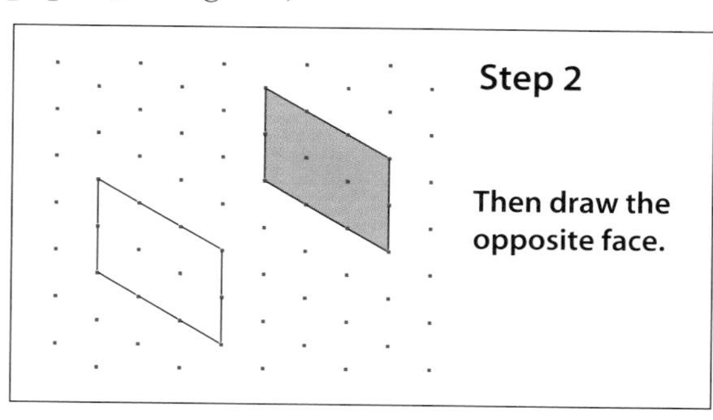

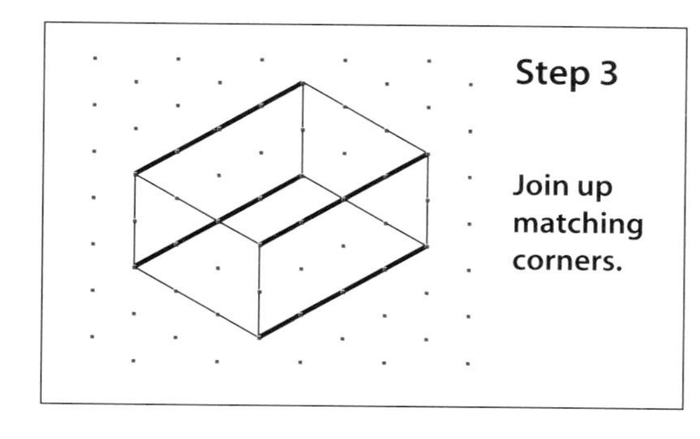

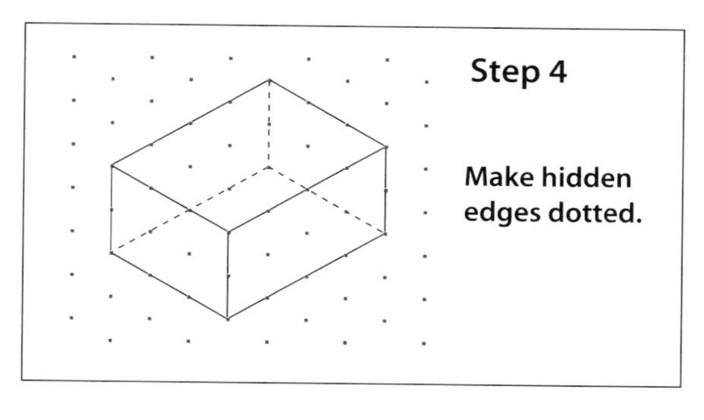

Exercise 5

1 Make drawings of these solids on squared paper.

(a)

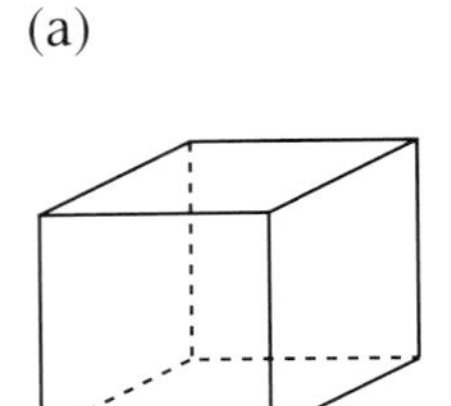

(b)

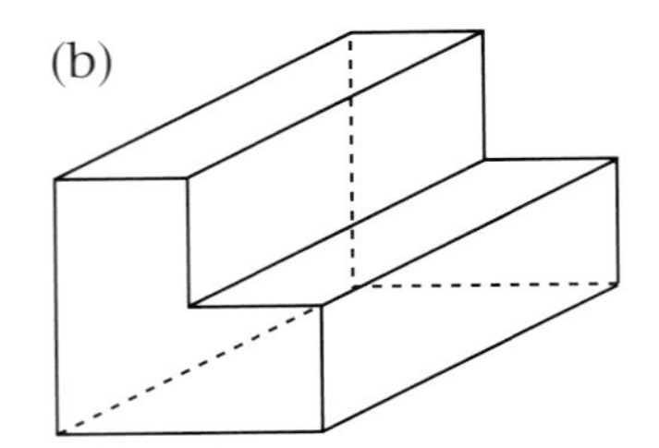

(c)

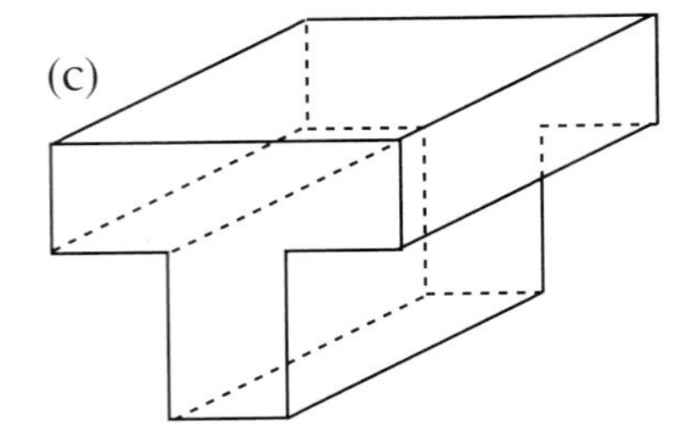

2 Make drawings of these solids on isometric paper.

(a)

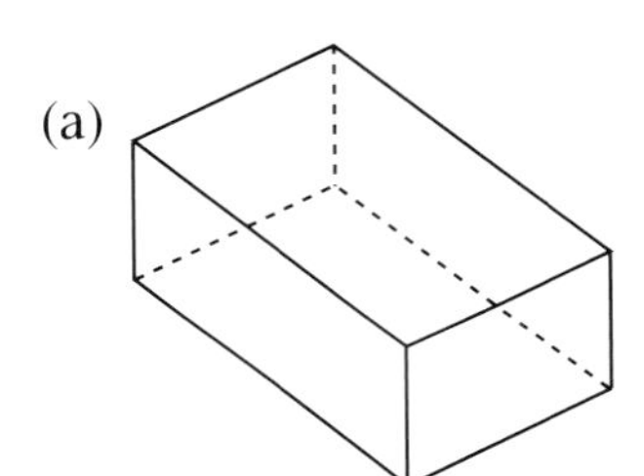

(b)

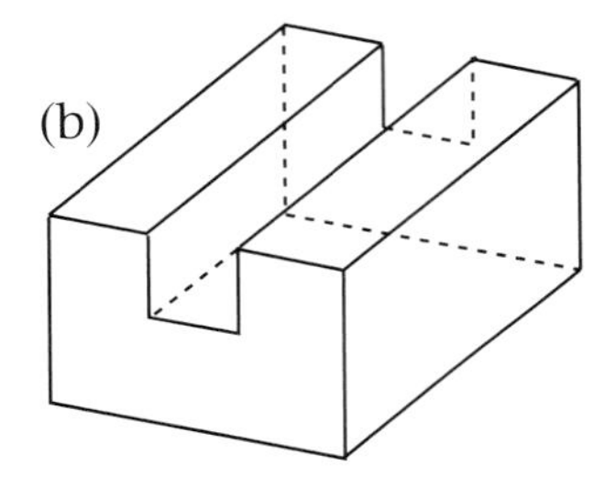

(c)

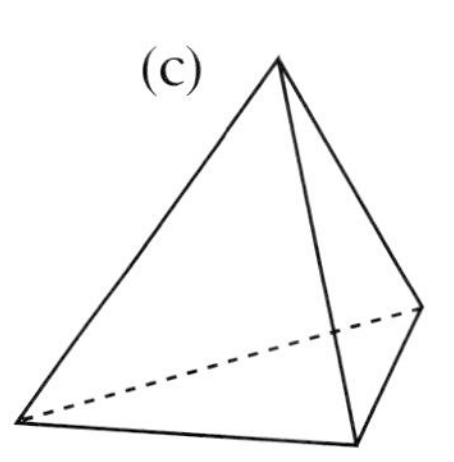

3 Draw some 3-D shapes of your own. Perhaps you could draw the shapes you made in exercise 4.

Key facts

A face is a flat surface.

An edge is where two faces meet.

A vertex is a corner where edges meet.

Properties of polyhedra

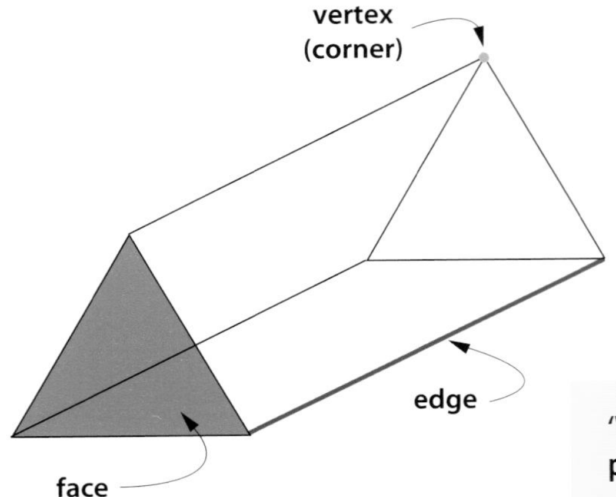

This prism has 5 faces, 9 edges and 6 vertices.

'Vertices' is the plural of 'vertex'.

Exercise 6

1 Copy the table for these polyhedra and complete it.

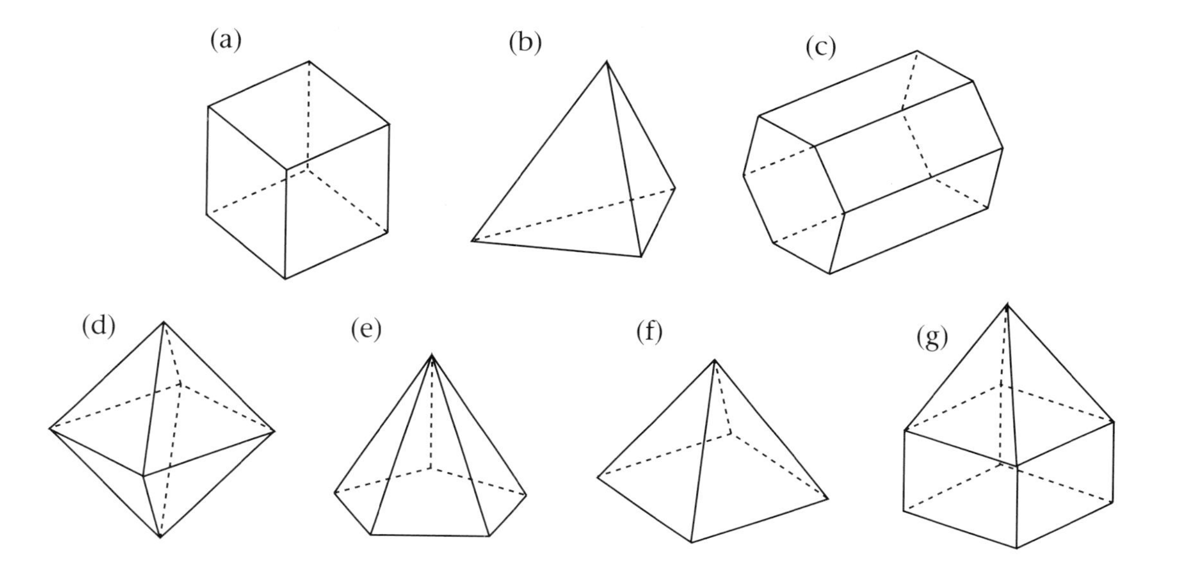

Shape	Number of faces	Number of vertices	Number of edges
(a)	6	8	12
(b)	4		
(c)			
(d)			
(e)			
(f)			
(g)			

2 There is a relationship between the number of faces, vertices and edges which works for all the shapes.

Describe this relationship.

The relationship is known as Euler's Rule. Leonhard Euler (1707–1783) was a famous Swiss mathematician.
He enjoyed studying shapes and he proved his rule in 1752.

3 Use Euler's Rule to work out the missing numbers in the table.

Shape	Number of faces	Number of vertices	Number of edges
A	10	12	
B	9		15
C		8	16

4 A cube has 12 edges, 6 faces and 8 vertices.

An octahedron has 12 edges, 8 faces and 6 vertices.

They have the same number of edges but the number of faces is swapped with the number of vertices.

When this happens, each shape is known as the **dual** of the other.

So a cube is the dual of an octahedron, and vice-versa.

Describe the dual of a triangular prism. Make a sketch of both shapes.

5 A tetrahedron has 4 faces and 4 vertices. It is its own dual, that is a **self-dual**.

Find other polyhedra which are self-duals.

Can you make any general statements?

Looking at 3-D shapes

So far in this chapter you have seen solids (3-D shapes) shown like this:

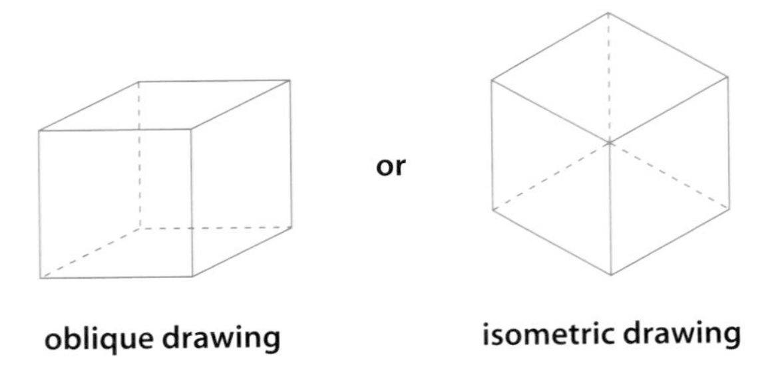

oblique drawing   isometric drawing

Although each drawing gives a good idea of what a cube looks like, some of the faces have been distorted.

Another way of showing shapes is to give views from different directions. This shape has five cubes: 2 red, 2 blue and 1 green.

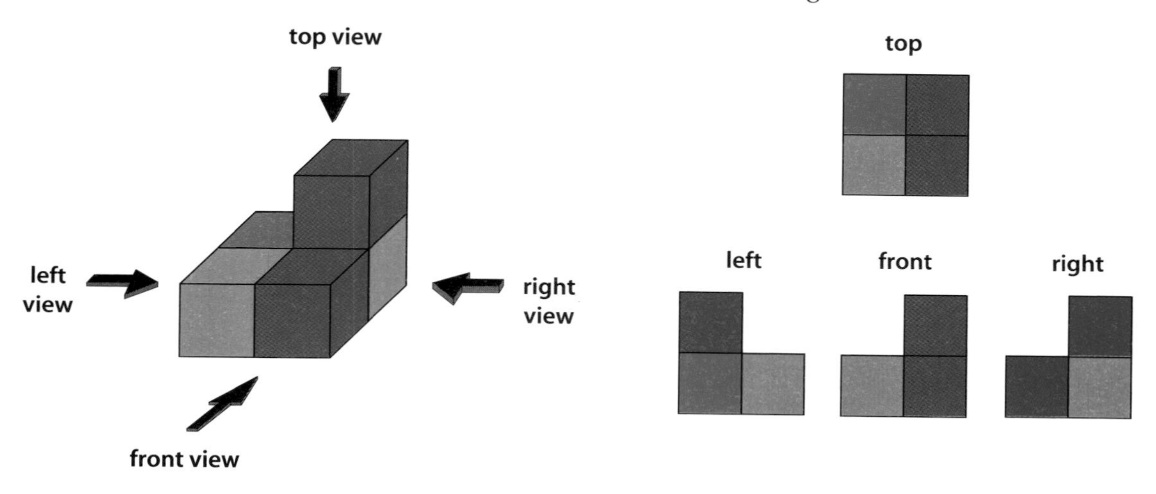

Exercise 7

Look at the views of various shapes shown in the examples.

See if you can make these shapes.
Compare your shapes with those of your friends.

1 This is an easy shape with only 5 cubes: 2 red, 2 yellow and 1 blue.

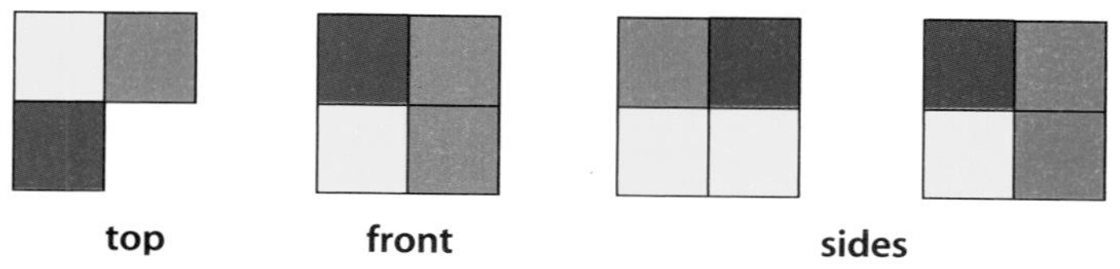

2 This is easier than it looks.
You need 12 cubes: 3 yellow, 3 red, 3 brown and 3 green.

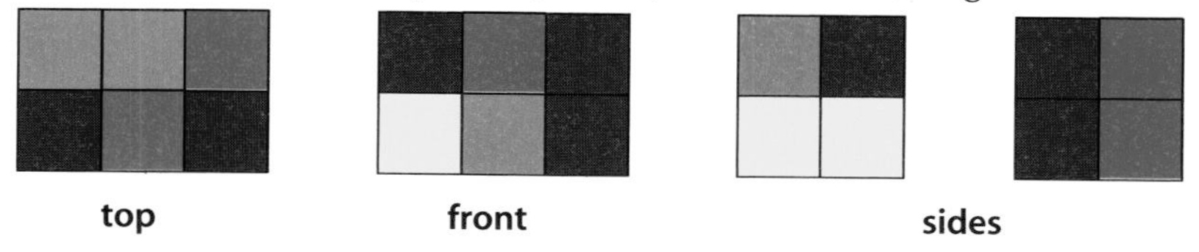

3 This is a shape with 10 cubes: 3 pink, 3 brown, 2 yellow and 2 green.

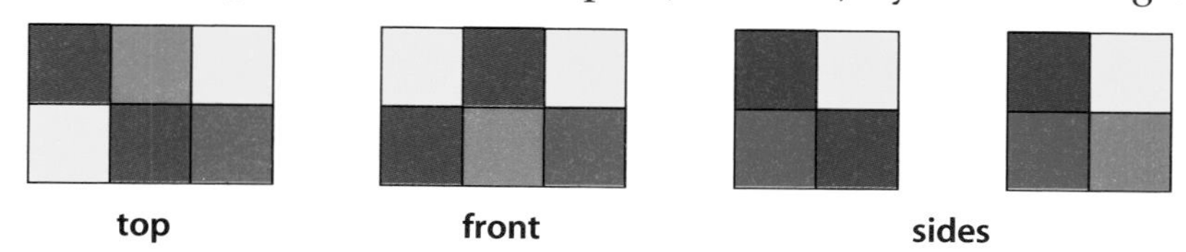

4 Try this 7-cube shape with 4 red, 2 orange and 1 yellow cube.

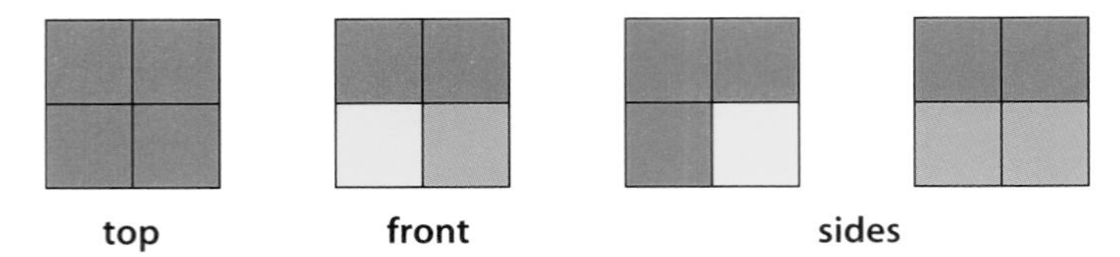

5 You need 9 cubes for this shape: 3 blue, 3 orange, 2 green and 1 yellow.

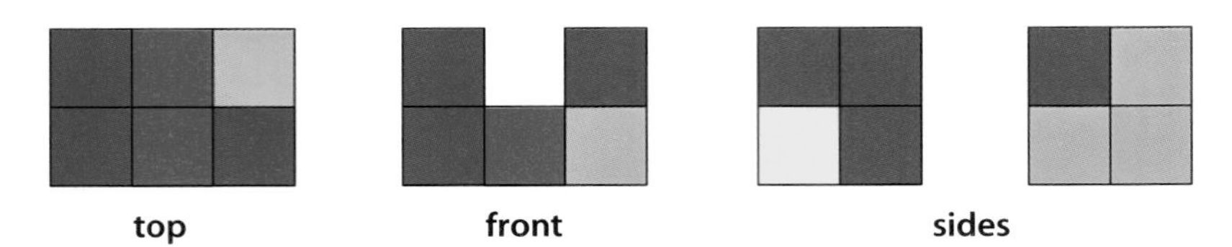

Assignment 3 Stacking boxes

You are given three boxes to stack against a wall.

The faces of the boxes must fit together exactly; they must not overlap.

Two possible ways are shown here.

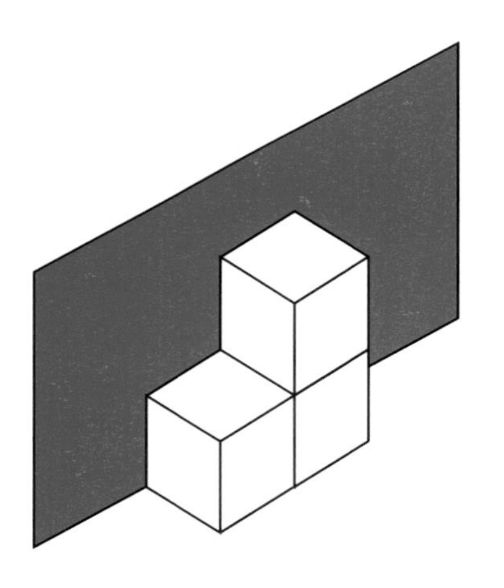

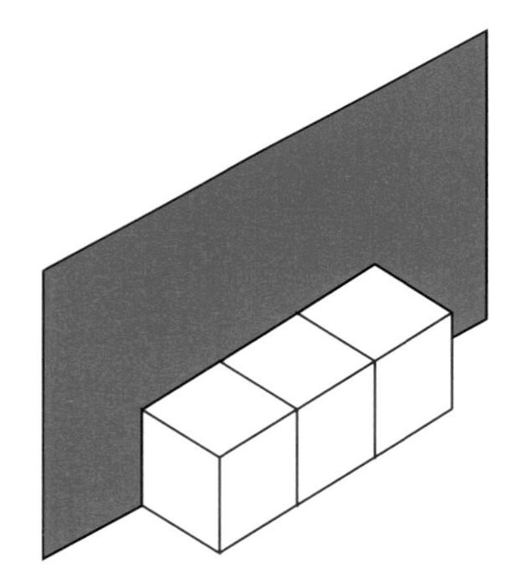

- How many ways can you find to stack the boxes?
- What about two boxes? ten boxes? any number of boxes?
- Extend your ideas.

Key fact

A plane of symmetry cuts a solid into 2 equal parts which are mirror images of each other.

Symmetry of 3-D shapes

In Chapter 6 you worked on the symmetries of 2-D (flat) shapes.
A 3-D shape also has symmetry.

Planes of symmetry

Look at the cuboid below. It can be sliced into two equal parts in three ways, as shown. The two parts are mirror images of each other.

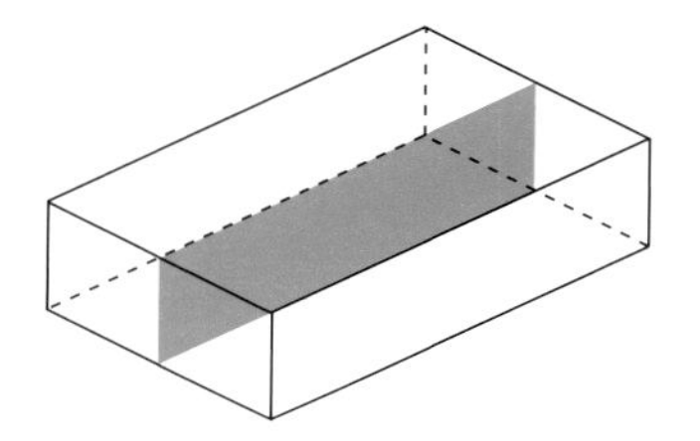

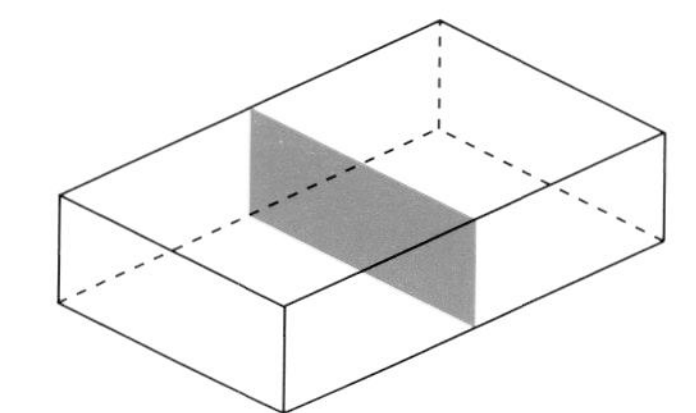

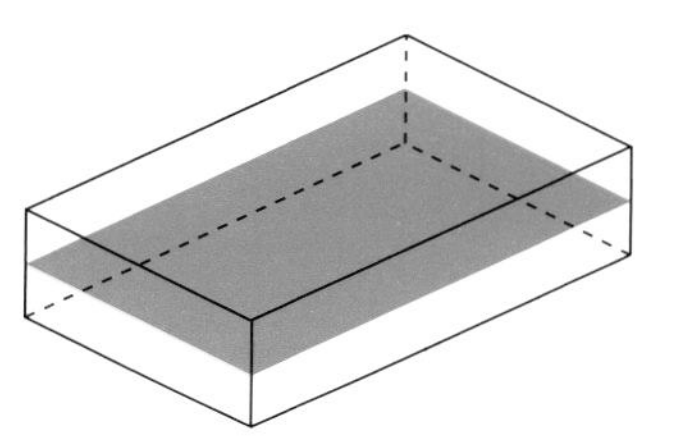

Each cut surface is a **plane of symmetry**.

This cuboid has 3 planes of symmetry.

The cuboid is symmetrical about each plane.

A plane is a flat surface.

Exercise 8

1 How many planes of symmetry has each of these solids?

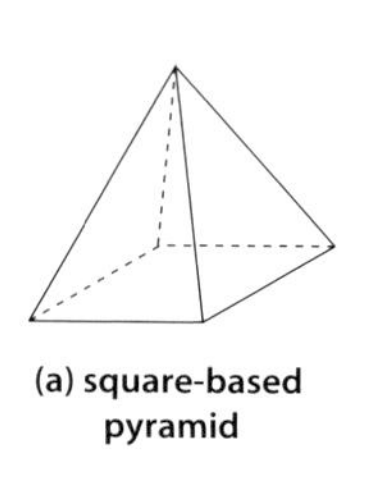

(a) square-based pyramid

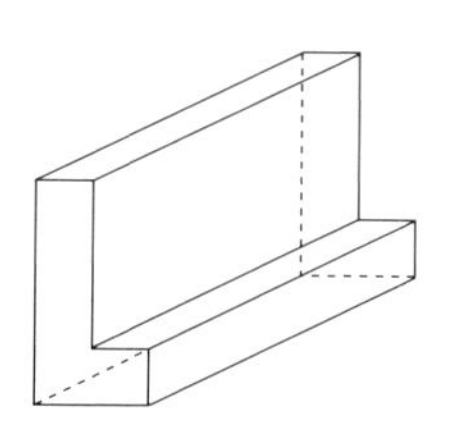

(b) L-shaped prism

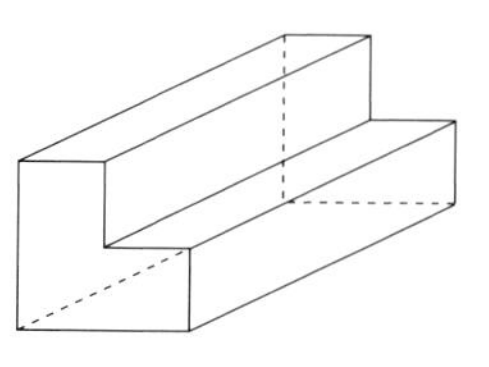

(c) L-shaped prism

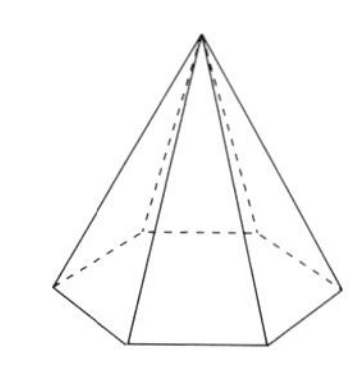

(d) regular hexagonal-based pyramid

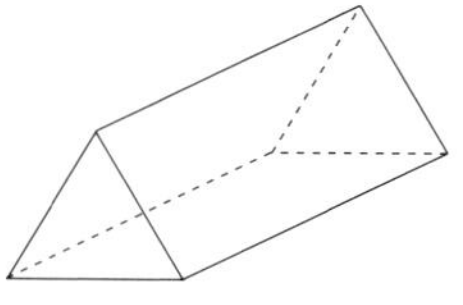

(e) equilateral triangular prism

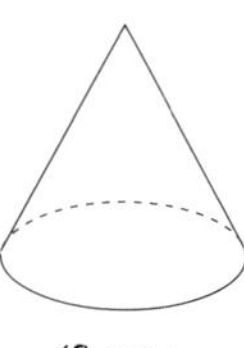

(f) cone

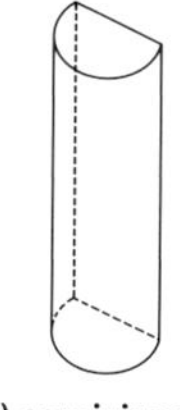

(g) semicircular prism

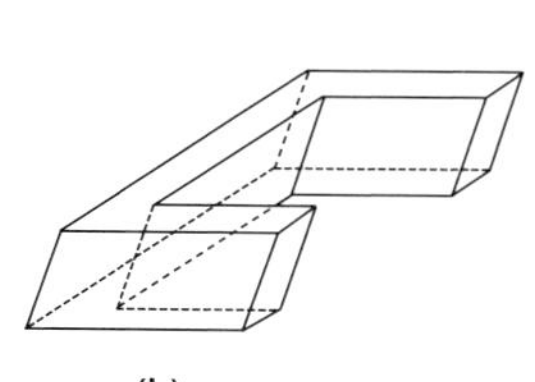

(h)

2 How many planes of symmetry has each of these objects?

(a) the tray of a matchbox (b) a chair

(c) a rectangular table (d) a shoe

(e) a pair of sunglasses (f) a can of soup

3 Imagine you slice through a prism like this.
If the slices are parallel to the end faces, the 'new' faces would all be the same. These faces are called **cross-sections**.

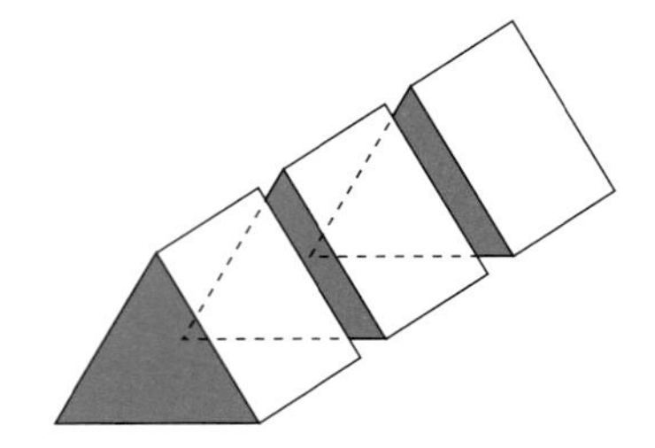

The cross-sections of several prisms are shown below.

(a) 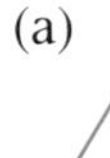

(b) 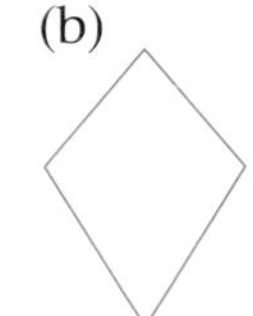

(c)  

(d)  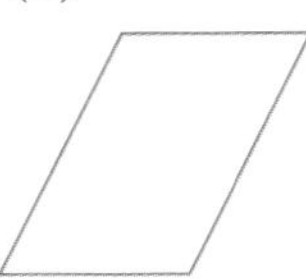

(e) 

(f) 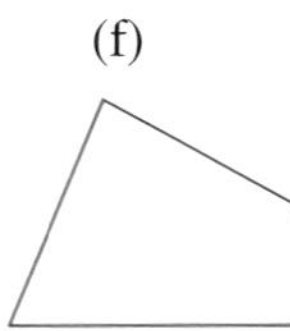

(g) 

(h) 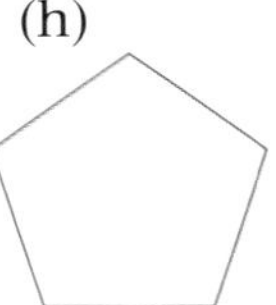

(i) 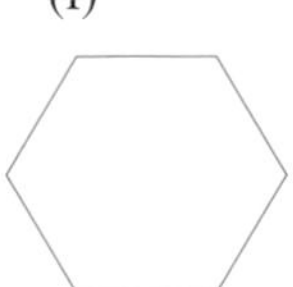

(j) 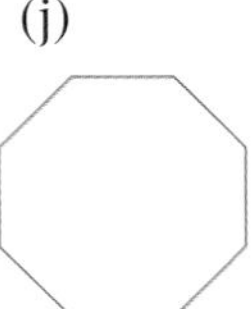

(k) 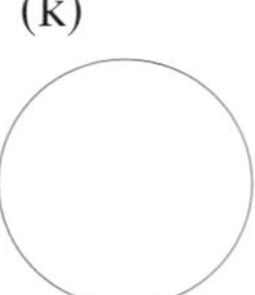

(l) 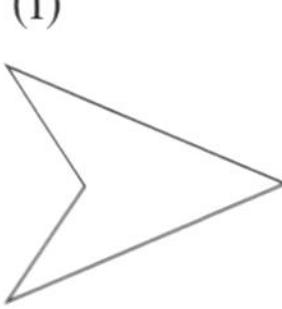

Copy and complete the table. You will have to think very carefully because you are not shown all of each shape.
The first two have been done for you.

Type of prism	Number of planes of symmetry
(a) Equilateral triangular	4
(b) Kite-faced	2
(c)	

Review Exercise

1 Copy this table. Write in the correct letter for each shape.

Name of polyhedron	Shape
cube	(c)
tetrahedron	
pentagonal-faced prism	
cuboid	
square-based pyramid	
dodecahedron	

(a)

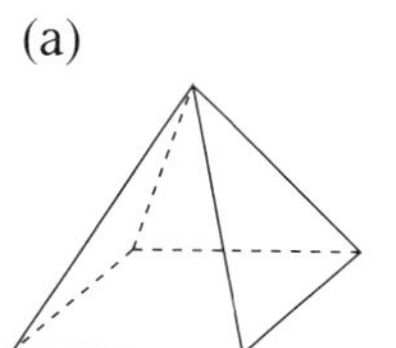

(b)

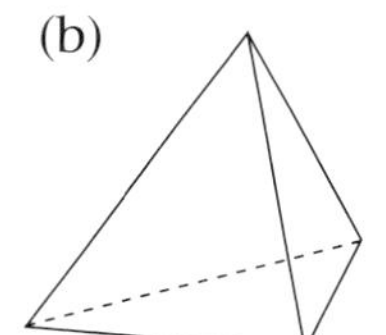

(c)

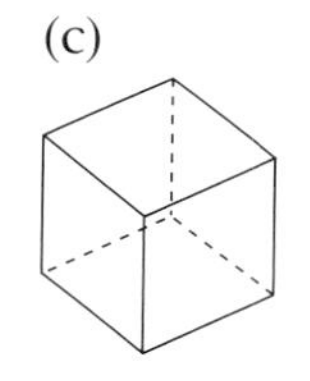

(d)

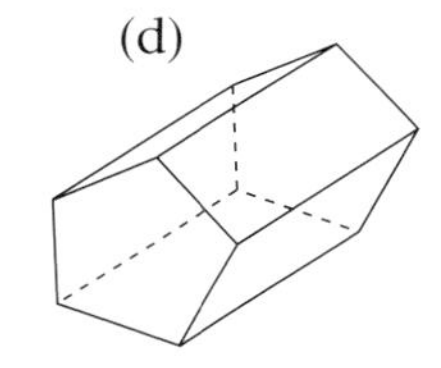

(e)

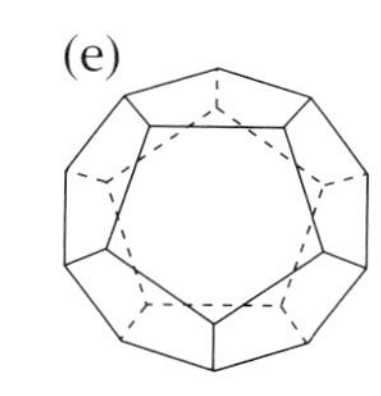

(f)

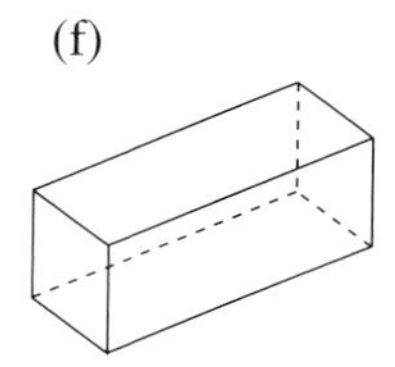

2 Which of the shapes in question 1 are regular polyhedra?

3 Name the polyhedra that can be made with the following nets.

(a)

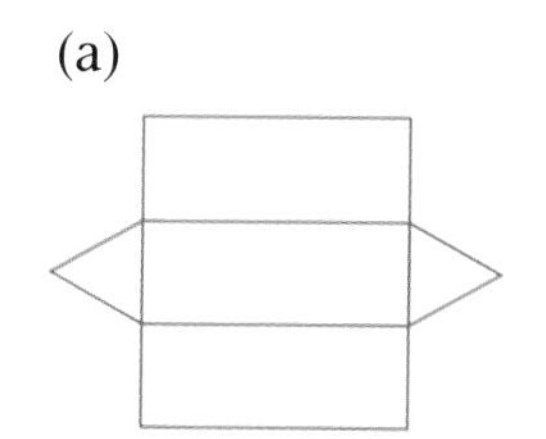

(b)

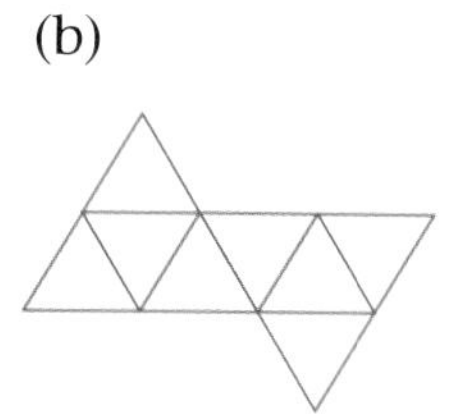

(c)

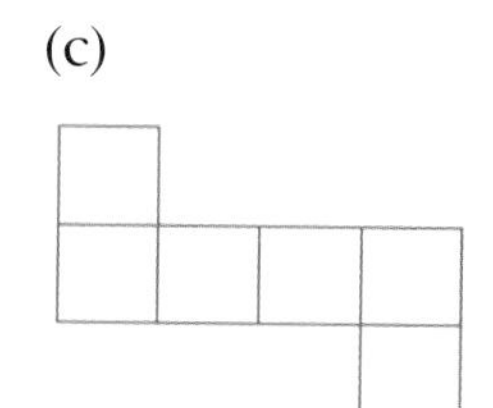

(d)

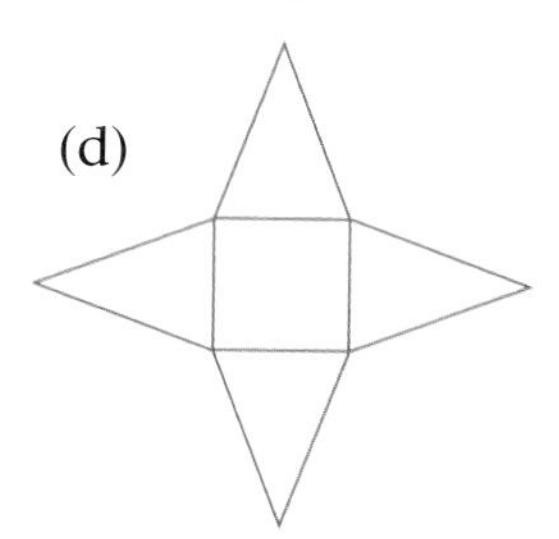

4 Use squared paper to draw a 2-D representation of this shape.

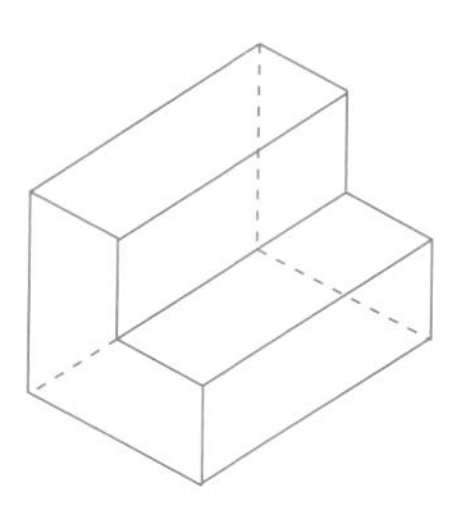

5 For each of these shapes, write down (i) the number of faces (ii) the number of vertices (iii) the number of edges.

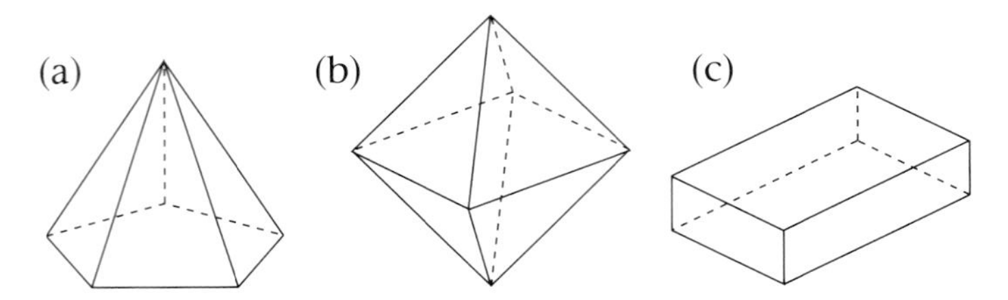

6 Which of the drawings shows the view from above this shape?

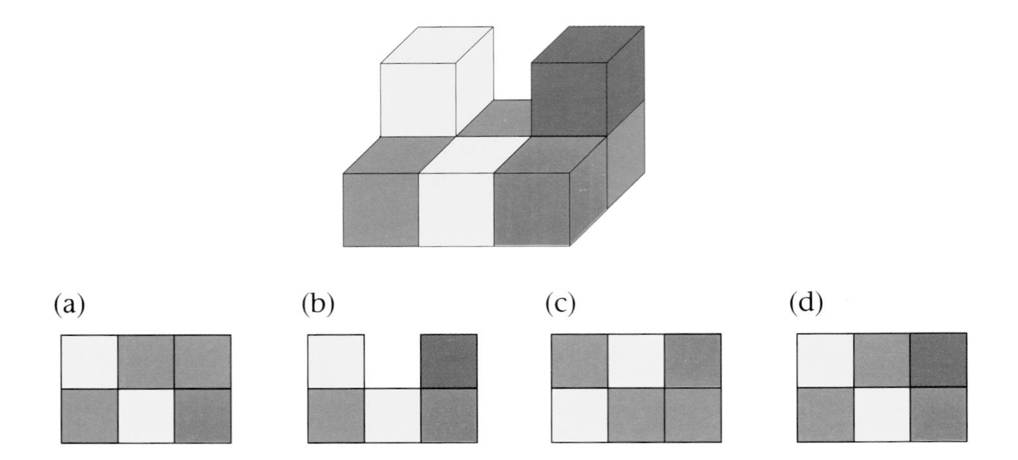

7 How many planes of symmetry has each of these shapes?

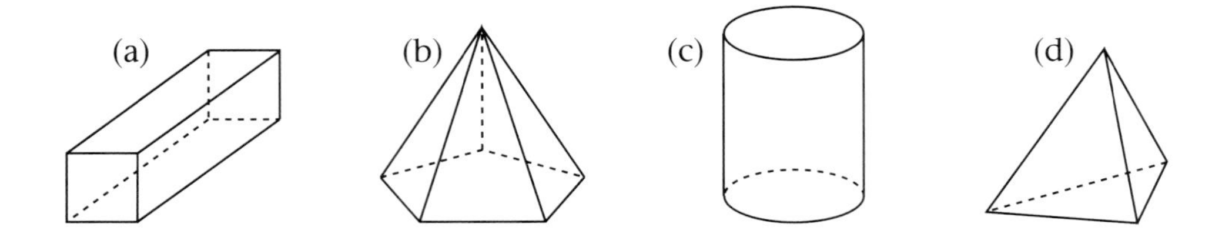

8 A polyhedron has 10 faces and 24 edges.
How many vertices does it have?

9 A polyhedron has 8 faces, 12 vertices and 18 edges.
What is it?

Probability

In this chapter you will learn:

- → **about probability**
- → **how to use a probability scale**
- → **how to write probabilities**

Starting points

Before starting this chapter you will need to know:

- a little about chance and when things are more or less likely to happen
- how to carry out a tally
- how to draw a bar chart
- how to make the net of a cube

Exercise 1

1 Draw a line about 10 cm long, similar to the one below.
Use it to show the chance of the following things happening.

(a) You will spend some money today.

(b) It will snow tomorrow.

(c) You will get a phone call tonight.

(d) You will have homework to do this evening.

(e) There will be a fire drill today.

(f) The address of a house or flat is an even number.

2 This story is taken from *Oliver Twist* by Charles Dickens.

For the next eight or ten months, Oliver was the victim of a systematic course of treachery and deception. He was brought up by hand. The hungry and destitute situation of the infant orphan was duly reported by the workhouse authorities to the parish authorities.

(a) Copy the following table. Record in your table how many times each of the letters listed occurs in the passage.

Remember to tally in 5s.

𝍸

Letter	*Tally*	*Frequency*
e(E)	II	
u(U)		
a(A)	I	
s(S)		
z(Z)		

(b) Draw a bar chart to show your results. Give your chart a title.

(c) Which letter do you think occurs most often in the whole book?

(d) Write the five letters in order of the chance of finding them in the book.

More than a guess

Red sky at night is a shepherd's delight.
Red sky in the morning is a shepherd's warning.

This well-known saying is believed by many people to be true.

It means that if you see a red sky at night you are *likely* to have good weather the next day.

It is often difficult to find words to describe the **probability** accurately. People may disagree on which word is most suitable.

The words '*possibly, sometimes, maybe, perhaps, usually, likely*' all seem to have the same meaning but some seem more definite than others.

certain

no chance
impossible

Assignment 1 Describing probability

Copy the scale from the side of this page. Write the following words where you think they should go on the scale.

- possibly
- sometimes
- maybe
- perhaps
- usually
- likely

Exercise 2

> **It is important to be able to justify your comments.**

1 How likely is it that there will be a thunderstorm tomorrow?
Why did you decide this?

2 Estimate the likelihood that you will go swimming within the next week.
Write down why you decided on this estimate.

3 How likely is it that you will have chicken to eat today?
How did you decide?

4 If you drop a piece of buttered bread, how likely is it that the bread will land with the butter side down?
Why did you give this answer?

5 If an egg falls off a table, how likely is it the egg will break?
Why did you decide this?

6 What is the likelihood that you will watch more than three hours television today?
What things did you consider before making your estimate?

7 Look back at all the answers you have given to the questions in this exercise.
Which event do you think is the most likely to happen?

8 Which of your answers for questions 1 to 5 do you think is the least likely to happen?

Probability scales

Key fact
A probability of 1 means something is certain to happen.

Key fact
A probability of 0 means something has no chance of happening.

Key fact
Probability can be written as a fraction, decimal or percentage.

(a) It is **certain** that the year after 1999 will be the year 2000.

(b) The probability that an apple will fall downwards if you drop it on this planet is **1** or **100%**.

(c) It is **impossible** for the word zebra to be the first word in the English dictionary.

(d) The probability that an African elephant can fly by flapping its ears is **0** or **0%**.

As all probabilities lie between 0 and 1, or 0% and 100%, we can show these probabilities on a **probability scale**.

Exercise 3

1 Write down three events which you consider have a probability of 0 or 0%. Explain your answers.

2 Write down three events which you consider have a probability of 1 or 100%. Explain your answers.

3 What kind of events have a probability of 0.5 or 50%?

4 Write down three events which have a probability of 50%.

Uncertainty

When a new electric light bulb is switched on it is **very likely** that it will light up. This is shown on the probability scale below.

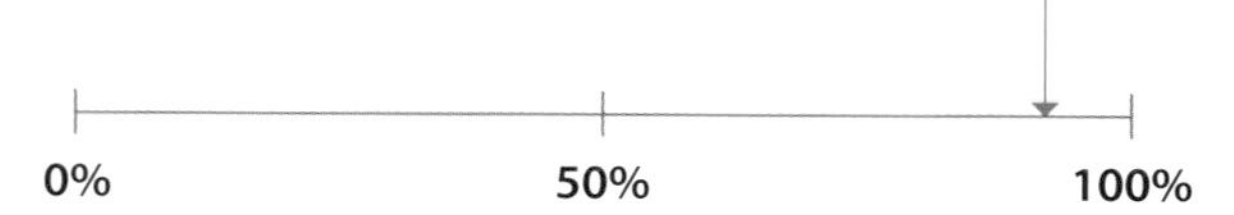

The probability that you will pass a police car on the way home from school is less certain.

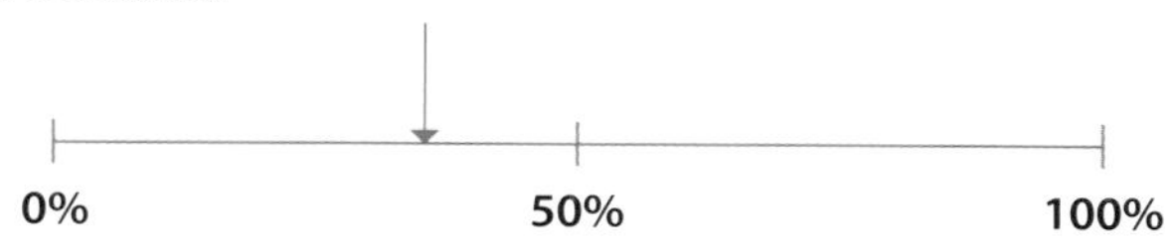

Exercise 4

1 Write these events in order of likelihood of happening. Start with the most likely to happen.

(a) You will get more than 70% in the next maths test.

(b) You will come to school on Saturday.

(c) The first person you meet when you leave this room will be male.

Show the probabilities on a probability scale.
Explain your answers.

2 These things were dropped onto the floor.

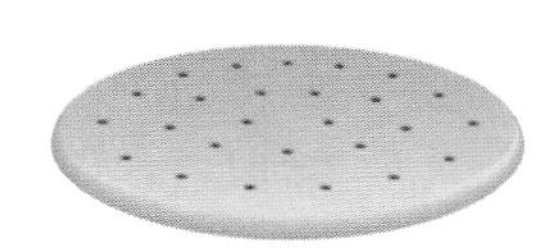

Write these down in order of the probability of their being damaged.
Start with the most certain to be damaged.

How did you decide?
Show your results on a probability scale.

3 Write these events in order of likelihood. Start with the *least* likely.

(a) Meeting a friend whilst on holiday 100 km from home

(b) Meeting a friend outside their home

(c) Meeting a friend at the shops

(d) Meeting a friend at school

(e) Meeting a friend whilst visiting the USA

Show your results on a probability scale.

4 If you step outside your classroom now you would be likely to meet one of the following persons.

(a) a male teacher

(b) a male pupil

(c) a female teacher

(d) a female pupil

(e) the caretaker

(f) a parent

Write, in order of likelihood, who you might meet.
Name the most likely first.
Show your results on a probability scale.

5 The following table shows the number of gold, silver and bronze medals won by some countries in the 1992 Summer Olympic Games.

Country	*Gold*	*Silver*	*Bronze*
Italy	6	5	8
France	8	5	16
Germany	33	21	28
China	16	22	16
Canada	6	5	7
Great Britain	5	3	12

Using this information write down these countries in order of the likelihood of their winning the most gold medals at the next games.
Start with the most likely.
Show your results on a probability scale.

Experimental probability

Assignment 2 A fair dice

A dice is **fair** if all the numbers are equally likely to be thrown.

You will need a fair, six-faced dice that is numbered from 1 to 6.

Before you begin, draw a probability scale. Show on it your estimate for the probability of throwing a 3.

Make a table like this for your results. Include a column for the tally you will make after each throw.

Score	*Tally*	*Frequency*
1		
2		
3		
4		
5		
6		

Throw the dice 60 times. Record the results on your table.

- Complete the table of results.
- Draw a bar chart of the results.

  Write down what the results tell you about the probabilities.

Remember to group your tallies in 5s. 卌

- Compare your results with your neighbour's.

  Are they the same? What do you notice?

- Look back at the probability scale you drew before you did the experiment.

  Do you want to change your estimate?

- Explain why you think this is the probability of throwing a 3 with a fair dice.

Assignment 3 A biased dice

- Make a net for a cube, to make your own dice. Use light-weight card and make every edge 4 centimetres. Cut it out.
- Write numbers on the faces to make it into a dice.

> A dice is **biased** if the numbers are not all equally likely.

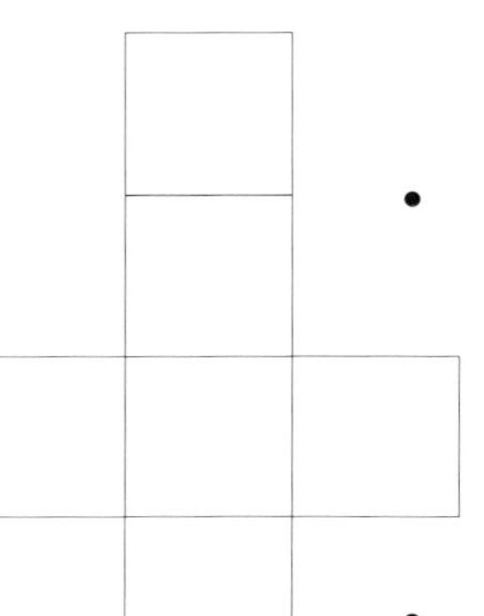

net of a dice

- Stick a small piece of modelling clay to the middle of one face which will be on the inside of your dice. (Remember which number you attached it to.) Stick your cube together.
- Exchange your completed cube with your neighbour.

> Remember to use tabs.

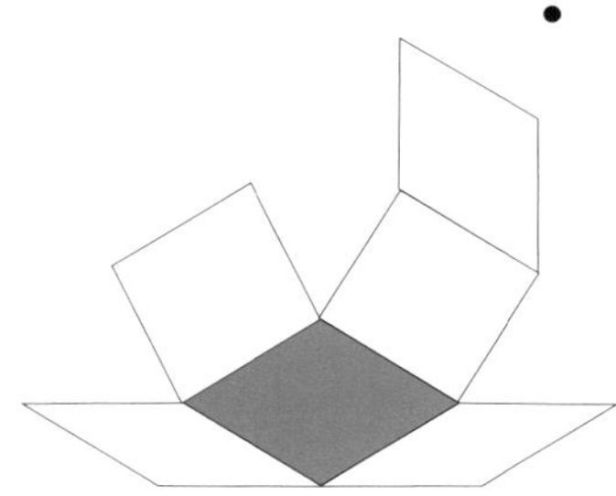

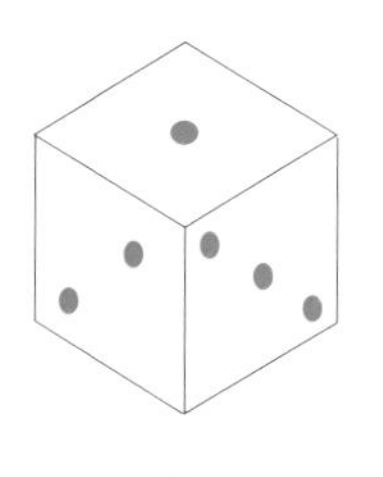

dice

- Repeat the experiment of Assignment 2, rolling the new dice 60 times. Record the results in the same way as before.
- Draw a bar chart of your results. What do your results show?
- What do you think would have happened to your results if you had rolled the dice fewer times?
- How different might your results have been if you had done 1000 rolls?
- Compare your results with your neighbour's.
- Which face has the modelling clay attached to it? Write down how you know.
- Check with your neighbour to see if you are correct.

Assignment 4 Length of words

This assignment investigates the likelihood of finding words of various lengths in a story.

- If possible use one of your own stories in this assignment. Use a passage containing 100 words.
- Write down which length of word you think will occur most often. Make a table to record your results.

Length of word (number of letters)	*Tally*	*Frequency*
1		
2		
3		
	Total number of words	

- Use a tally to record each word of your story. Draw a bar chart to show your results.
- Which length of word occurs:

  (a) most often

  (b) least often?

- Estimate the probability of finding words of these lengths in your story.
- Show these probabilities on a probability scale.
- Was your prediction correct?
- Compare your results with your neighbours'. If their results are different write down why you think this is so.
- Why is the probability of finding an eight-letter word not the same for all books?

Outcome of events

Key fact

A fair event is when all outcomes are equally likely.

If you spin one coin there is a 50% chance of it landing to show a head and a 50% chance of it showing a tail. It could show either heads or tails.

The coin is said to be **fair** if heads and tails are equally likely to occur. The coin is not weighted or **biased**.

Biased coins, for example 2-headed coins, give unfair results.

If you spin two coins there are four possible ways for them to land.

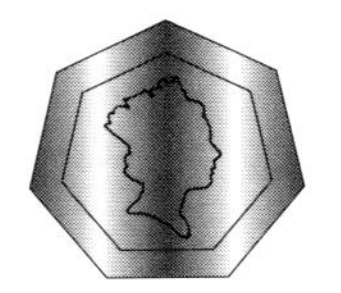

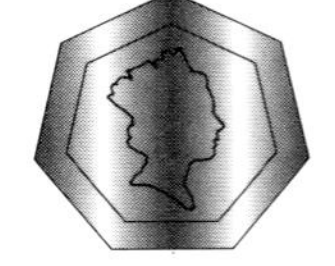

Exercise 5

1 You have a packet of chocolate biscuits and a packet of plain biscuits in the cupboard. Write down all the different ways of choosing two biscuits.

2 A snack bar sells three different types of fruit: apples, oranges and pears. You want to buy two pieces of fruit.

Write down all the possible choices you might make.
Set out your list clearly.

1st choice	*2nd choice*
apple	**orange**
apple	**pear**
apple	**...**

Complete the list.

'Outcome' is another way of saying 'arrangement'.

3 The canteen has four kinds of drink on sale.

If two people both want a drink, how many different arrangements are possible? Write them all down.

1st person	*2nd person*
orange	blackcurrent
orange	cola

Complete the list.

4 There are three brown, two black and two blue socks in the cupboard. You choose two socks without looking.

(a) Write down all the possible outcomes.

(b) How many arrangements are there?

1st sock	*2nd sock*

(c) How many of the arrangements are pairs of the same colour?

(d) For which colour are you most likely to get a pair?

Dice can mean one or more than one dice.

5 There are six faces on a dice. Write down all the possible outcomes when you roll two dice.
Be systematic in your recording and make sure you list them all. The list is started below.

1st dice	*2nd dice*
1	1
1	2
1	3
1	4

(a) How many arrangements are there altogether?

(b) How many of the arrangements are doubles? (A double means the same number is on both dice.)

Games involving probability

Exercise 6

1 At the school fête, the wheel of fortune is a popular game. The visitors buy a ticket and choose a number on the wheel. The arrow indicates the winning number when the wheel stops.

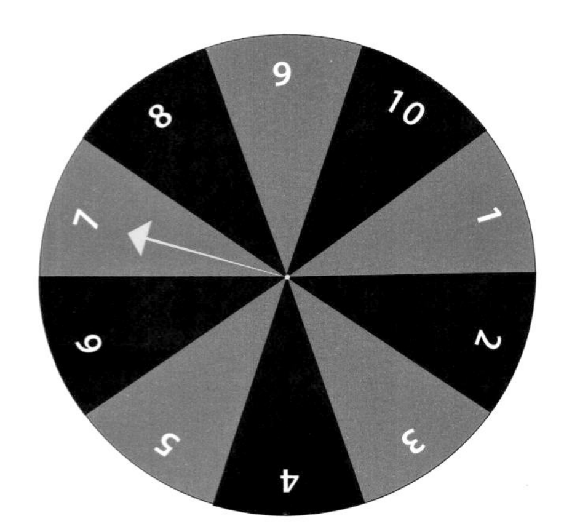

Draw a line 10 cm long and mark it as a probability scale similar to this one.

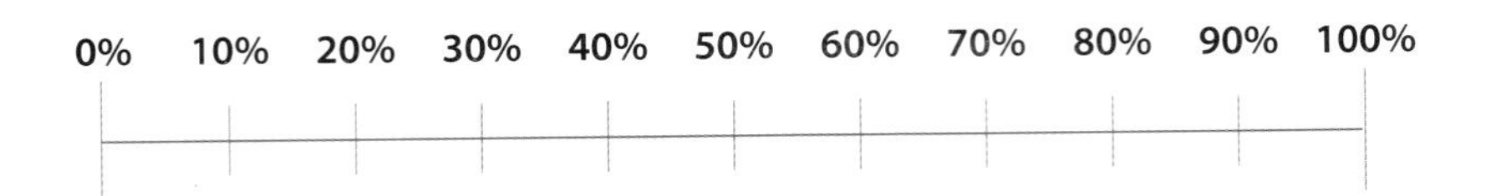

Show on the probability scale, when using this wheel of fortune, the probability that:

(a) the winning sector is red

(b) the winning number is 3

(c) the winning sector is black with an odd number on it

(d) the winning number is 6 or 10

(e) the winning sector is either red or 2

2 Another stall involves rolling a coin onto a board. The board is marked out in squares. Some are winning squares and some losing squares. If your coin lands completely within a winning square, not touching any lines, you win a prize.

The boards in the diagram are being used at the fair.
W is a winning square and **L** is a losing square.

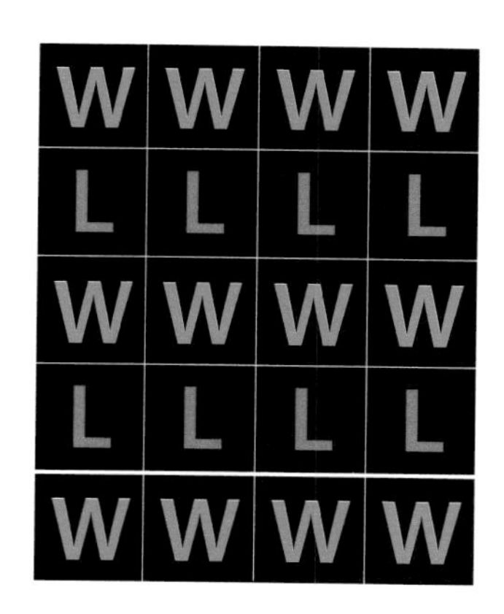

Board A

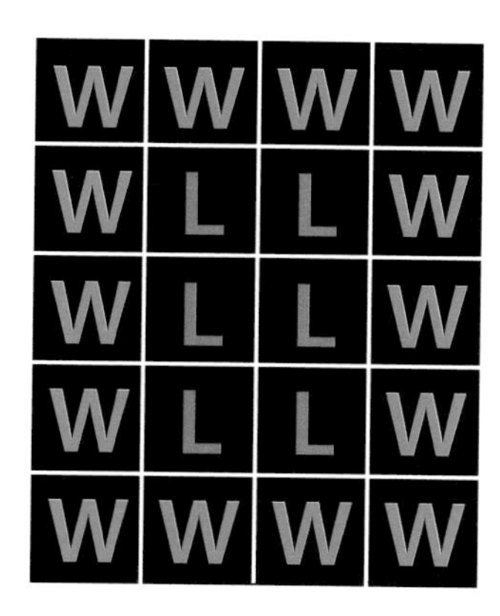

Board B

Draw a line 10 cm long and make it into a probability scale.

(a) On which board would you be most likely to lose?
Explain your answer.

(b) On the probability scale show the probability of winning on board A.

(c) On the scale, show the probability of losing on board B.

(d) Explain how you decided on the probabilities in parts (a) and (b).

3 Make a board to play 'roll-a-coin'.

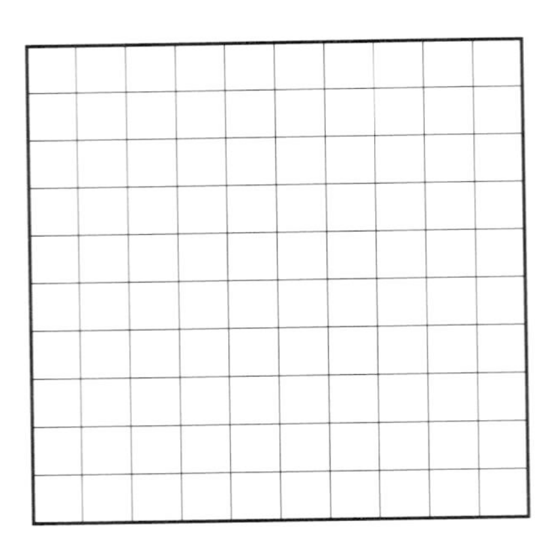

Make sure the squares are big enough for your coin.

Make the board from a square piece of card.

Your board should have 100 small squares on it.

Make sure the small squares are large enough for the coin that you decide to use.

Use **W** and **L** to stand for winning and losing squares.

Design the board so that if you were the stall holder you would make a profit.

Make sure that the visitors would enjoy playing the game and think that they had a good chance of winning.

- Explain why you decided to design it in your way and what would be the probability of winning?
- How much would you charge?
- How much would a winner receive to ensure you made a profit?

Try out your board and play the game.

Write down what you discovered and any problems you had.

Assignment 5 Starting with a six

In many board games you have to roll a six to start. When you need a six it seems to take ages to throw one. If the dice is fair then a six should be as easy to roll as any other number.

You are going to investigate how many throws it takes to roll a six. You need a fair dice.

- Estimate the highest number of throws that might be necessary to throw a six.
- Make a table in which to record your results.

Number of throws to roll a six	*Tally*	*Frequency*
1		
2		
3		
4		
5		

- Count how many throws it takes you to get your first six.
  Put a tally mark in the correct row of your table. Continue in this way until you have thrown 50 sixes.
- Complete the table of results.
- Draw a bar chart of the results.
- Compare your results with your neighbour's.
- What did you discover? Are the results as you had expected them to be?

Calculating probability

Sometimes it is possible to calculate the likelihood of an event.

Probabilities are written as fractions, decimals or percentages.

A fair coin has two faces. When you spin the coin there is an **equal likelihood** of heads or tails showing on the top.

In two throws you would expect each to appear once.

This means that one in two throws would be a 'head', and one in two throws would be a 'tail'.

The probability of throwing a head is:

p(head) = $\frac{1}{2}$ or 0.5 or 50%

p(head) is a short way of writing the probability of throwing a head.

The probability of throwing a tail is:

p(tail) = $\frac{1}{2}$ or 0.5 or 50%

Exercise 7

Give your answers as fractions, decimals and percentages.

1 If you choose a card from an ordinary pack, what is the chance that it is:

(a) a red card (b) a black card (c) a club (d) a heart?

A tetrahedron has four faces.

2 A fair dice can be made from a tetrahedron. The score is the number on the bottom. If you throw this dice, what are the following probabilities?

(a) p(1) the probability of scoring 1

(b) p(4) the probability of scoring 4

3 There were 100 tickets sold for a raffle. If you bought 10 tickets, what would the probability be of you winning first prize?

4 Small tiles with letters on them were used to make the word

The tiles were put into an envelope and Ben took one out, without looking. What is the probability that Ben chose:

(a) a letter 'L' (b) a letter 'K' (c) any vowel (d) any consonant?

5 In a lake there were five different kinds of fish. There were two perch, three bream, one roach, two carp and two tench. An angler uses bait that is attractive to all of them. Work out the following probabilities.

(a) p(carp) the probability of catching a carp

(b) p(perch) the probability of catching a perch

(c) p(tench) (c) p(roach) (d) p(bream)

6 There are ten snooker balls in a bag: six are red, two are white, one is black and one is pink. If you choose a ball without looking, work out the following probabilities.

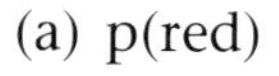

(a) p(red) (b) p(black)

(c) p(white) (d) p(pink)

(e) p(white or red) (f) p(pink or black)

(g) p(any colour except white)

7 You are offered a bag of sweets. In the bag are four red sweets, seven blue sweets, three green sweets and six orange sweets. You take a sweet without looking in the bag. What is the probability that the sweet will be:

(a) red (b) orange (c) blue

(d) a sweet you like, if you dislike red ones

(e) a sweet you dislike, if you like blue ones and orange ones?

8 The probability of rain on any day in February is 25%.

(a) On how many days would you expect rain?

(b) How many days would you expect to be dry?

FEBRUARY						
S	M	T	W	T	F	S
			1	2	3	4
5	6	7	8	9	10	11
12	13	14	15	16	17	18
19	20	21	22	23	24	25
26	27	28				

9 Amy rolls a dice 60 times.

(a) How many times should she expect to get a six?

(b) How many times should she expect to get a number greater than four?

Calculation or estimation?

At an athletics meeting, the running track has four lanes.
There is someone in each lane for the 100 m sprint.

The probability of the athlete in lane 2 winning is *not* $\frac{1}{4}$.
There is *not* an equal likelihood of the runner in each lane winning.
The athlete in lane 1 may be more experienced than the athlete in lane 2. Personal bests and previous performance are useful indicators to help predict the probability of winning.

Exercise 8

In the following questions the outcomes are not equally likely.

1 Why is one person more likely to live to 100 than their neighbour?

2 Two films are released at the same time. Why is one more likely to be successful than the other?

3 Why is one car driver more likely to have a road accident than another?

Review Exercise

1 Write down an example of an event that has a probability of 0.

2 Write down an example of an event that has a probability of 1.

3 Give an example of an event that has a probability of 50%.

4 There are two red pens, three blue pens and five pencils in a pencil case. You remove something from the pencil case without looking.

(a) Write down the order of likelihood of getting a blue pen, a red pen or a pencil. Start with the most likely.

(b) Show the probabilities on a probability scale.

(c) Giving your answers as fractions, decimals and percentages work out:

(i) p(a red pen)
(ii) p(a blue pen)
(iii) p(a pencil)
(iv) p(a red pen or a pencil)

5 You have a day off school. You enjoy three sports: bowling, ice-skating and table tennis. You have time to do only two of these. Write down all the possible arrangements of two of these sports in which you could take part.

6 Why is one person more likely than another to grow to over six feet tall?

Coordinates

In this chapter you will learn:

- → **to use coordinates to specify locations in the first quadrant**
- → **to use coordinates to specify locations in all four quadrants**
- → **to use negative numbers**

Starting points

Before starting this chapter you will need to know how to:

- recognise and name simple polygons
- recognise line symmetry
- read a simple decimal scale.

Assignment 1 The telephone game 1

This is a game for two players.

Rules

- Draw a simple picture on a sheet of paper. Do not let your partner see your picture.
- Now imagine that you are making a telephone call to your partner. Describe your picture without actually saying what it is. Give your instructions, step by step.
- Your partner has to make a copy of your picture, following your instructions. It is important that you cannot see each other's pictures.
- When your partner has finished, it is your turn to make a copy of their picture, following their instructions.

The winner is the one who has made the best copy of the other person's picture.

Treasure hunt

Lines are always numbered from the bottom left-hand corner.

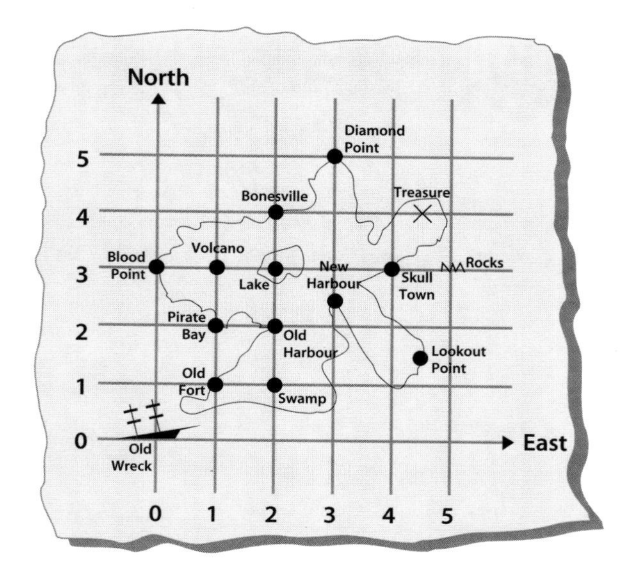

Look at this treasure map. To find a place on the map, you must know the distance east and the distance north from the old wreck.
These distances are written in brackets, like this.

Key fact
The first number is east, the second number is north.

Diamond Point is at (3, 5).
3 east ↗ ↖ 5 north

Exercise 1

1 Write down the east and north distances of the following places. The first one has been done for you.

(a) Pirate Bay (1, 2) (b) Old Harbour
(c) Skull Town (d) New Harbour
(e) Bonesville (f) Lake
(g) Old Wreck (h) Treasure

2 Write down what you would expect to find at each of these points.
(a) (5, 3) (b) (1, 1) (c) (1, 3) (d) (2, 1) (e) (0, 3) (f) $(4\frac{1}{2}, 1\frac{1}{2})$

3 Design your own treasure map.
- Include at least eight places.
- Below your map, write the place names with their (east, north) distances.

Coordinate codes

Sometimes it is useful to describe the exact position of a **point** on a grid.

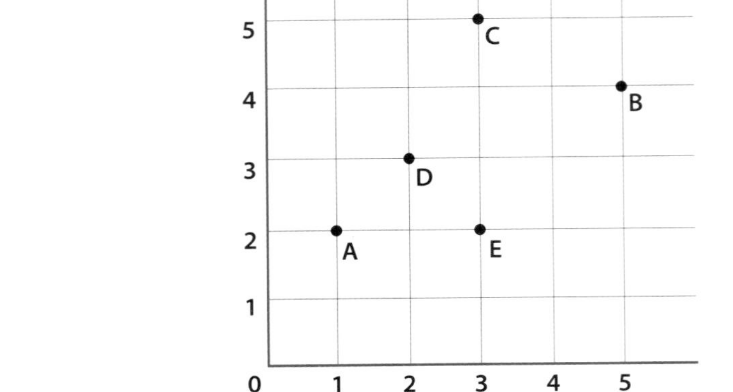

A grid is made from lines set out to form a square lattice, as shown.

You can do this by numbering the grid lines and describing the position of a point on the grid using **coordinates**.

When writing coordinates you always start at 0 and count across first, and then up.

The position of A from 0 is one square across and two squares up.

The **coordinates** of A are (1, 2).

You always put brackets round the coordinates.

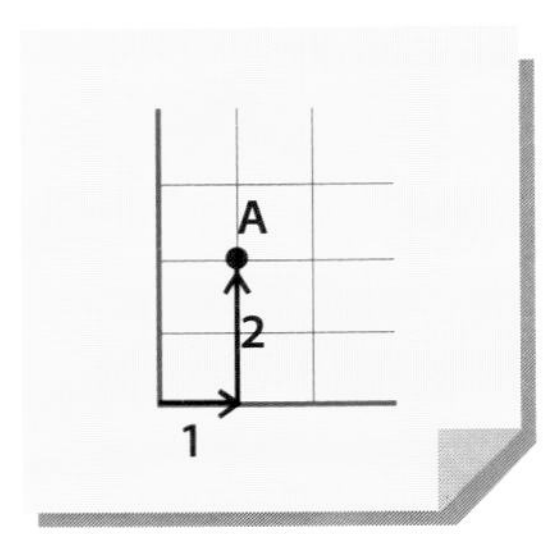

Exercise 2

1 (a) What are the coordinates of B?

(b) What are the coordinates of C?

(c) What are the coordinates of D?

(d) What are the coordinates of E?

Use this grid to answer the rest of the questions in this exercise.

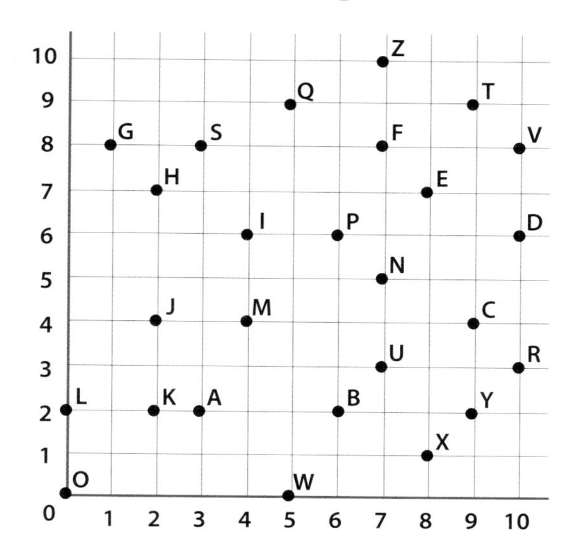

2 What are the coordinates of these points?

(a) A (b) S (c) O (d) Z (e) P

(f) Q (g) R (h) E (i) G (j) W

3 What letters have these coordinates?

(a) (6, 2) (b) (9, 9) (c) (7, 3) (d) (4, 4) (e) (10, 8)

(f) (8, 1) (g) (9, 4) (h) (7, 5) (i) (9, 2) (j) (0, 2)

4 Use the alphabet on the grid to solve this coded message.

(9, 9) (2, 7) (4, 6) (3, 8)
(5, 9) (7, 3) (8, 7) (3, 8) (9, 9) (4, 6) (0, 0) (7, 5)
(9, 9) (3, 2) (2, 2) (8, 7) (3, 8)
(3, 2)
(6, 2) (4, 6) (9, 9)
(0, 2) (0, 0) (7, 5) (1, 8) (8, 7) (10, 3)
(9, 9) (0, 0)
(9, 4) (0, 0) (4, 4) (6, 6) (0, 2) (8, 7) (9, 9) (8, 7)

5 Write your own coded message.

- Use the letters in the grid.
- Give it to a friend to solve.

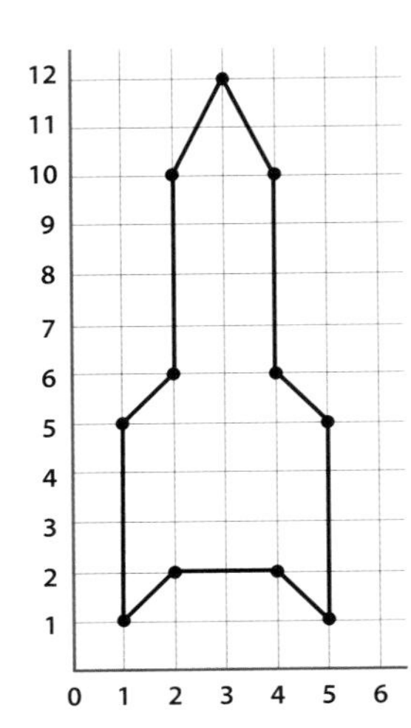

Coordinate pictures

If you were to join one coordinate to another, then another, in order, you could draw a picture.

(1, 1) (2, 2) (4, 2) (5, 1) (5, 5) (4, 6) (4, 10) (3, 12) (2, 10) (2, 6) (1, 5) (1, 1)

These coordinates give a rocket.

Exercise 3

1 Draw a grid numbered from 0 to 6 along each axis, as shown.

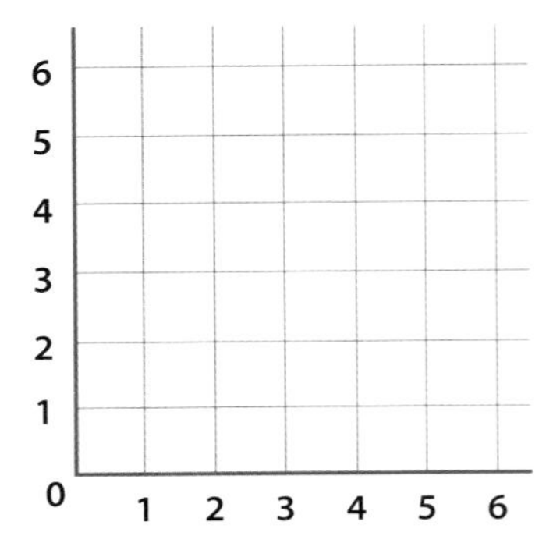

As you plot each point, join it to the previous one.

(a) Plot the following points and join them up, in order.
(2, 1), (6, 3), (5, 5), (1, 3), (2, 1)

(b) Write down the name of this shape.

2 Draw a grid numbered from 0 to 6 along each axis, as shown.

(a) Plot the following points and join them up, in order.
(6, 1), (6, 9), (2, 9), (2, 7), (1, 5), (2, 5), (2, 4), (3, 5), (2, 3), (2, 2), (3, 1), (6, 1)

(b) What should you put at the point (3, 7)?

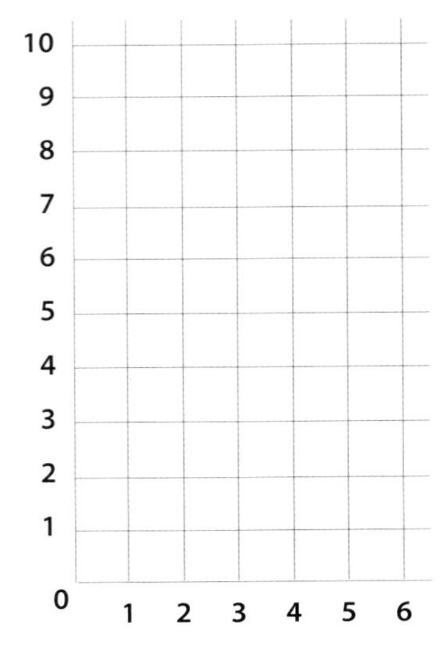

3 The picture on this grid is not complete. The dotted line should be a line of symmetry.

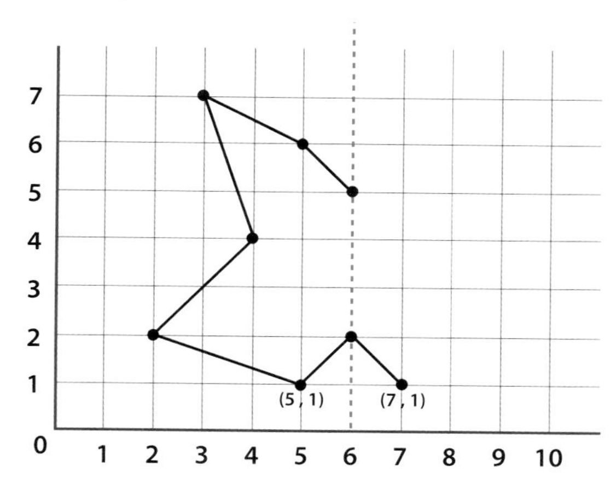

(a) Copy the picture and complete it by plotting points and joining them up.

(b) Write down the coordinates of each point on the diagram.

4 Draw and number another grid the same size as the one in question 3.

(a) Plot these points and join them up, in order.

(2, 5), (1, 7), (2, 7), (3, 5), (2, 5)

The vertices are the corners.

(b) Plot these points and join them up, in order.

(8, 5), (9, 7), (10, 7), (9, 5), (8, 5)

(c) Plot these points and join them up, to form a kite.

(3, 5), (4, 4), (3, 0), (2, 4), (3, 5)

(d) Draw another kite on the grid so that the shapes form a symmetrical pattern.

(e) Write down the coordinates of the vertices of the kite you have just drawn. Write them next to the points.

More coordinate pictures

Exercise 4

1 Copy the axes onto squared paper. Plot each set of points and join them up, in order. If you come to a star, stop and start again on the next set.

(a) (1, 5), (0, 5), (1, 7), (4, 7), (4, 8), (5, 8), (5, 7), (6, 7), (7, 5), (6, 5), (6, 0), (1, 0), (1, 5), (6, 5)

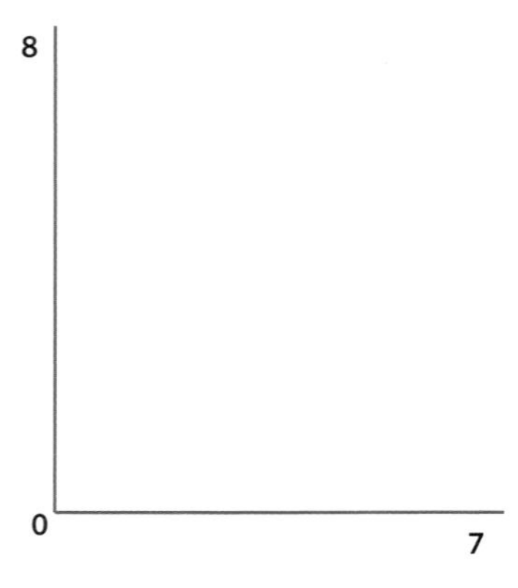

(b) (4, 0), (4, 1), (0, 1), (2, 3), (1, 3), (3, 5), (2, 5), (4, 7), (3, 7), (5, 9), (7, 7), (6, 7), (8, 5), (7, 5), (9, 3), (8, 3), (10, 1), (6, 1), (6, 0), (4, 0)

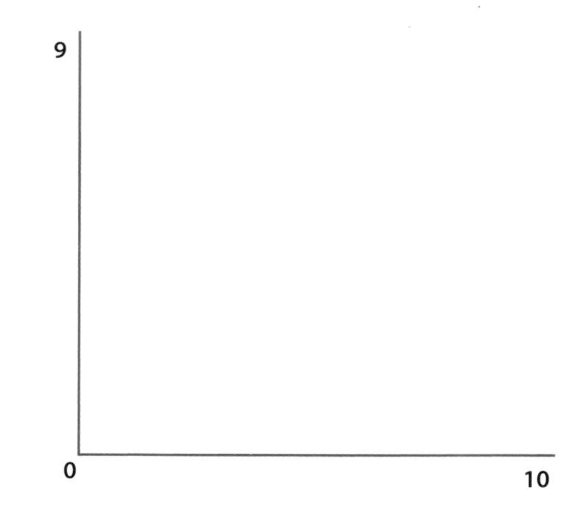

(c) (1, 0), (1, 2), (0, 2), (1, 4), (1, 6), (7, 6), (7, 0), (1, 0) ★

(1, 1), (2, 1), (3, 2) ★

(3, 4), (3, 5), (4, 5), (4, 4), (3, 4)

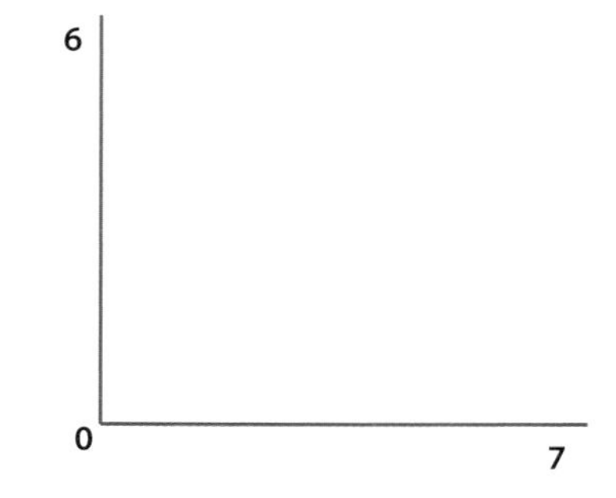

2 Copy these axes onto squared paper. Plot each set of points separately and join them up, in order. Stop at each star and start on the next set.

(a)

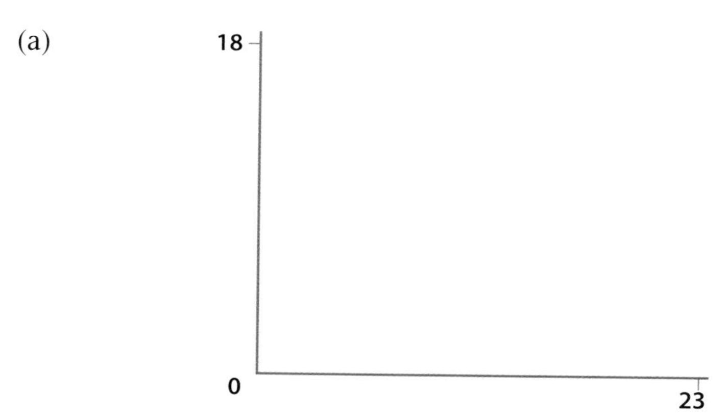

(4, 15), (5,15) ★

(7,17), (6,16), (6,13), (7,12), (10,12), (11,13), (11,14), (10,16) ★

(7, 1), (9, 1), (9, 7), (10, 7), (10, 1), (12, 1), (12, 7), (17, 7), (17, 1), (19, 1), (19, 7), (20, 7), (20, 1), (22, 1), (22, 11), (23, 8), (22, 13), (19, 16), (11, 16), (9, 18), (5, 18), (3, 16), (1, 11), (1, 8), (4, 8), (4, 10), (3, 10), (3, 9), (2, 9), (2, 11), (4, 12), (6, 10), (7, 7), (7, 1) ★

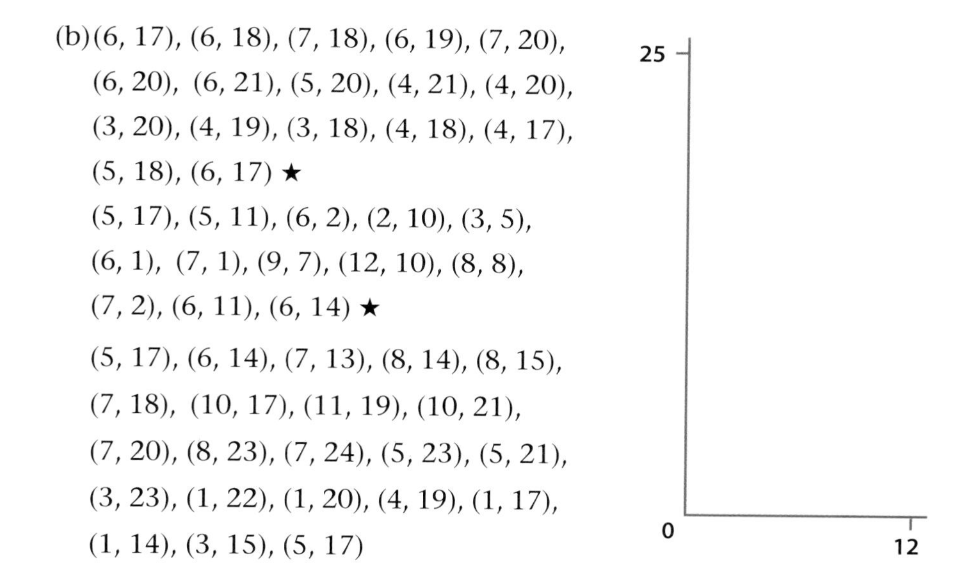

(b) (6, 17), (6, 18), (7, 18), (6, 19), (7, 20), (6, 20), (6, 21), (5, 20), (4, 21), (4, 20), (3, 20), (4, 19), (3, 18), (4, 18), (4, 17), (5, 18), (6, 17) ★

(5, 17), (5, 11), (6, 2), (2, 10), (3, 5), (6, 1), (7, 1), (9, 7), (12, 10), (8, 8), (7, 2), (6, 11), (6, 14) ★

(5, 17), (6, 14), (7, 13), (8, 14), (8, 15), (7, 18), (10, 17), (11, 19), (10, 21), (7, 20), (8, 23), (7, 24), (5, 23), (5, 21), (3, 23), (1, 22), (1, 20), (4, 19), (1, 17), (1, 14), (3, 15), (5, 17)

Assignment 2 Telephones again

- With a partner, play the telephone game again.
  How can you use what you have learnt so far in this chapter?

Assignment 3 Rhino

You can use Sheet 10X for this assignment.

- Load the program 'Rhino' from Microsmile® 'The first 30'.
- Play the game for 10–15 minutes.
  – What is the best coordinate to start on?
  – What is the lowest number of moves it takes to finish the rhino?
- Write a brief report.
  – Explain the challenge.
  – Describe what you did.
  – Say what you have noticed.
  – Write down your conclusions.

Labelling the grid lines

Key fact
The across coordinate is called the ***x*-coordinate.**
The up coordinate is called the ***y*-coordinate.**

The two numbered lines are known as **axes.**

The numbered line going across the page is labelled the ***x*-axis.**
The numbered line going up the page is labelled the ***y*-axis.**

The coordinates of P are (3, 2).
3 is the x-coordinate of P.
2 is the y-coordinate of P.

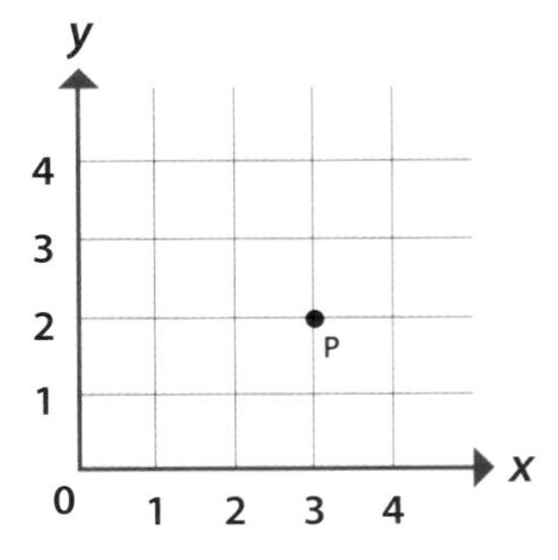

Exercise 5

1 In this diagram, the x-axis is numbered from 0 to 6 and the y-axis is numbered from 0 to 5. ABCD is a rectangle.

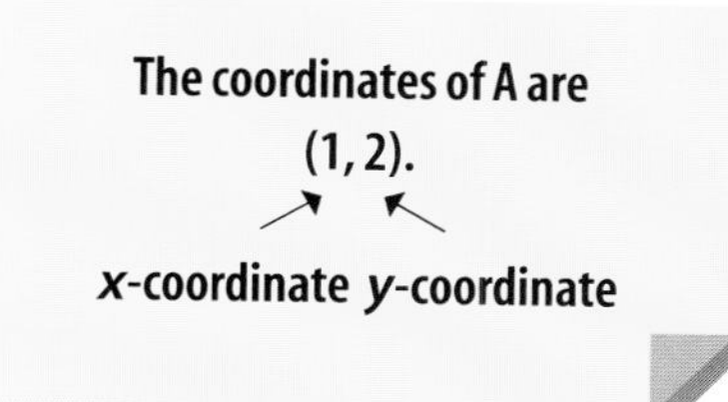

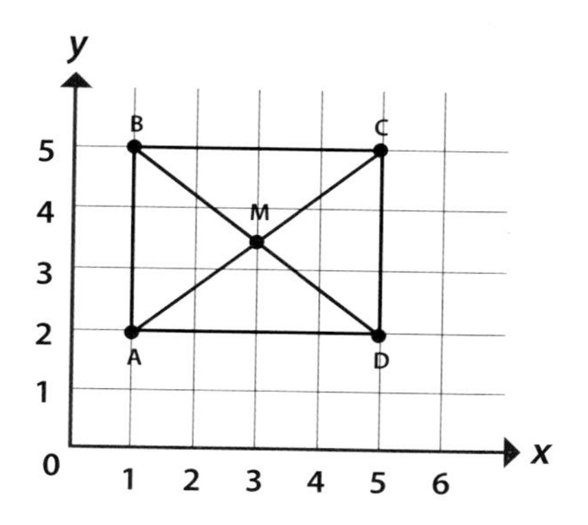

The diagonals of the rectangle cross at the point M.

(a) What is the x-coordinate of M?

(b) The y-coordinate is halfway between 3 and 4. What is the y-coordinate of M?

(c) Write down the coordinates of M.

2 Draw axes so that the x-axis is numbered from 0 to 6 and the y-axis is numbered from 0 to 6. Label the axes x and y.

(a) Plot the points A(0, 4), B(1, 6), C(3, 6), D(5, 4), E(3, 2) and F(1, 2). Join them up to make a six-sided shape. What is the name for a six-sided shape?

(b) This shape has a line of symmetry. Draw the line of symmetry on your diagram.

(c) Write down the coordinates of six points on the line of symmetry. What do you notice about the y-coordinate for each of your points?

3 Draw x and y axes numbered from 0 to 8. Label the axes x and y.

(a) Plot the points A(1, 4), B(2, 6), C(7, 6), D(8, 4) and E($4\frac{1}{2}$, 0). Join them up to make a five-sided shape. What is the name for a five-sided shape?

(b) This shape has a line of symmetry. Draw this line on your diagram.

4 A fairground game involves taking three shots at a target like the one here. Hits in the rings are scored as shown, for example, a hit in the outer ring scores 1 point.

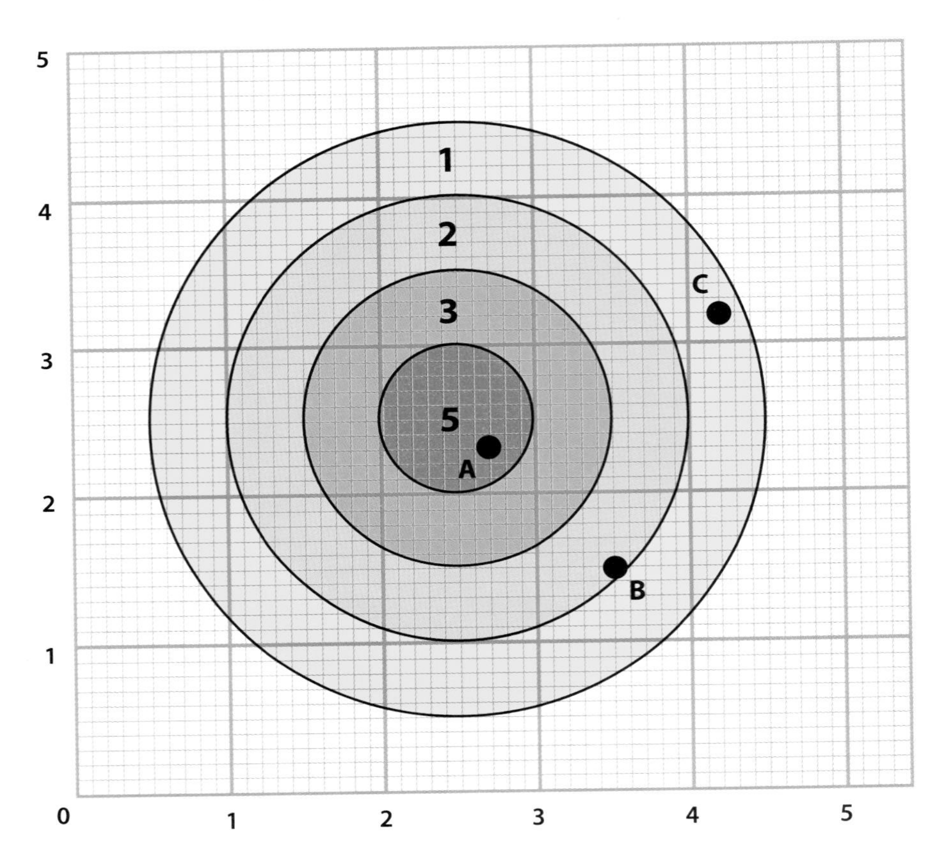

Three shots hit the target at A, B and C.
The coordinates of A are (2.7, 2.3).

(a) What are the coordinates of shots B and C?

(b) How many points are scored by:

(i) shot D(2.0, 1.0) (ii) shot E(2.9, 2.1) (iii) shot F(3.1, 3.7)?

Two more people fired at the target.
The first person scored hits at (1.2, 1.9), (2.7, 3.4) and (3.1, 2.6).
The second person scored hits at (2.3, 2.1), (2.9, 3.8) and (1.2, 4.2).

(c) What was the first person's total score?

(d) What was the second person's total score?

The four quadrants

This shape is drawn in the first quadrant.

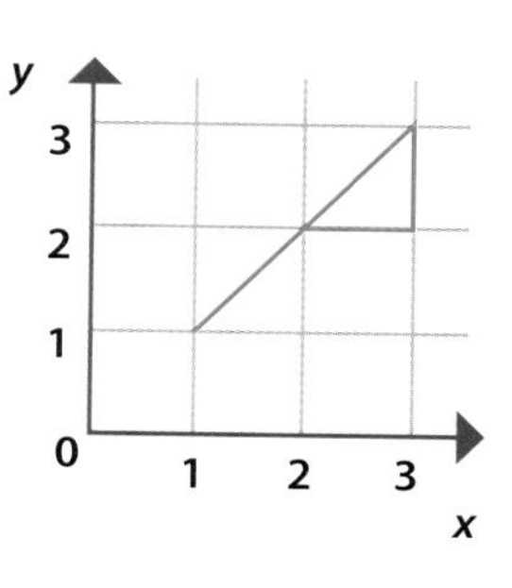

Key fact
Quadrants occur when an area is divided into four regions.

The completed pattern fits into the four quadrants like this.

The axes now form a cross.

Note:
Axes is the plural of axis.

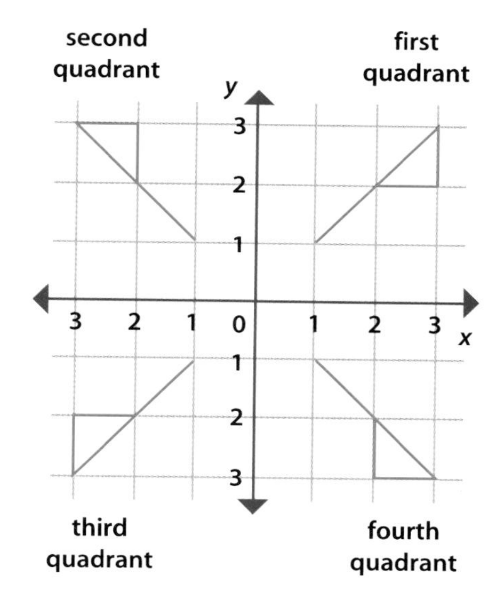

Key fact
The origin (0, 0) is a reference point when plotting coordinates.

The point where the two axes meet is called the **origin**. It has coordinates (0, 0). To distinguish the numbers on the x-axis to the left of the origin, negative signs are used.

To distinguish the numbers on the y-axis that are below the origin, negative signs are used.

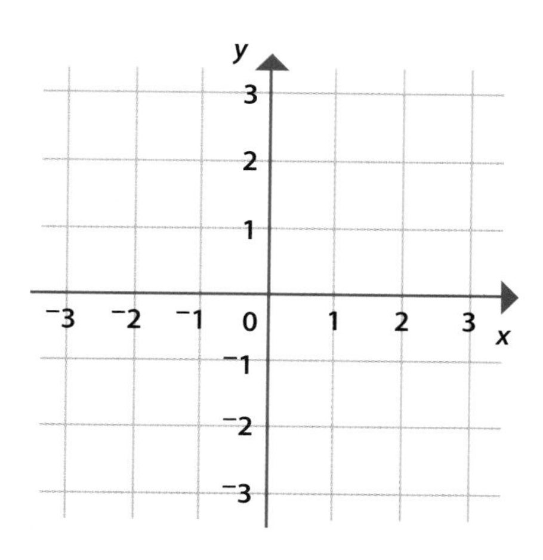

Exercise 6

1 Copy the table and complete it for this shape.

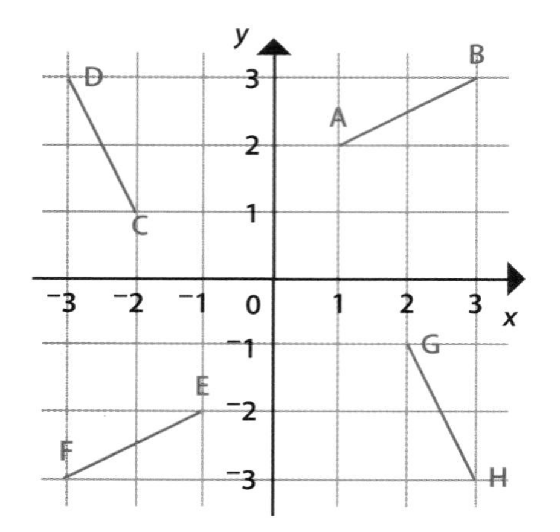

Point	Coordinate
A	(1, 2)
B	(, 3)
C	($^-$2,)
D	(, 3)
E	(, $^-$2)
F	($^-$3,)
G	(,)
H	(,)

2 Look at this grid.

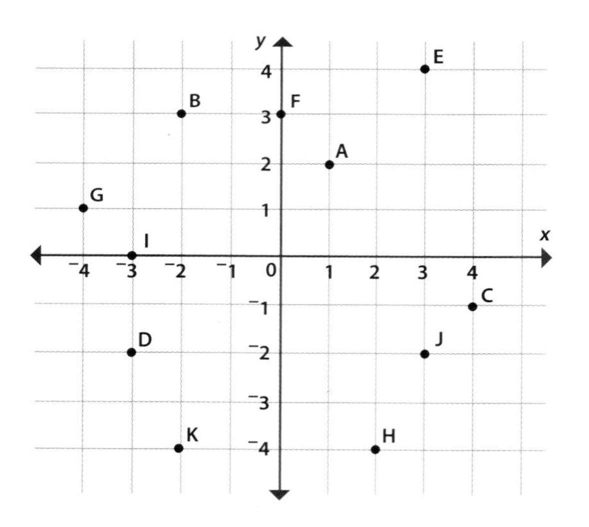

Copy this table and complete it for the points in the grid.

Point	Coordinates
A	(1, 2)
E	
	(3, $^-$2)
G	
	($^-$2, 3)
D	

Point	Coordinates
	($^-$2, $^-$4)
F	
	(4, $^-$1)
	($^-$3, 0)
H	

3 Draw a grid numbered from $^{-}6$ to 6 along each axis. Plot each of the following sets of points and join them, in order. Write down the names of the polygons formed.

(a) (2, 2), (6, 4), (5, 6), (1, 4), (2, 2)

(b) (2, 0), (6, $^{-}1$), (5, 3), (2, 0)

(c) ($^{-}4$, $^{-}1$), ($^{-}2$, 1), ($^{-}4$, 5), ($^{-}6$, 1), ($^{-}4$, $^{-}1$)

(d) ($^{-}6$, $^{-}3$), ($^{-}5$, $^{-}6$), ($^{-}2$, $^{-}5$), ($^{-}3$, $^{-}2$), ($^{-}6$, $^{-}3$)

(e) (1, $^{-}1$), (4, $^{-}2$), (4, $^{-}5$), (1, $^{-}6$), ($^{-}1$, $^{-}3$), (1, $^{-}1$)

4 Draw a grid numbered from $^{-}6$ to 6 along each axis.

(a) Plot the points P(4, 1), Q($^{-}1$, 6) and R($^{-}6$, 1).

P, Q and R are three vertices of the square PQRS.

(b) Plot the point S and write down its coordinates.

(c) What are the coordinates of the midpoint (centre) of the square?

Review Exercise

1 Copy this table and complete it, for the points on the grid.

Point	Coordinates
A	(1, 2)
D	
	(3, 1)
	(2, 0)
F	
	$(3\frac{1}{2}, 3)$
E	

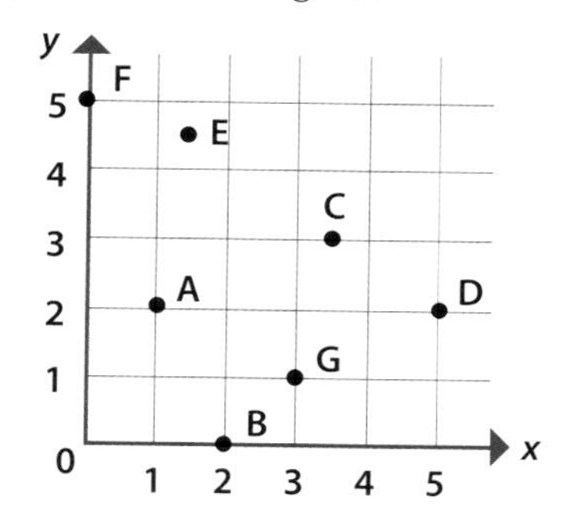

2 Copy this grid.

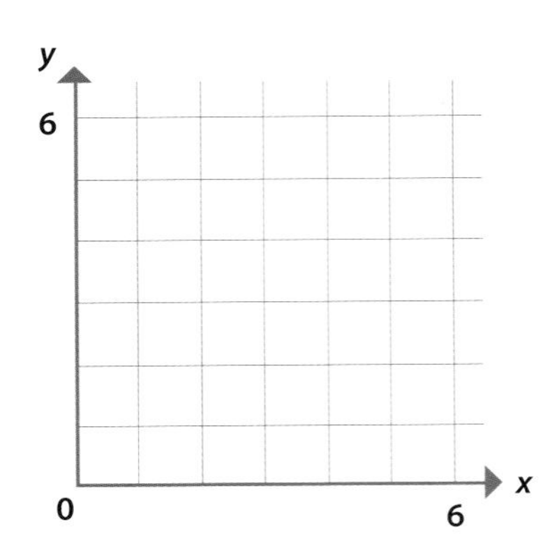

(a) Plot the following points and join them up, in order.

(1, 2), (1, 1), (5, 1), (6, 2), (1, 2), (4, 6), (4, 2), (6, 3), (4, 5)

(b) What picture have you drawn?

3 Draw a set of axes. Number the x-axis from 0 to 8, and the y-axis from 0 to 7.

(a) Plot the following points and join them up in order.
(3, 1), (5, 2), (7, 1), (7, 5), (5, 7), (1, 7), (2, 5), (1, 3), (3, 1)

(b) What is the mathematical name for the polygon you have drawn?

(c) This polygon has a line of symmetry. Draw it in.

4 Copy this table and complete it for the points in the grid.

Point	Coordinate
P	$(1, {}^{-}2)$
S	
	$({}^{-}2, 3)$
Q	
	$(0, {}^{-}1)$
	$({}^{-}1, 1\frac{1}{2})$
V	

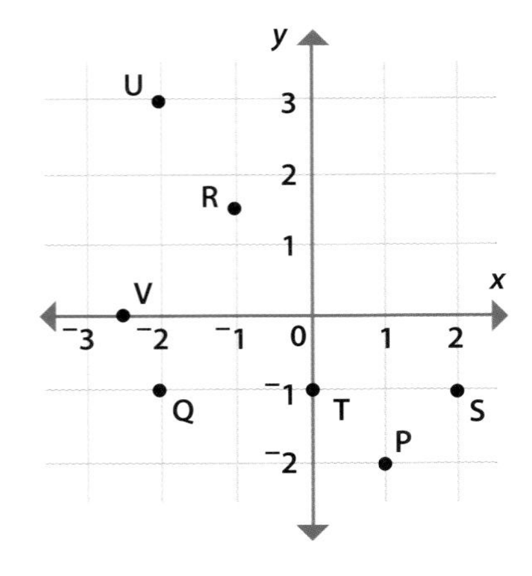

5 Draw a grid numbered from $^{-}3$ to 3 along each axis.

(a) Plot these points and join them up, in order.

$(^{-}3, ^{-}2)$, $(^{-}3, ^{-}1)$, $(^{-}2, ^{-}1)$, $(^{-}2, 2)$, $(^{-}3, 2)$, $(^{-}3, 3)$, $(^{-}1, 3)$, $(^{-}1, 1)$, $(1, 1)$, $(1, 3)$, $(3, 3)$, $(3, 2)$, $(2, 2)$, $(2, ^{-}1)$, $(3, ^{-}1)$, $(3, ^{-}2)$, $(1, ^{-}2)$, $(1, 0)$, $(^{-}1, 0)$, $(^{-}1, ^{-}2)$, $(^{-}3, ^{-}2)$

You should find that this shape has line and rotational symmetry.

(b) Draw the lines of symmetry on your diagram.

(c) What is the order of rotational symmetry?

(d) Write down the coordinates of the centre of rotation.

6 Draw a grid numbered from $^{-}3$ to 3 along each axis.

(a) Draw your own shape with rotational symmetry of order 2.

(b) Write down the coordinates that would enable someone else to copy your shape.

Assignment 4 Four in a line

This is a game for two players or two teams.
One team uses ✘s and the other used ❍s.

Rules

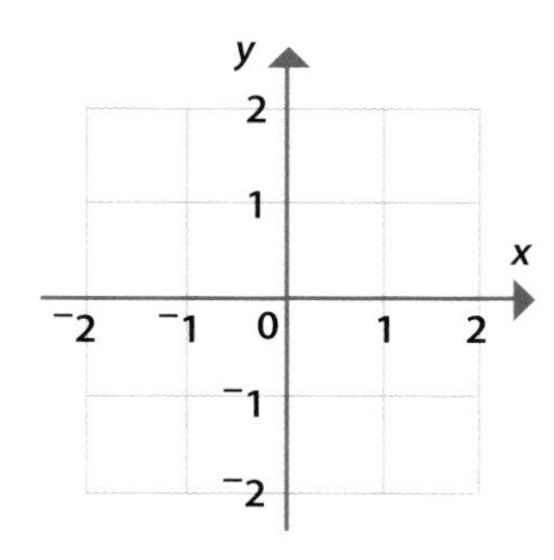

- Decide who goes first.
- Team A chooses a coordinate. Team B puts a ✘ on the grid at that point. This point is then 'owned' by Team A.
- Team B chooses a coordinate. Team A puts a ❍ on the grid at that point. This point is then 'owned' by Team B.
- The winner is the first to obtain four points in a line. It can be a horizontal, vertical or diagonal line.

CHAPTER 11

Perimeter, area and volume

In this chapter you will learn:

- → **how to find the distance round the outside of a shape**
- → **how to measure the space inside 2-D and 3-D shapes**

Starting points

Millimetre, centimetre, metre and kilometre are metric units for length.

Before starting this chapter you will need to know:

- common metric and imperial units
- how to use some different measuring instruments.

Inch, foot, yard and mile are imperial units for length.

Exercise 1

1 Look at the lengths marked on the diagram of the hand.

Which of these lengths on *your* hand do you think is the nearest in length to one inch?

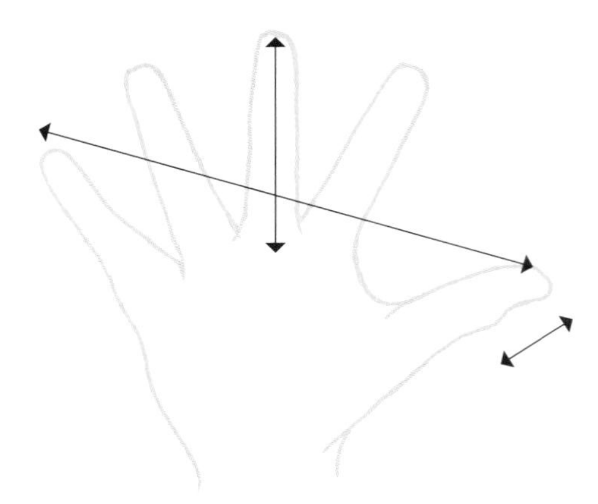

2

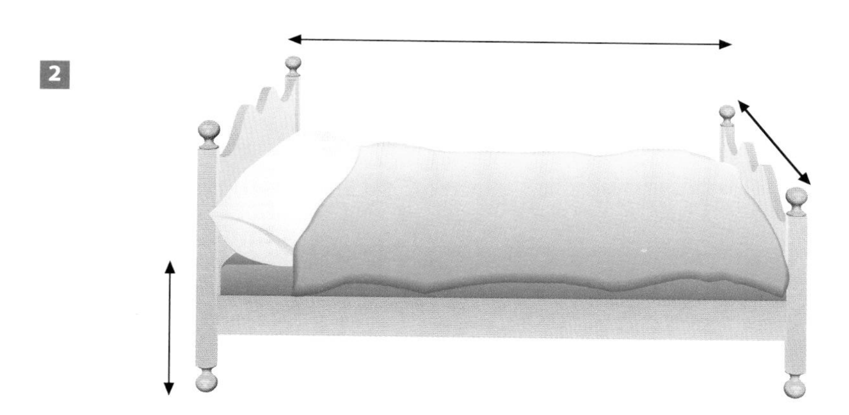

(a) Which of these lengths do you think is the nearest to 1 foot?

(b) Which length is nearest to 1 yard?

3 Which of these lengths do you think is nearest to 1 metre?

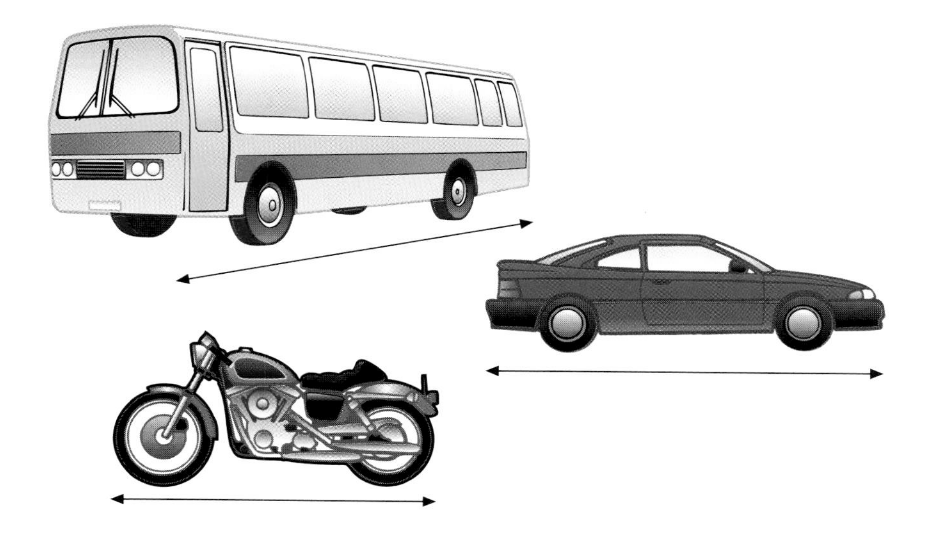

m means metre
cm means centimetre
mm means millimetre
km means kilometre

4 (a) How many centimetres are there in 1 m?

(b) How many millimetres are there in 1 m?

(c) How many centimetres are there in 1 km?

5 (a) How many inches are there in 1 ft?

(b) How many inches are there in 1 yd?

ft means foot
yd means yard

6 Measure these lines.
Write down your answer and say which units you used.

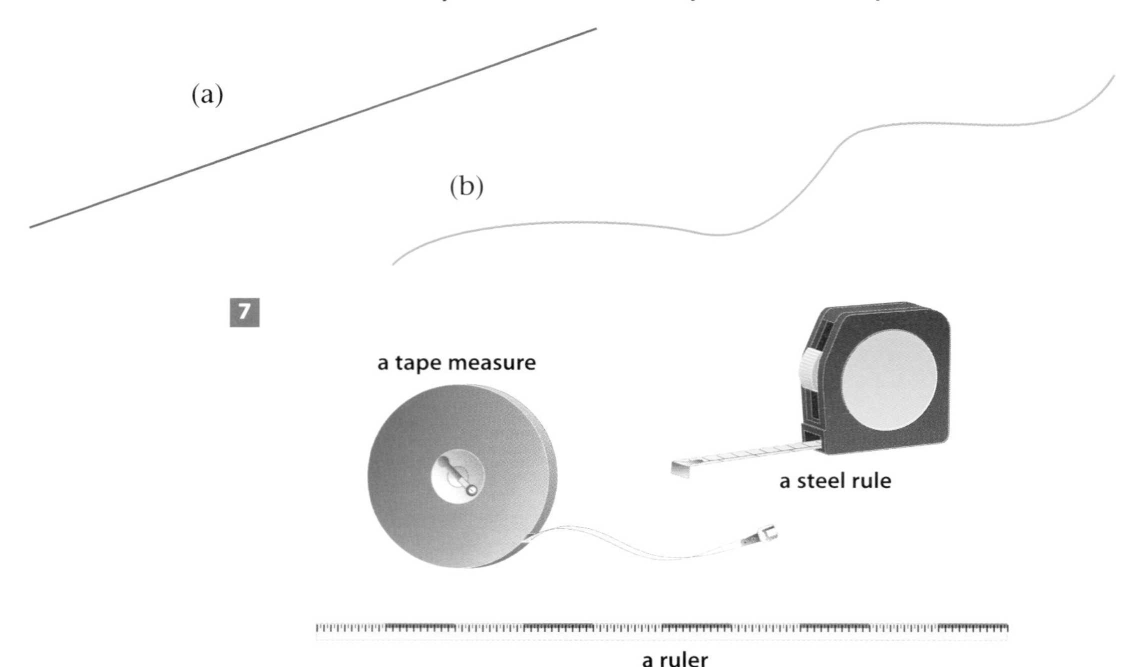

7 Which of the measuring instruments shown here would you use to measure the following? In each case, write down the units you would use.

(a) a football pitch (b) this book (c) your classroom

(d) your height (e) your bedroom

Assignment 1 How long is a mile?

In Roman times a mile was measured as 1000 paces.

We know that a mile is 1760 yards or 1609 metres.

- How long was a Roman pace?
  Does your answer seem reasonable?
- Look at books on the Romans in your library. Write down what you find out about the Roman pace.

Perimeter

Key fact

The perimeter of a shape is the distance round the outside.

Exercise 2

1 Measure the perimeter of this rectangle, in centimetres, and write down your answer.

2 Find the perimeters of these shapes.

(a)

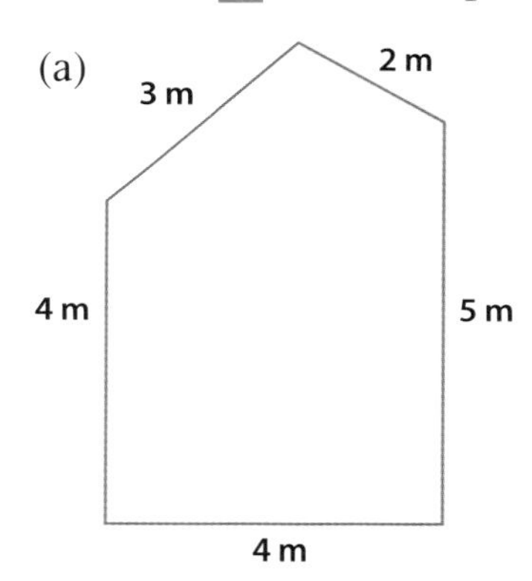

(b)

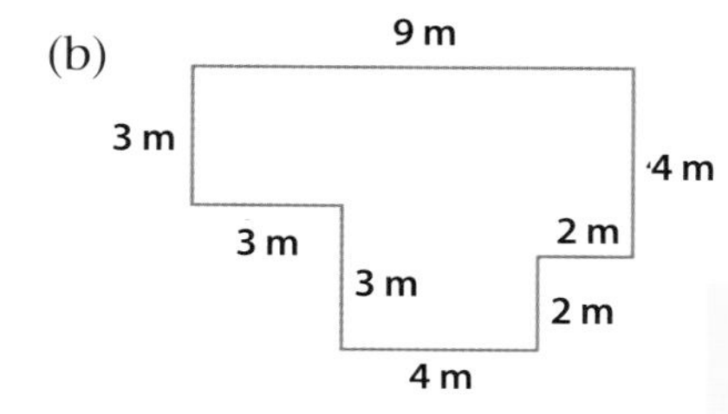

Do not measure the shapes; they are *not* to scale.

3 (a) Find the perimeter of this book, in millimetres.
Write down your answer.
(b) What is the perimeter of this book in centimetres?

4 This shape has some measurements missing.
Work out the perimeter of the shape, explaining how you reached your answer.

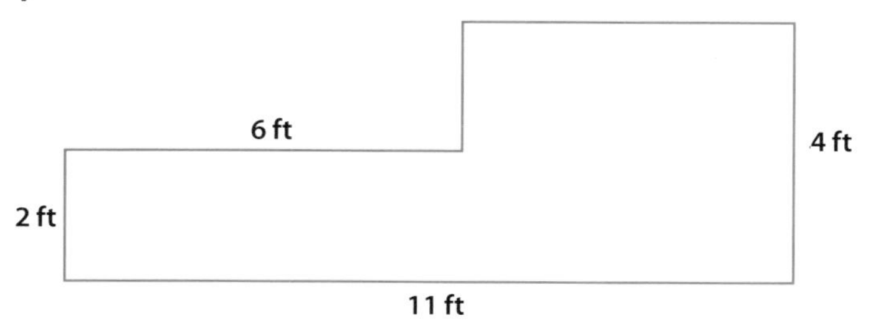

5 Measure the perimeter of your bedroom.
Describe any difficulties you had to overcome.

6 Describe clearly how you would measure the perimeter of your school field.

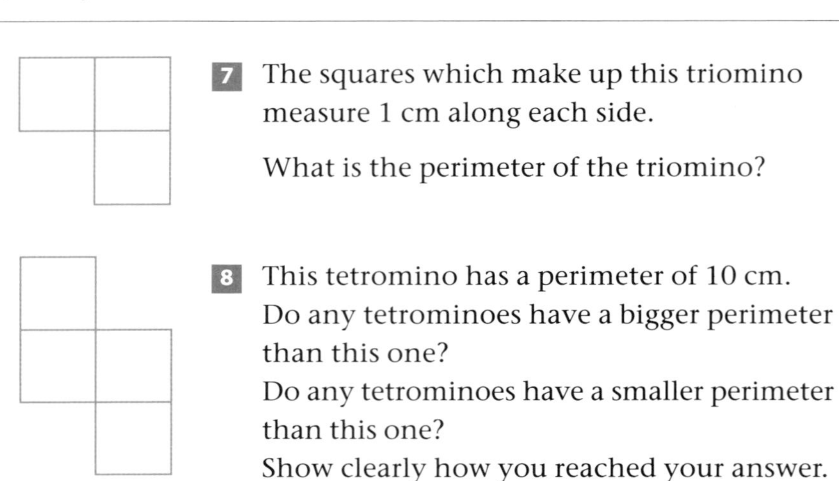

7 The squares which make up this triomino measure 1 cm along each side.

What is the perimeter of the triomino?

A triomino is a shape made from three squares that are joined at the edges.

8 This tetromino has a perimeter of 10 cm.
Do any tetrominoes have a bigger perimeter than this one?
Do any tetrominoes have a smaller perimeter than this one?
Show clearly how you reached your answer.

Tetro means four. Pento means five.

9 There are 12 different pentomino shapes. Work out what they look like.

- Which pentomino has the largest perimeter?
- Which one has the smallest perimeter?

Assignment 2 Perimeters

For this assignment the squares can be joined together in any way as long as they touch.

Copy this table and fill in the largest and smallest perimeter for each number of squares. Draw the shapes to show how you found your answer.

- Describe any patterns you notice in the numbers.

Number of squares	Largest perimeter	Smallest perimeter
1	4	4
2		
3		
4		
5		
6		

Write down any patterns or rules you have worked out to find the largest perimeter and the smallest perimeter.

Area

Exercise 3

1 Copy this square and the rectangular tile onto centimetre-squared paper.

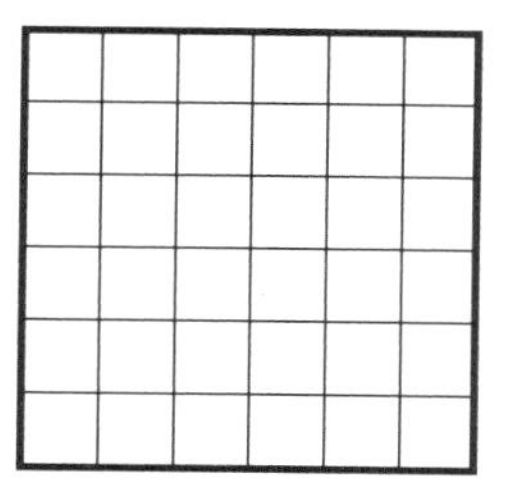

How many of the rectangular tiles can you fit into the square?
Draw them on your diagram.

2 Copy these shapes onto centimetre-squared paper.
How many of the rectangular tiles in question 1 can you fit in each of these shapes? Record your answers on centimetre-squared paper.

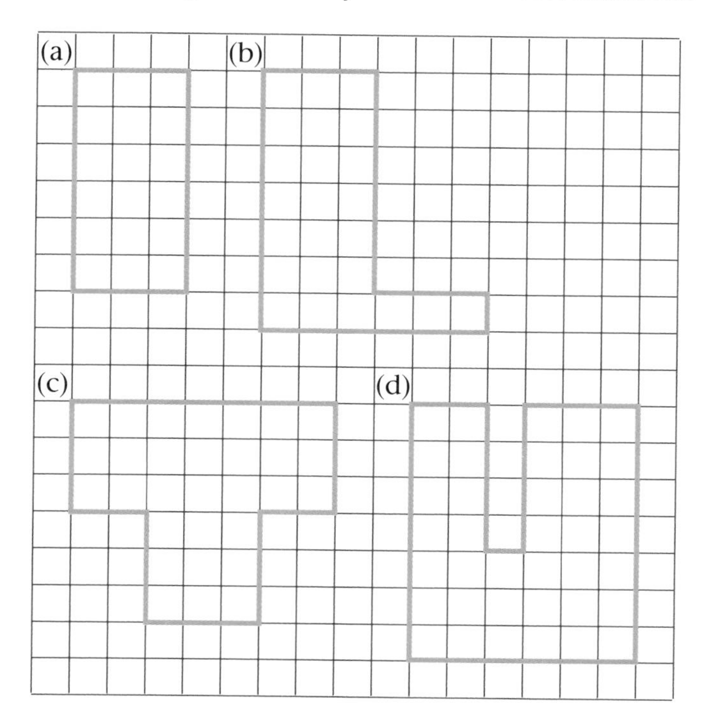

3 How many equilateral triangles like this one can you fit into the following shapes?

An equilateral triangle has all sides equal.

Record your answers on triangular dotted paper.

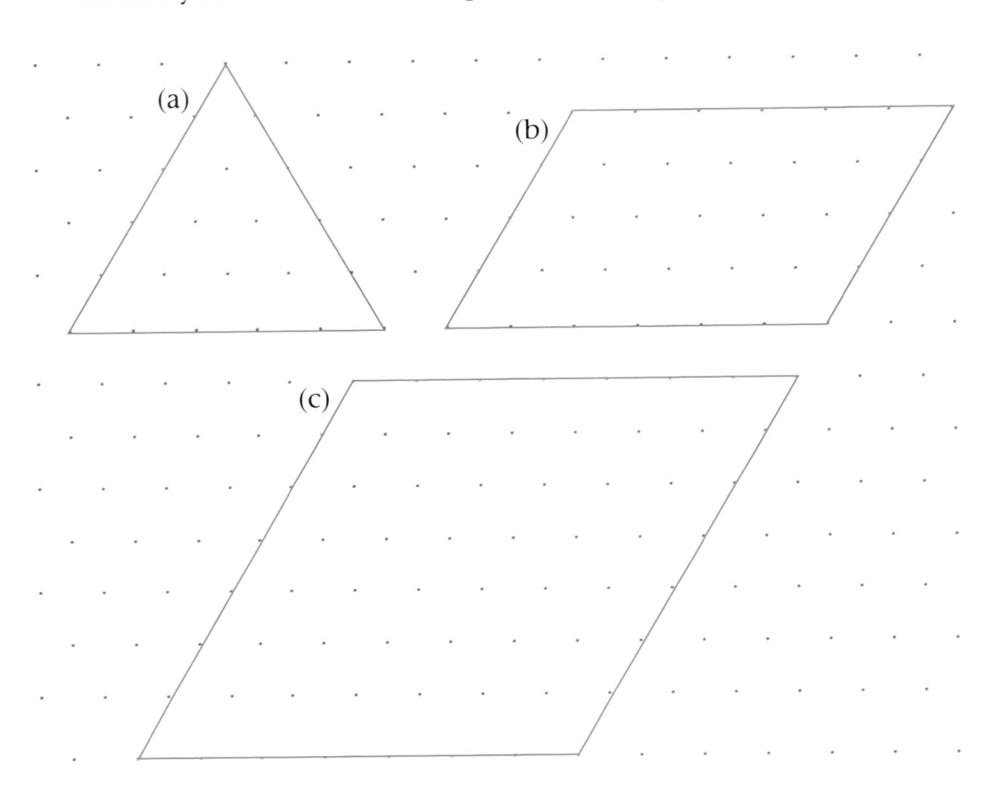

4 Copy this shape onto centimetre-squared paper. Divide your shape into four parts which are all the same size and shape.

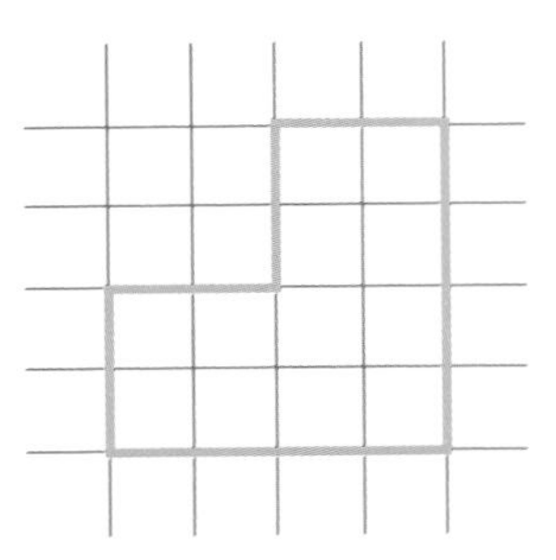

5 Copy the outline of this grid onto centimetre-squared paper.

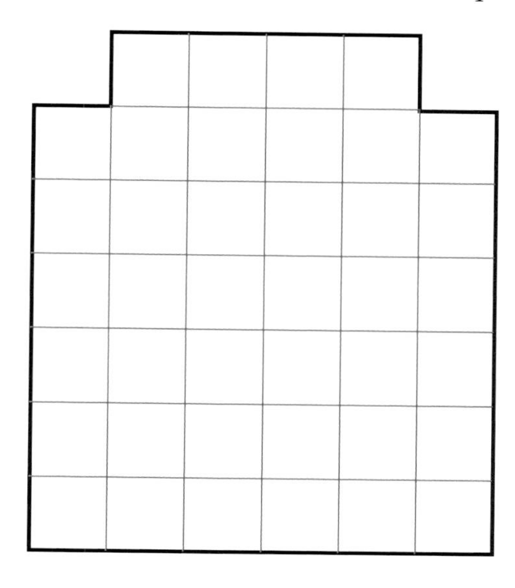

- Show how you can cover the grid with tiles like this one.

- Copy the outline of this grid and use the same tiles to cover it.

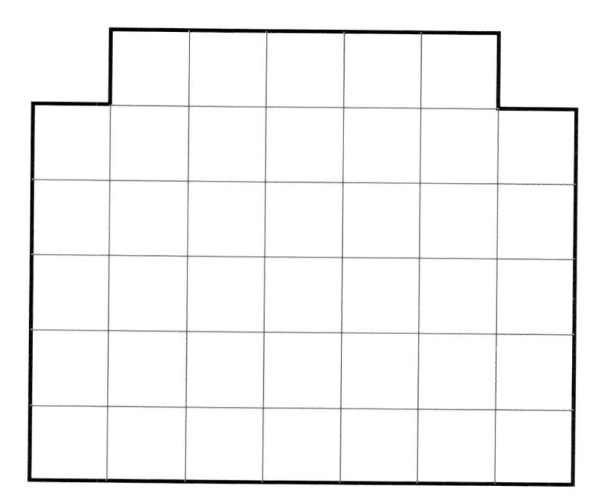

- Write down what you discover.
- Both of these outlines surround 40 squares. Draw other outlines which surround 40 squares. Try to cover each of them with the same tiles. Write down what you discover.

6 The diagram shows the same parallelogram covered with different shapes.

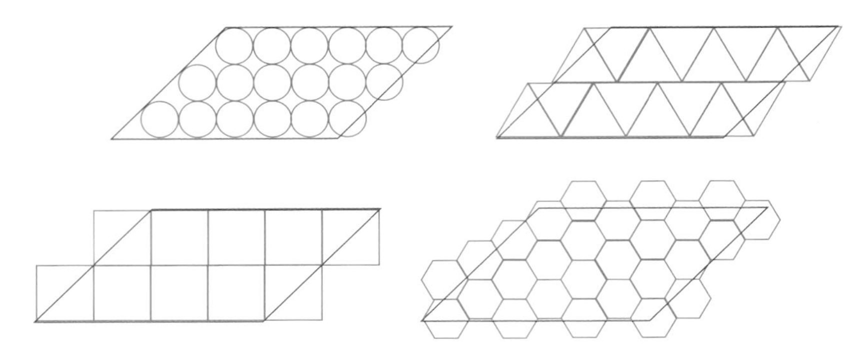

Key fact

Area is the space inside the perimeter of a 2-D shape.

- How many of each shape do you think you would need to fill the parallelogram without leaving any gaps?
- Do all the shapes cover the parallelogram without leaving gaps?
- Which shape did you find hardest to work out?
- Which shape did you find easiest?

7 This diagram shows the outline of the floor of a room.
How many square carpet tiles that measure 1 m by 1 m are needed to cover the floor completely?

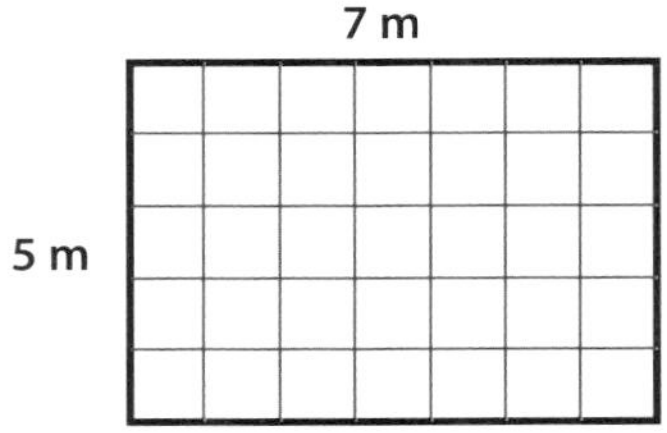

7 m

5 m

8 Here are the outlines of the floors of other rooms.
How many carpet tiles are needed to cover them completely?

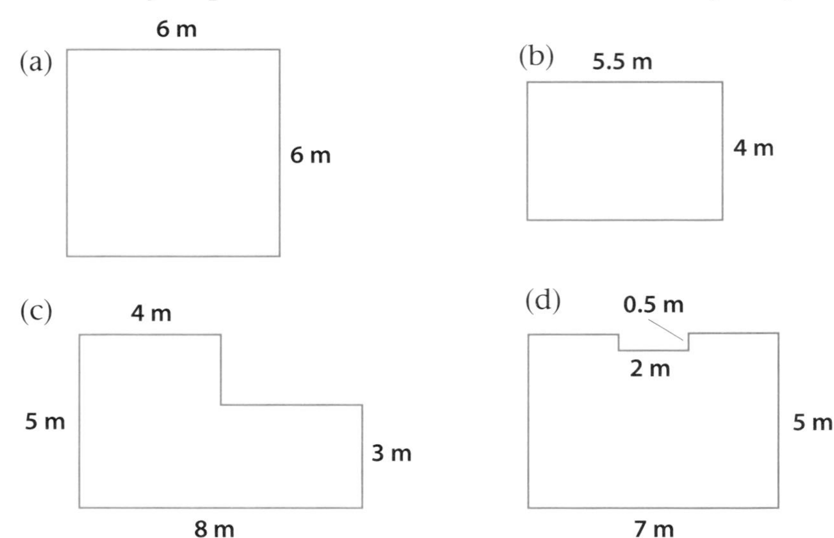

> **Area is measured in squares. The length of the side of the square is one unit. If the square measures 1 cm by 1 cm, the unit of area is 1 square centimetre, which is written 1 cm^2.**

9 Write down the units that are used to measure the area of each of these rectangles.

(a) 4 m by 3 m

(b) 4 ft by 3 ft

(c) 4 km by 3 km

(d) 4 mm by 3 mm

10 This triangle has an area of 6 squares.

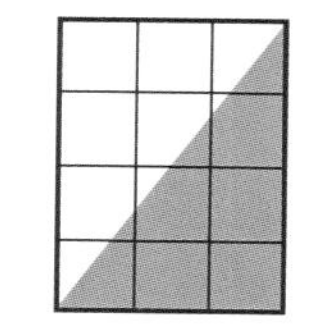

An easy way to work this out is to find the area of the rectangle, which is 12 squares, and halve this, because the triangle cuts the rectangle in half.

Write down the area of each of these triangles.

(a) (b) (c) (d) (e)

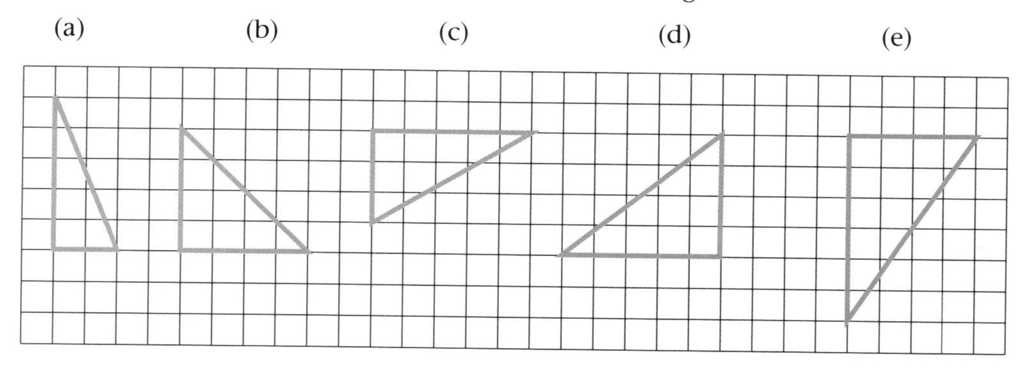

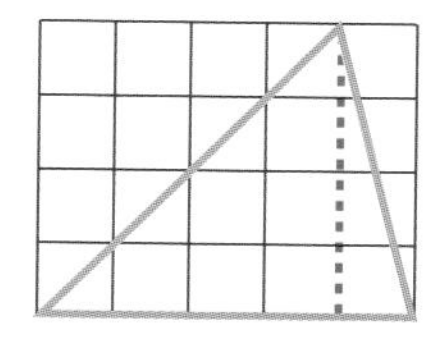

11 This triangle has an area of 10 squares. One way to work this out is to split the triangle into two right-angled triangles and then work out the area of each, as before.

Write down the area of each of these triangles. Use drawings to explain how you reached your answer.

(a) (b) (c) (d)

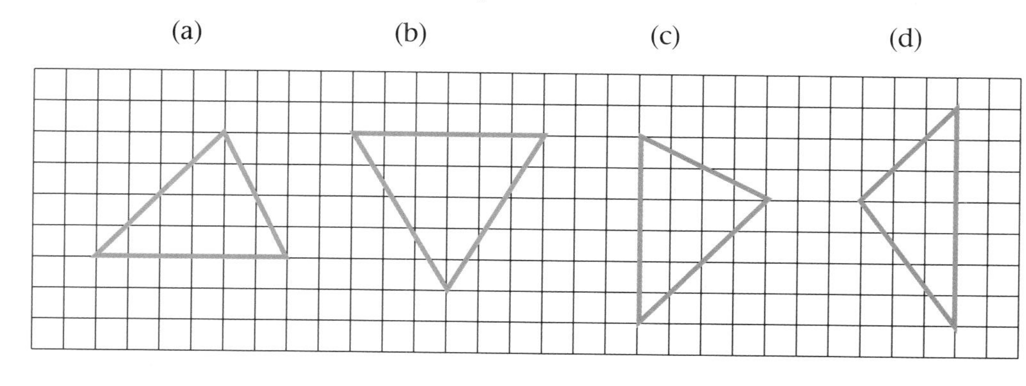

12 Work out the area of each of these shapes.
Use drawings to explain how you reached your answer.

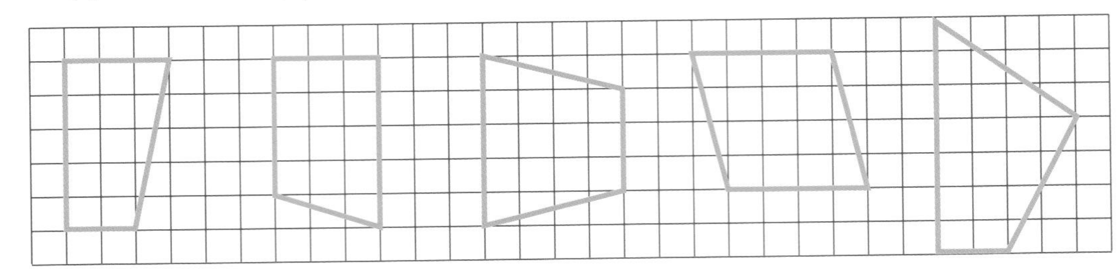

Assignment 3 Areas

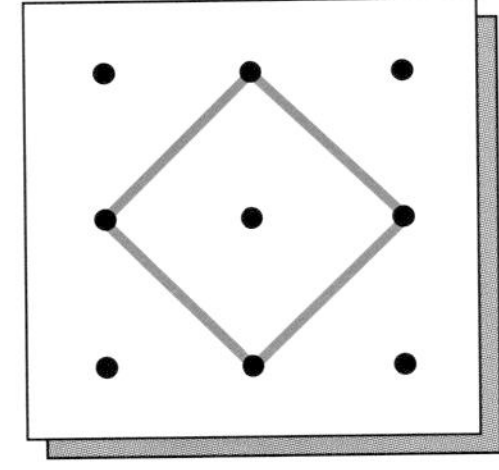

You will need a 3 by 3 pinboard.

This polygon has an area of 2 cm².

- How many more polygons can you find with an area of 2 cm²? Record your answers on squared dotted paper.
- How many shapes can you find with an area of 3 cm²?
- How many shapes with an area of 4 cm² can you find on a 3 by 3 pinboard?
- How many shapes with an area of 2 cm² can you find on a 4 by 4 pinboard?
- Choose two areas and find how many shapes you can draw on a 4 by 4 pinboard with these areas.

Assignment 4 Area and perimeter

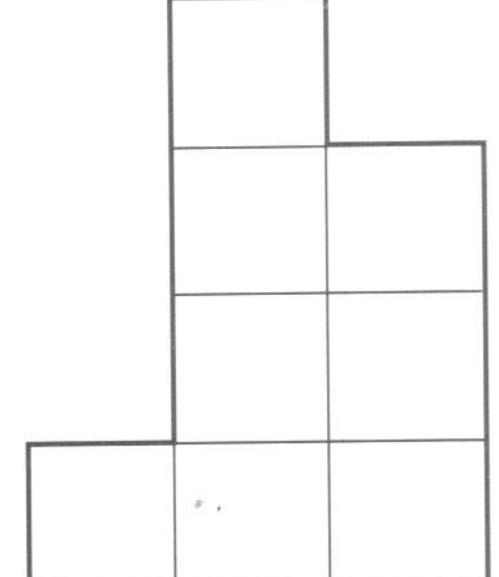

These two shapes both have a perimeter of 14 cm but the first shape has an area of 8 cm² and the second shape has an area of 6 cm².

- Use centimetre-squared paper. Find as many shapes as possible with a perimeter of 14 cm. Write down the area of each shape.
- Choose another length for the perimeter. Draw all the shapes you can find and write down their areas.
- Comment on your answers.

Volume

Exercise 4

1 How many matchboxes like the one shown will fit into the box?

2 cm 2 cm 3 cm

2 cm 6 cm 12 cm

Write down how you worked out your answer.

2 How many matchboxes like this one will fit into this box?

Write down how you worked out your answer.

1 cm 2 cm 3 cm

8 cm 5 cm 15 cm

> **Key fact**
> Volume is the amount of 3-D space inside a 3-D shape.

3 What units are used to measure the volume of each of these boxes?

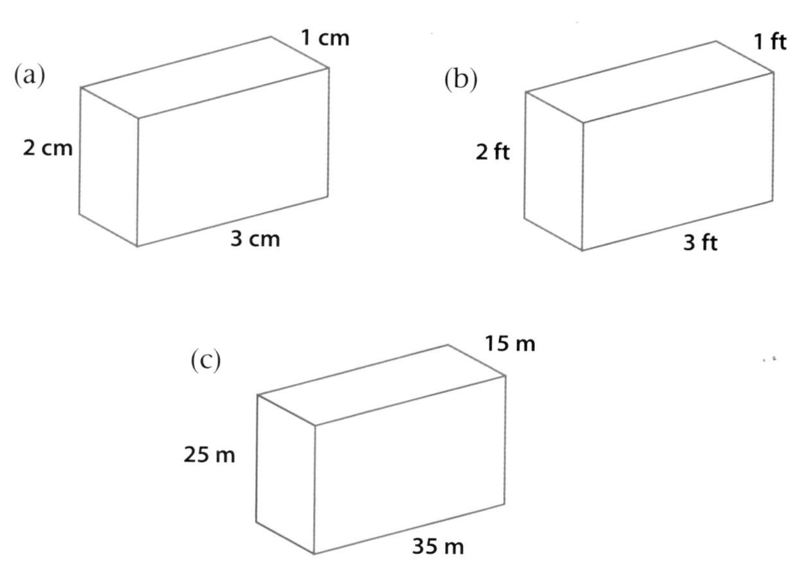

Volume is measured in cubes. The length of the edge of the cube is one unit. If the cube measures 1 cm by 1 cm by 1 cm the unit of volume is the cubic centimetre, written as cm³

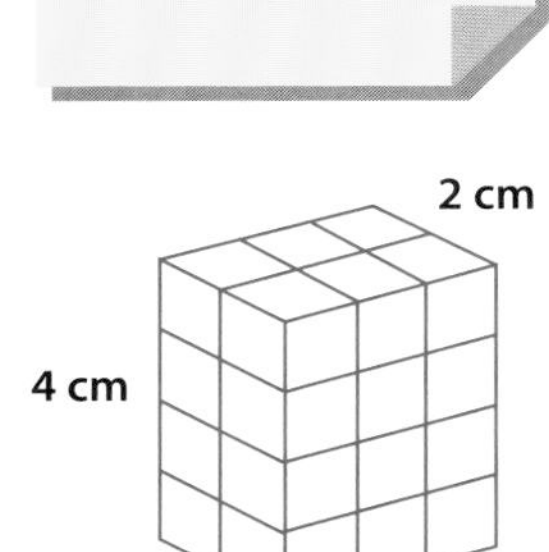

4 (a) How many of these 1-cm cubes will fit in the bottom of this box?

(b) How many layers of cubes will fit in the box?

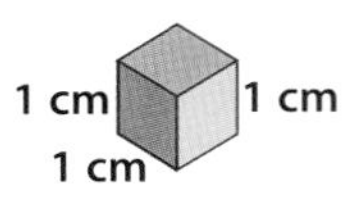

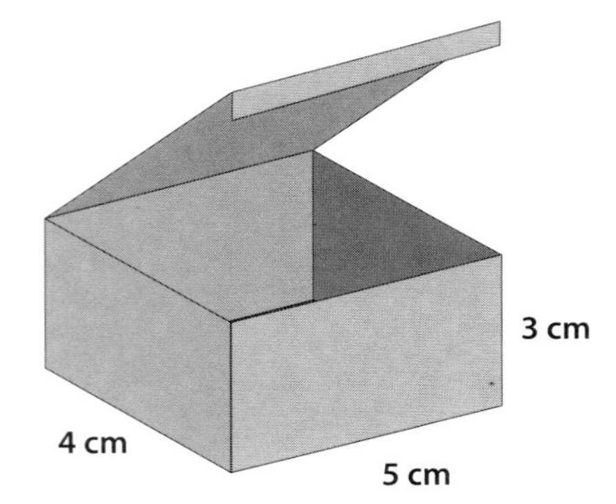

5 Explain how you could use your answers in question 4 to work out the number of cubes which will fit into the box.

6 This shape holds 24 centimetre cubes. The volume of the shape is 24 cm³.

Work out the volume of each of these shapes, showing clearly how you worked out your answers.

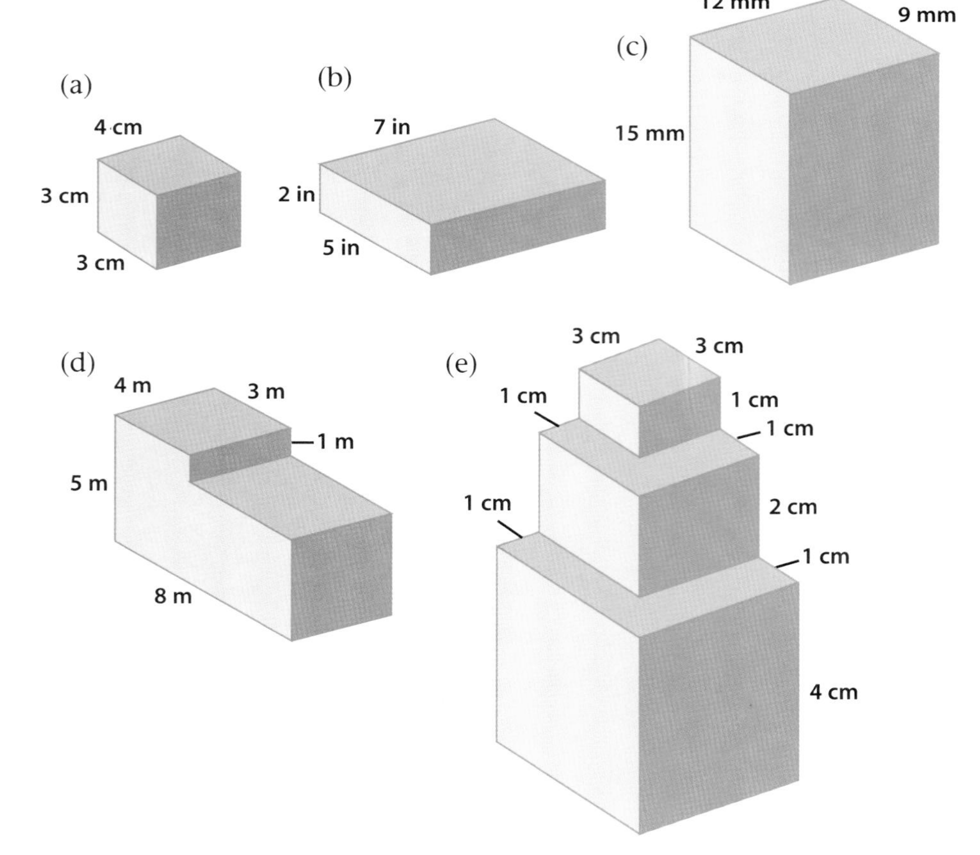

Assignment 5 A tower of cubes

This tower is made from 10 cubes. It has 36 faces that can be seen on the outside of the shape.

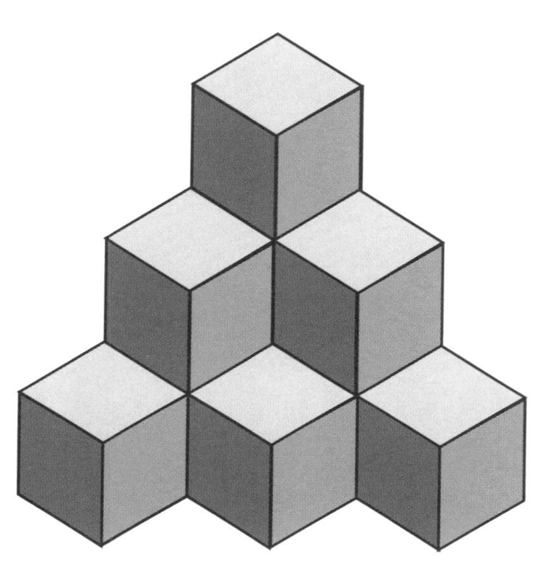

- What arrangement of the 10 cubes leaves the smallest number of faces showing?
- Make shapes with different numbers of cubes.
- Find the arrangement which shows the smallest number of faces for your shapes.
- Record your shapes on triangular dotted paper like this.

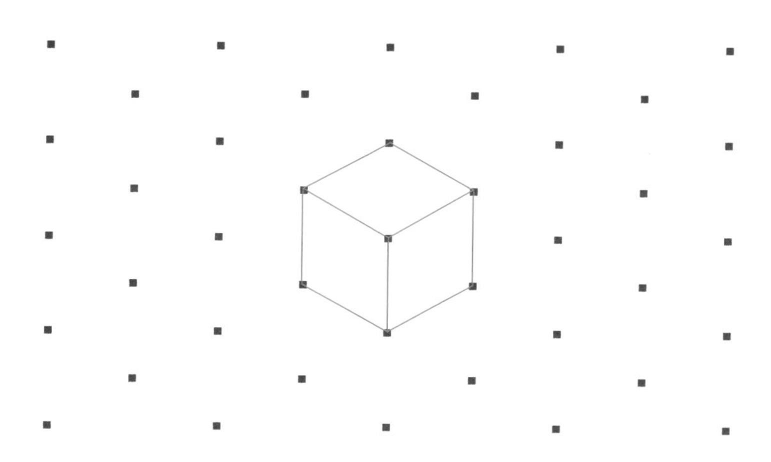

Review Exercise

1 Write down the perimeter of each of these shapes.

(a) (b)

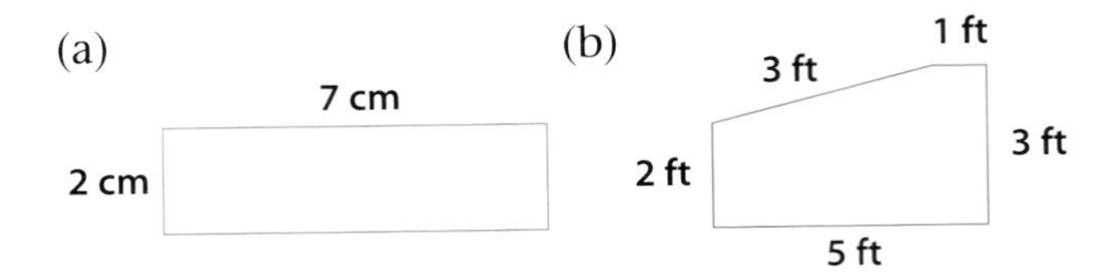

(c) (d)

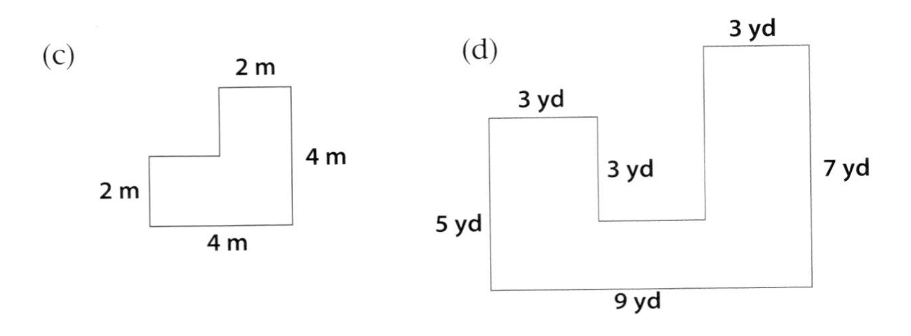

2 Work out the area of each of these shapes.
Write down how you worked out your answers.

(a) (b)

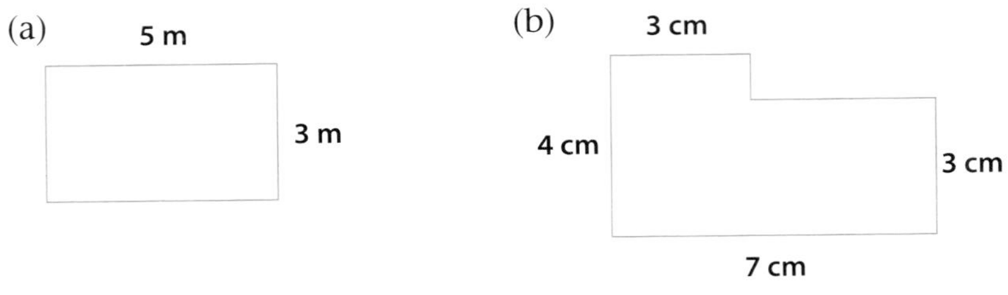

(c) (d)

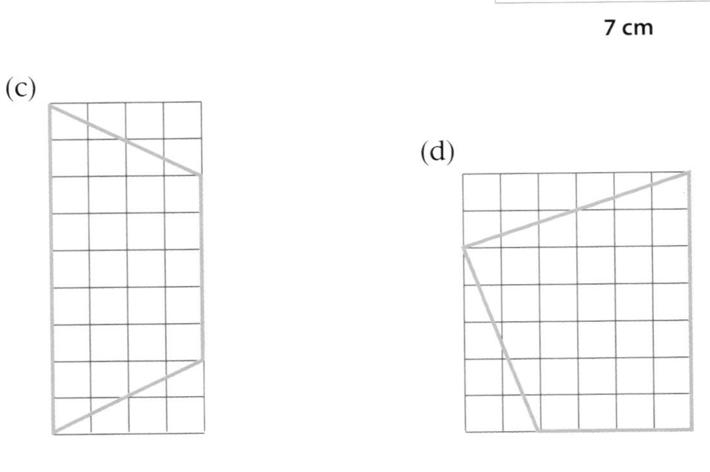

3 Work out the volume of each of these shapes.
Write down how you worked out your answers.

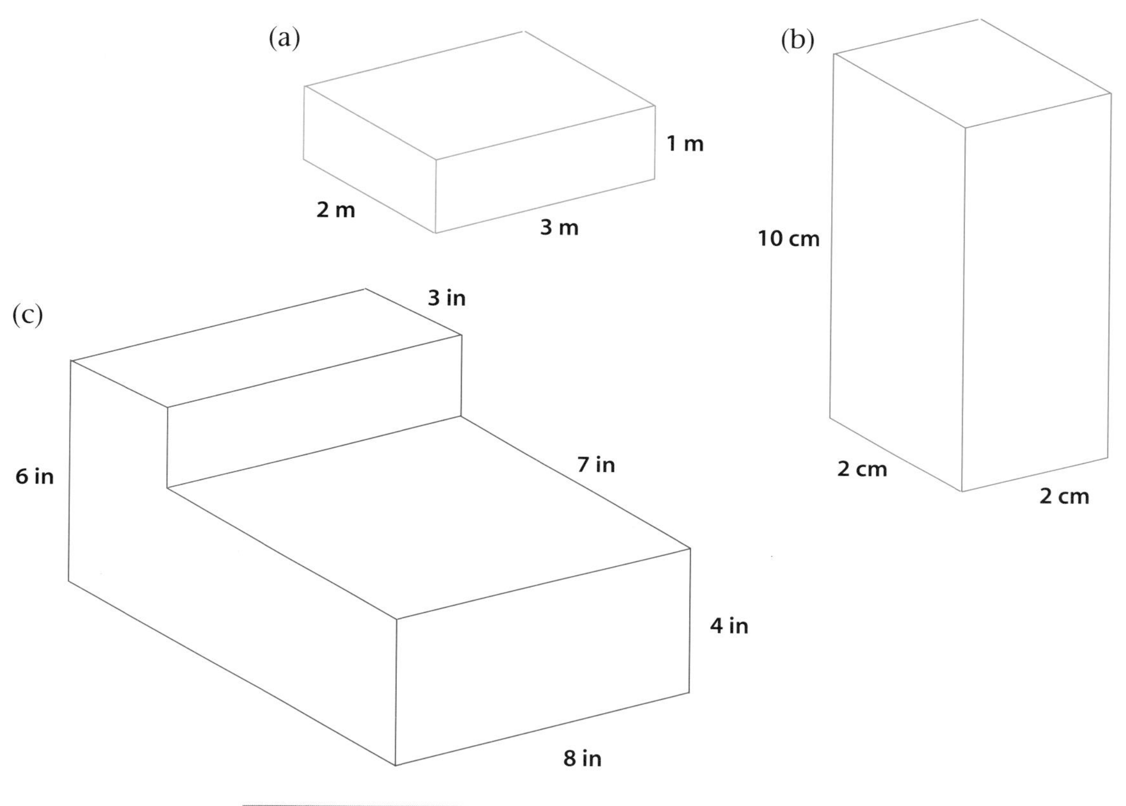

Assignment 6 Rabbit runs

A rabbit breeder has to build some runs for his rabbits. He has a supply of fencing in pieces that are all 1m long. Each rabbit needs 1 m^2 of space.

The breeder realises that four pieces of fencing in the shape of a square make a run which is just big enough for one rabbit.

As his rabbits do not have to be kept separate, he can keep two rabbits in a run made of six pieces of fencing.

He decides that it is easiest to join the fencing at right angles for the corners and he wants to use as little fencing as possible.

- How many pieces of fencing will he need to make a hutch for
  (a) three rabbits (b) four rabbits?
- Is it possible to calculate how much fencing he would need for 100 rabbits? Explain your answer.

Patterns with numbers

In this chapter you will learn:

- → **the names for different types of numbers**
- → **some facts about numbers**
- → **how to recognise properties of numbers**
- → **how to use number facts**

A number sequence is a list of numbers which are found by following a rule.

Starting points

Before you start this chapter you will need to:

- recognise odd and even numbers
- make simple number sequences
- continue simple number sequences.

Exercise 1

1 Copy the following set of numbers. Draw a loop round each odd number. Draw a box round each even number.

15	29	14	51
64	74	101	245
321	428	1465	2004
3457	2220	13 578	246 801

2 Copy the following sentences and complete them. Use either 'odd' or 'even'.

(a) 18 is the ninth ________ number.

(b) 24 follows an ________ number.

(c) ________ numbers are in the two times table.

(d) When we put things into pairs we always have an ________ number.

(e) When numbers are ________ they cannot be divided exactly by two.

3 Copy the following sentences, and complete them using either 'odd' or 'even'. Use numbers to give an example to support each answer. The first one has been done for you.

(a) An even number + an even number = an even answer.
4 + 6 = 10

(b) An odd number + an odd number = an ________ answer.

(c) An even number × an even number = an ________ answer.

(d) An odd number × an even number = an ________ answer.

(e) An odd number × an odd number = an ________ answer.

(f) An even number – an odd number = an ________ answer.

(g) An odd number – an odd number = an ________ answer.

4 Copy the following sequences.
Continue each of them for another three terms.

(a) 1, 3, 5, 7, ...

(b) 20, 18, 16, 14, ...

(c) 1, 10, 100, 1000, ...

(d) 3, 6, 9, 12, ...

(e) 100, 90, 80, 70, ...

(f) 1, 2, 4, 8, ...

(g) 1, 4, 7, 10, ...

A term is the name given to any number in a sequence.

Number patterns

Recognising patterns

Work with a partner.

- Cover the six cards that are on the next page.
- Uncover the first one for about two seconds.
- Cover it up again.
- Write down the number of dots.

Do the same for all the other cards.

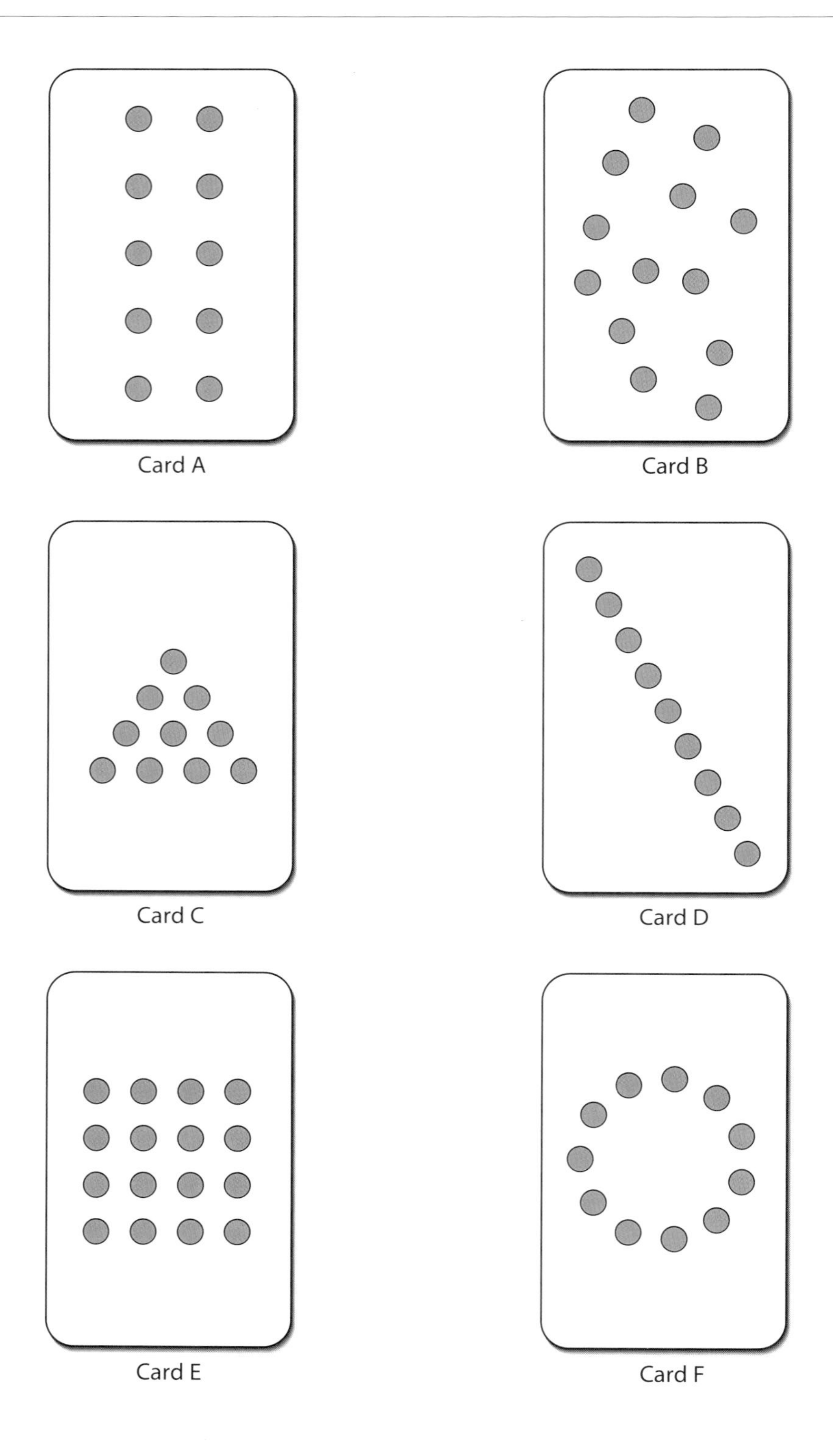
Card A
Card B
Card C
Card D
Card E
Card F

Now go back and carefully count the number of dots on each card to see how well you did.

Which numbers did you recognise most easily? Were the patterns of dots familiar? Have you seen any of them before?

Think about the patterns used on dice, dominoes and playing cards. Are these patterns often the same for any particular number? Why?

Key fact
A rectangle number can be shown as a rectangular pattern of dots.

Rectangle numbers

Six dots can be drawn to form a rectangle.

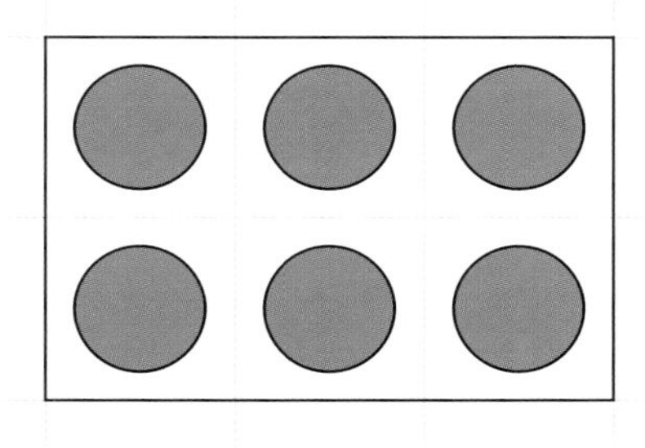

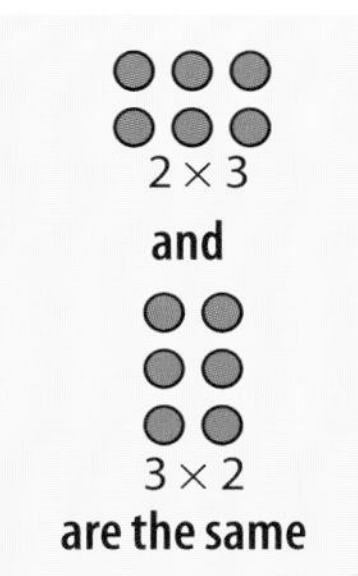

6 is a rectangle number.

6 can be arranged as 2 rows of 3 dots.

For short we say 2 by 3

or 2×3

Exercise 2

Use squared paper to help you with this exercise.

1 Make drawings of the following rectangle numbers. Underneath each pattern, describe it as we did above, e.g. 2×3.

(a) 8 (b) 15 (c) 10 (d) 21 (e) 12

2 You may have noticed that there are several ways to draw part (e). Find as many different ways of drawing these rectangle numbers as you can.

(a) 12 (b) 20 (c) 24

This pattern shows a row of dots in a 1×7 arrangement. It is not considered to be a rectangle.

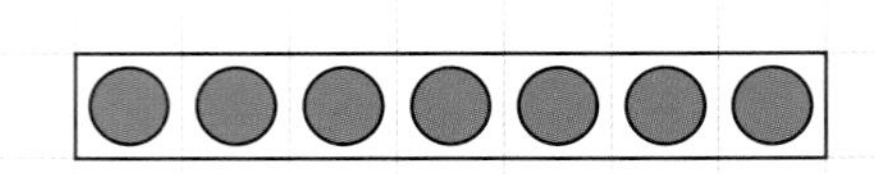

Key fact
A number that divides exactly into another number is called a factor of that number.

Factors

Consider the number 8.

We can see that $8 = 2 \times 4$
and $8 = 1 \times 8$

2 and 4 are **factors** of 8.

1 and 8 are also factors of 8.

Factors of 8: 1, 2, 4, 8

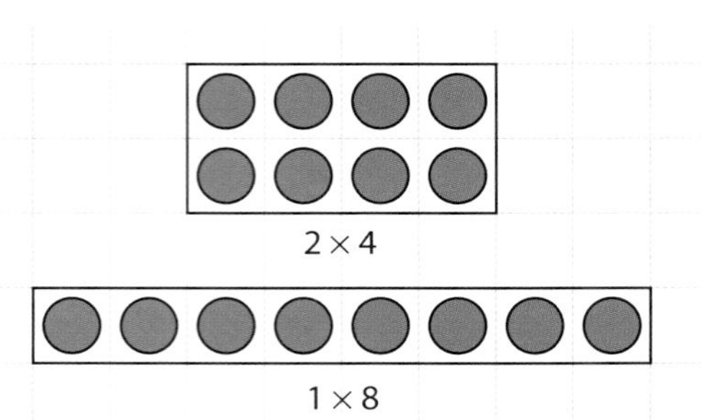

Note that factors come in pairs.
We can use one factor of a number to find another factor.

Exercise 3

1 Find the factors of the following numbers. Use your tables, or draw rectangle patterns.

(a) 6 (b) 12 (c) 15 (d) 20 (e) 32

(f) 36 (g) 25 (h) 11 (i) 63 (j) 72

2 (a) Write down the factors of 12.
(b) Write down the factors of 20.
(c) What factors are in both lists?
These are the **common factors** of 12 and 20.
(d) What is the biggest number that is in both lists?
This is the **highest common factor** (HCF) of 12 and 20.

Dividing by one factor always gives another factor.

3 Use the same steps as in question 2 to find the HCF of each of these pairs of numbers.

(a) 32 and 72 (b) 27 and 36 (c) 25 and 45

(d) 24 and 60 (e) 18 and 32 (f) 48 and 64

Key fact
The highest common factor (HCF) of two numbers is the biggest number that will divide into both of them exactly.

Key fact
A square number can be shown as a square pattern of dots.

Square numbers

Some numbers can be drawn as a square pattern.

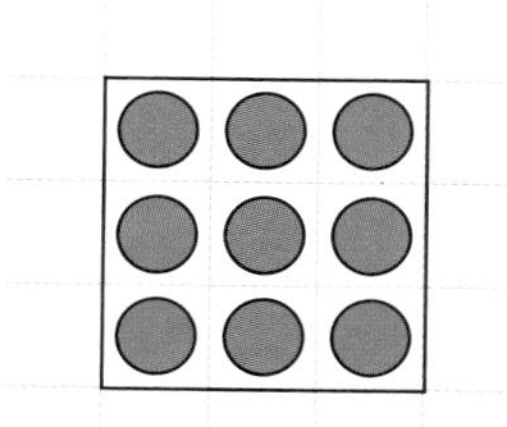

9 is a **square number**.

Note that $9 = 3 \times 3$

9 can be made by multiplying two **equal factors**.

$\mathbf{3 \times 3}$ is often written as $\mathbf{3^2}$.
We read it as **'3 squared'**.

Key fact
A triangle number can be shown as a triangular pattern of dots.

Triangle numbers

Some numbers can be arranged into a triangular pattern.

This diagram shows the first four triangle numbers.

1

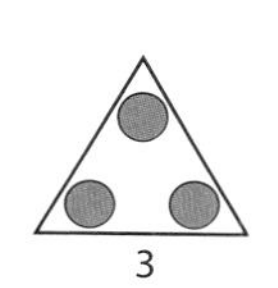
3

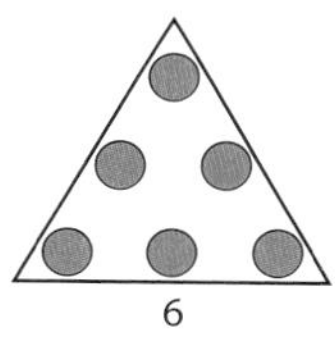
6

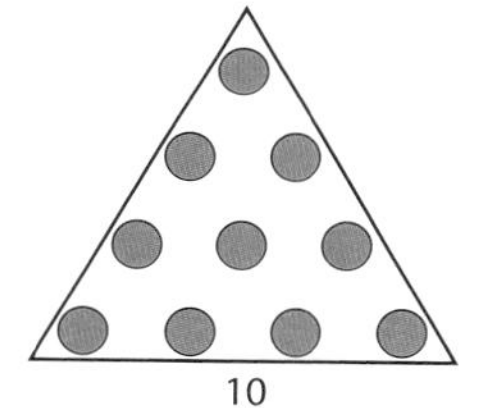
10

Example

6 is a **triangle number**.

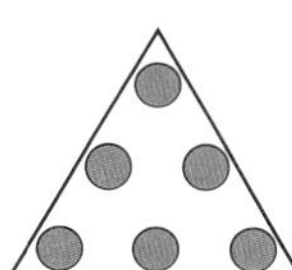

Exercise 4

1 Copy this table and continue it to find the first ten square numbers.

1st square number	$1^2 = 1 \times 1 =$	1
2nd square number	$2^2 = 2 \times 2 =$	?
3rd square number	$3^2 = 3 \times 3 =$	?
4th square number		

2 (a) Copy this table and continue it to find the first ten triangle numbers.

1st triangle number	$1 =$	1
2nd triangle number	$1 + 2 =$	3
3rd triangle number	$1 + 2 + 3 =$	?
4th triangle number	$1 + 2 + 3 + 4 =$	?

(b) Explain why this method gives triangle numbers.

3 This diagram shows the square number 16.
It has been divided into different sized squares.

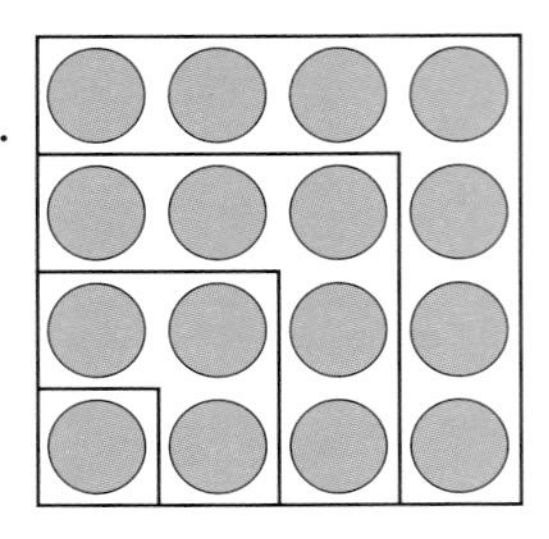

1 square (1^2)	=1 =	1
2 square (2^2)	= 1 + 3 =	4
3 square (3^2)	= 1 + 3 + 5 =	9
4 square (4^2)	= 1 + 3 + 5 + 7 =	16

(a) Can you see a connection between this number pattern and the diagram above? Try to explain the connection.

(b) Write down the next three lines of the number pattern.

(c) Use this pattern to work out the sum of the first ten odd numbers.

(d) Without writing them down, predict the sum of the first 25 odd numbers.

4

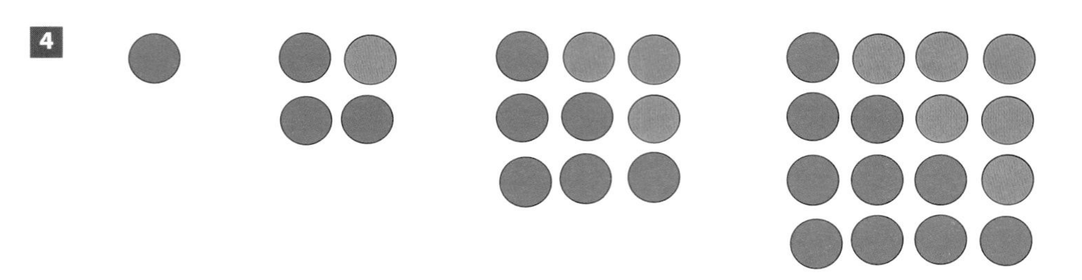

Copy this table and continue it to find the first ten square numbers.

	Number of red dots	Number of blue dots	Total number of dots
1st square	0	1	1
2nd square	1	3	4
3rd square	3	6	?
4th square	6	?	?
5th square	?	?	?

(a) Explain any patterns that you notice in the table.

(b) What is the connection between square numbers and triangle numbers?

(c) 100 is a square number. 100 is also the sum of two triangle numbers. Which two triangle numbers have a sum of 100?

(d) What is the sum of the 14th and 15th triangle numbers?

Assignment 1 Number chains

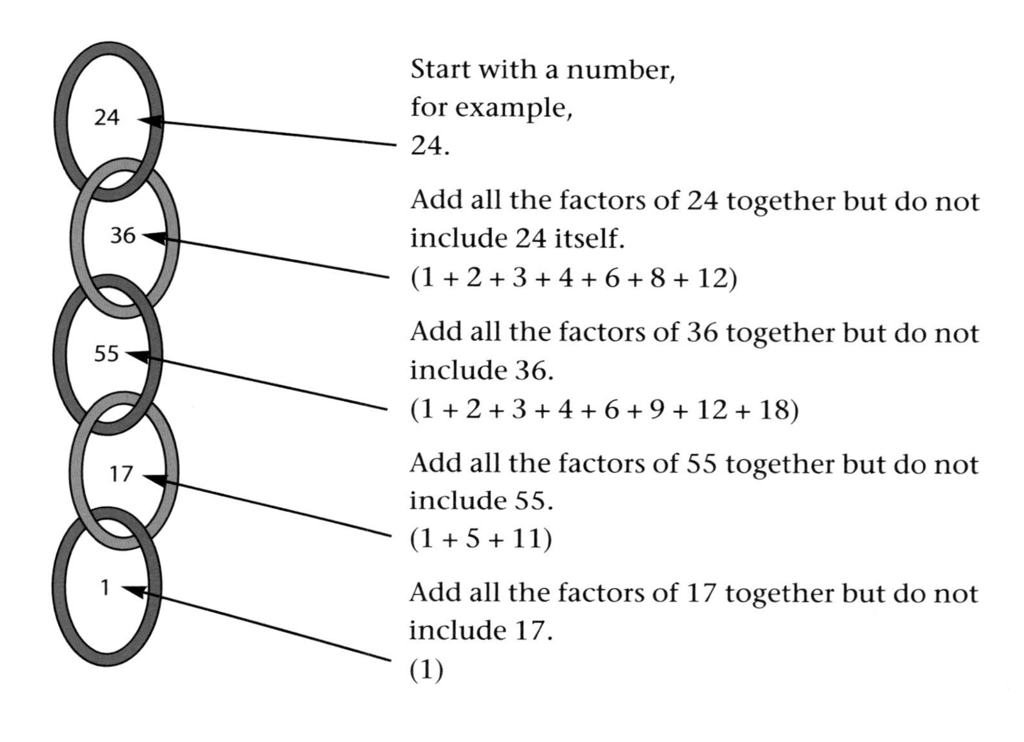

Start with a number,
for example,
24.

Add all the factors of 24 together but do not include 24 itself.
(1 + 2 + 3 + 4 + 6 + 8 + 12)

Add all the factors of 36 together but do not include 36.
(1 + 2 + 3 + 4 + 6 + 9 + 12 + 18)

Add all the factors of 55 together but do not include 55.
(1 + 5 + 11)

Add all the factors of 17 together but do not include 17.
(1)

Now start with 6.

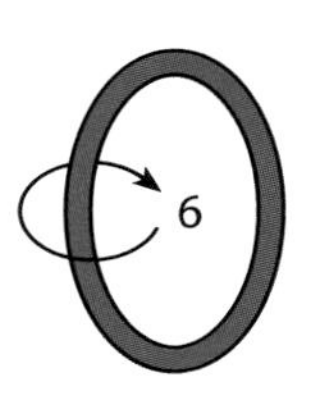

The factors of 6 add up to 6.

(1 + 2 + 3)

So if we start with 6 we get a loop.

- Make up your own number chains using the same rules.
  For a start, just use numbers below 50.
- What type of number always occurs before the 1? Why?
- Is there anything special about numbers which give longer chains?
- What number(s) below 50 gives the longest chain?
- Extend your ideas.

Patterns in the 100 number grid

You will need copies of the 100 number grid for Exercises 5 to 10.

The pattern of three

Exercise 5

1	2	3	4	5	6	7	8	9	10
11	12	13	14	15	16	17	18	19	20
21	22	23	24	25	26	27	28	29	30
31	32	33	34	35	36	37	38	39	40
41	42	43	44	45	46	47	48	49	50
51	52	53	54	55	56	57	58	59	60
61	62	63	64	65	66	67	68	69	70
71	72	73	74	75	76	77	78	79	80
81	82	83	84	85	86	87	88	89	90
91	92	93	94	95	96	97	98	99	100

1 Start with a blank 100 number grid. Shade the number 3. Miss the next two squares and shade 6.

Continue to shade every third square until the pattern is complete. The numbers shaded are called the **multiples** of 3 (up to 100).

2 How many threes are there in 100?

3 The 'pattern of three' numbers on the top row are 3, 6 and 9.

(a) Write down the 'pattern of three' numbers in the next row.
(b) For each of these numbers, find the digit sum.

Key fact
The multiples of a number can be obtained by multiplying it by the counting numbers 1, 2, 3, … .

4 (a) Write down six other multiples of 3.
(b) Find the digit sum for each number.

5 Try to think of a rule to check if a number is a multiple of 3.
Write it down in your own words.

To find the digit sum of a number, just add together its separate digits.
For example:
$12 \rightarrow 1 + 2 = 3$
Digit sum of $12 = 3$

Threes and sixes together

Exercise 6

1 Use the same 100 number grid that you used for Exercise 5. Draw a bold box around the number 6. It should look like this.

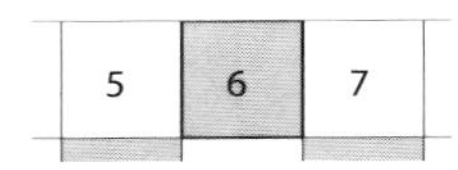

Draw a box around every sixth square until the pattern is complete. The boxed numbers are the multiples of 6.

2 How many sixes are there in 100?

3 'All the multiples of 6 are also multiples of 3.' Is this true?

4 Think of a rule to check if a multiple of 3 is also a multiple of 6.

5 Copy this table and complete it. ✔ means yes, ✘ means no.
Some have been filled in for you.

Number	Multiple of 3	Multiple of 6
32	✘	✘
63	✔	✘
66	✔	?
72		
84		
85		
87		
99		

The pattern of four

Exercise 7

1 Use a new 100 number grid.
Shade the number 4.

Shade every fourth square until the pattern is complete.

The shaded numbers are the multiples of 4.

1	2	3	4	5	6	7	8	9	10
11	12	13	14	15	16	17	18	19	20
21	22	23	24	25	26	27	28	29	30
31	32	33	34	35	36	37	38	39	40
41	42	43	44	45	46	47	48	49	50
51	52	53	54	55	56	57	58	59	60
61	62	63	64	65	66	67	68	69	70
71	72	73	74	75	76	77	78	79	80
81	82	83	84	85	86	87	88	89	90
91	92	93	94	95	96	97	98	99	100

2 How many fours are there in 100?

3 Write down the next six multiples of 4 after 100. What do you notice?

4 Is 140 a multiple of 4? Explain your answer.

5 Is 250 a multiple of 4? Explain your answer.

6 Copy the following numbers.
Draw a loop round each multiple of 4.

32 30 52 98 128

176 110 264 624 4566

The fives and tens

Exercise 8

1	2	3	4	5	6	7	8	9	10
11	12	13	14	15	16	17	18	19	20
21	22	23	24	25	26	27	28	29	30
31	32	33	34	35	36	37	38	39	40
41	42	43	44	45	46	47	48	49	50
51	52	53	54	55	56	57	58	59	60
61	62	63	64	65	66	67	68	69	70
71	72	73	74	75	76	77	78	79	80
81	82	83	84	85	86	87	88	89	90
91	92	93	94	95	96	97	98	99	100

1 Use a new 100 number grid. Shade the multiples of 5.
Draw bold boxes around the squares containing multiples of 10.

2 Explain how to tell if a number is a multiple of 5.

3 Explain how to tell if a number is a multiple of 10.

3 Copy the following set of numbers. Draw a loop round each multiple of 5. Draw a box round each multiple of 10.
Some numbers will have a loop *and* a box.

35 42 50 387

585 1700 2644 62 455

The pattern of nine

Exercise 9

1 Use a new 100 number grid. Shade the multiples of 9.

2 How many nines are there in 100?

3 How many nines are there in 200?

4 Write down the first ten multiples of 9.

$1 \times 9 = 9$ $2 \times 9 = 18$ $3 \times 9 = 27$

and so on. Describe any patterns that you notice.

5 Explain how to tell if a number is a multiple of 9.

6 Copy the following numbers. Draw a loop round each multiple of 9.

99 111 117 229 309

558 702 989 1234 5688

7 Look at this pattern.

$1 \times 9 = 10 - 1$ $2 \times 9 = 20 - 2$ $3 \times 9 = 30 - 3$ $4 \times 9 = 40 - 4$

Use this pattern to work out 99×9. Show your working.

Prime numbers

A formula is like a recipe which is used to work things out in maths.

People who study maths have always been fascinated by prime numbers. No one has ever been able to find a formula to work them out. In 1742 a famous German mathematician called Goldbach made an interesting conjecture. He said that every even number (except two) was the sum of two primes. Again, neither he nor anyone else has ever been able to prove that it works for all even numbers.

Exercise 10

Some of this has been started on the 100 number grid.

1 Use a new 100 number grid.

(a) Shade out 1. Put a box around 2.
Do not shade it. Shade the remaining multiples of 2.

(b) The next unshaded number is 3. Put a box around 3.
Do not shade it. Shade the remaining multiples of 3.

(c) Find the next unshaded number. Put a box around it.
Do not shade it. Shade the remaining multiples of this number.

(d) Continue to do this until there are no multiples left.

You should have 25 numbers remaining. These are the **prime numbers** below 100.

1	2	3	4	5	6	7	8	9	10
11	12	13	14	15	16	17	18	19	20
21	22	23	24	25	26	27	28	29	30
31	32	33	34	35	36	37	38	39	40
41	42	43	44	45	46	47	48	49	50
51	52	53	54	55	56	57	58	59	60
61	62	63	64	65	66	67	68	69	70
71	72	73	74	75	76	77	78	79	80
81	82	83	84	85	86	87	88	89	90
91	92	93	94	95	96	97	98	99	100

Key fact
1 is not a prime number.

2 Make a list of the prime numbers below 100.

3 Choose any five prime numbers. For each number, write down its factors. Explain what you notice.

Assignment 2 Goldbach's conjecture

Example

10	=	3	+	7
even number		prime		prime

A conjecture is a sensible guess.

Key fact
A prime number has exactly two different factors ... 1 and itself.

- Choose ten even numbers and write them down. Check to see if Goldbach's conjecture works.
- Some even numbers have more than one solution.

Example

18 = 5 + 13
18 = 7 + 11

Investigate.

- Which even number less than 50 has the most solutions?
- Is it possible to make odd numbers from two primes? Can it be done from three primes?
- Other ideas? Extensions?

Prime factors

Key fact
A prime factor is a factor which is also a prime number.

Factors of 45: 1 3 5 9 15 45

These are prime numbers. (3 and 5)

The **prime factors** of 45 are 3 and 5.

Note that the product of 3 and 5 is 15.

The product is the answer found by multiplying.

Writing numbers as a product of primes

Suggestion: Start with the smallest prime factor.

Example 1

Write 30 as the product of prime factors.

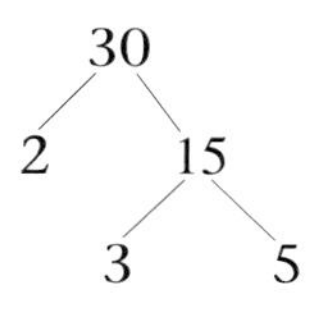

2 is the smallest prime factor.

3 is the next prime factor.
5 is also a prime factor.

You can make any number by multiplying all of its prime factors.

Example 2

Write 180 as the product of primes.

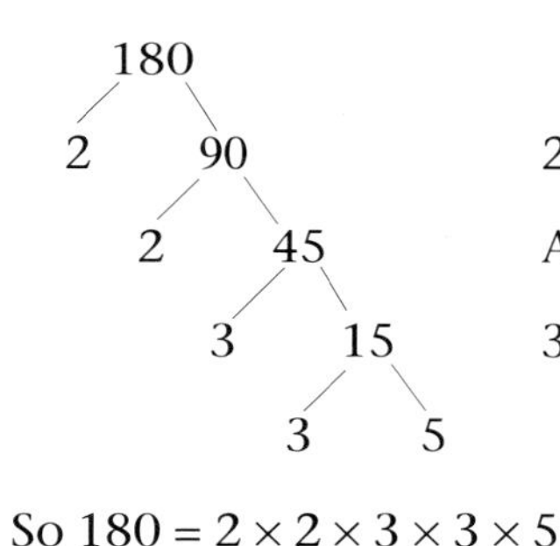

2 is the smallest prime factor.

Again, 2 is the smallest prime.

3 is the smallest prime factor.

So $180 = 2 \times 2 \times 3 \times 3 \times 5$

Sometimes prime factors are repeated.

Exercise 11

1 (a) Write down all the factors of 18.

(b) Which factors are prime numbers? Write them down.

(c) Now write 18 as a product of prime factors.

2 Find the prime factors of the following numbers.

(a) 12 (b) 33 (c) 49 (d) 29 (e) 75

3 Write each of the following as a product of prime factors.

(a) 12 (b) 24 (c) 30 (d) 72 (e) 660

4 Which of the following numbers are prime?

(a) 234 (b) 651 (c) 139 (d) 203 (e) 307

Remember: all squares are also rectangles.

Review Exercise

1 Copy this table and and complete it.

Number	Rectangle number	Square number	Triangle number	Prime number
1	✔	✔	✔	✘
3	✘	✘	✔	✔
4	✔	✔	✘	?
6	✔	?	?	
11				
22				
36				
56				
81				
89				
91				

2 Find all the factors of the following numbers.

(a) 10 (b) 16 (c) 30 (d) 48

(e) 56 (f) 75 (g) 84 (h) 100

3 (a) Write down the factors common to both 30 and 48.

(b) What is the highest common factor (HCF) of 30 and 48?

(c) Find the HCF of 56 and 84.

(d) Find the HCF of 36 and 63.

4 Copy this table and complete it.

Number	Multiple of 3	Multiple of 4	Multiple of 5	Multiple of 6	Multiple of 9
6	✔	✘	✘	✔	✘
20	✘	✔	?	?	?
36					
80					
81					
91					
120					
540					
810					
1215					

5 Write the following numbers as the product of prime factors.

(a) 20 (b) 55 (c) 63 (d) 126

(e) 182 (f) 525 (g) 1078 (h) 2520

6 It is fairly easy to calculate square numbers. For example, the 5th square number is found by calculating 5×5.
Consider the triangle numbers ...

1st triangle number = 1
2nd triangle number = 3
3rd triangle number =
4th triangle number =

Try to find a rule (called a formula) for working out the triangle numbers. Use this formula to find the 100th triangle number.

Assignment 3 Perfect? Nearly perfect!

In Assignment 1, you may have found out something special about 6. The factors of 6 are 1, 2, 3, 6.
Adding together all the factors of 6, but not including 6 itself, gives $1 + 2 + 3 = 6$ which is the number you started with!

A number which is equal to the sum of its own factors, excluding itself, is a **perfect number**.

12 is not a perfect number.

$12 \neq 1 + 2 + 3 + 4 + 6$

But, if just one factor is left out ...

$12 = 1 + 2 + 3 + 6$ (leaving out the 4)

12 is a nearly perfect number.

- There is one more perfect number below 50. Can you find it?
- Nearly perfect numbers are easier to find. How many are there below 50?

The symbol ≠ means 'is not equal to'.

Extensions

- What about 'almost nearly perfect' numbers? Try leaving out two factors.
- Try going beyond 50.

Number	Factors (not including itself)
2	1
3	1
4	1, 2
5	1
6	1, 2, 3
7	1
8	1, 2, 4
9	1, 3
10	1, 2, 5
11	1
12	1, 2, 3, 4, 6
13	1
14	1, 2, 7
15	1, 3, 5
16	1, 2, 4, 8
17	1
18	1, 2, 3, 6, 9
19	1
20	1, 2, 4, 5, 10
21	1, 3, 7
and so on.	

Angles

In this chapter you will learn:

- → **how to measure angles**
- → **how to bisect an angle**
- → **about the angles in a triangle**

Starting points

Before starting this chapter you will need to know:

- what we mean by 'right', 'acute' and 'obtuse' angles
- how to draw a right angle and triangle with equal angles.

Exercise 1

1. Use a ruler and compasses to construct a right angle.
2. Use a ruler and compasses to draw an equilateral triangle with all the sides 10 cm long.
3. Copy this table.

Angle	Acute angle	Right angle	Obtuse angle
a			
b			
c			

Look at the following angles. Put a tick in the correct column of your table, for each angle.

(a)

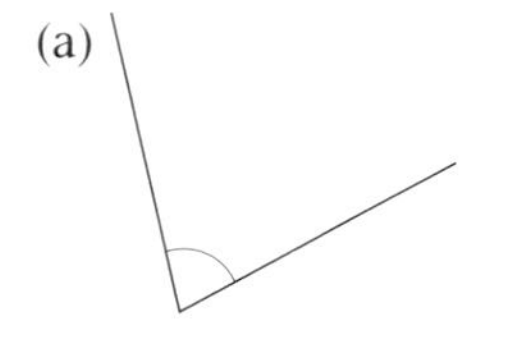

(b)

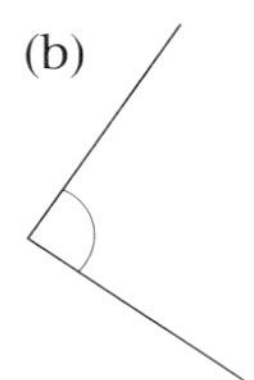

(c)

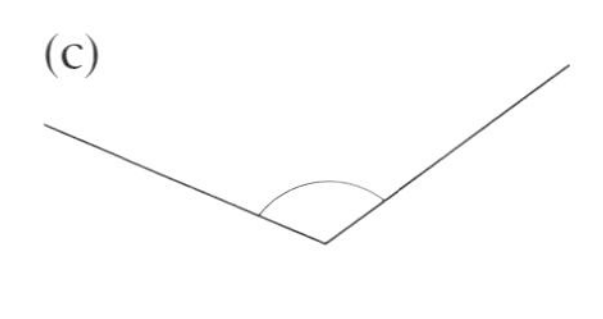

Bisecting an angle

Key fact

'Bisect' means 'cut in half'.

Assignment 1 Bisect an angle

You will need tracing paper, a sharp pencil, a ruler and compasses.

- Using tracing paper, draw two lines which cross. Make sure each line is more than 5 cm long. Mark the point where the lines cross A.

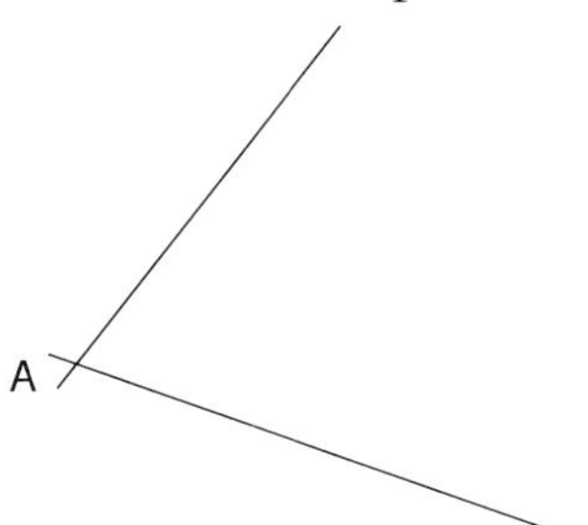

- Open your compasses to about 5 cm and put the point carefully on point A. Draw small arcs to cross each of the lines at B and C.

- Put your compass point carefully on point B. Draw another arc. Repeat this at point C.

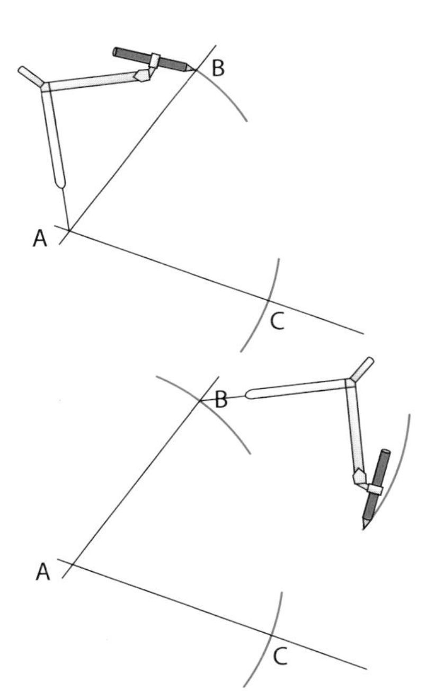

- Label the point where the arcs cross D. Join points A and D.
- Fold your tracing paper along the line joining A to D.
- Write down what you notice.

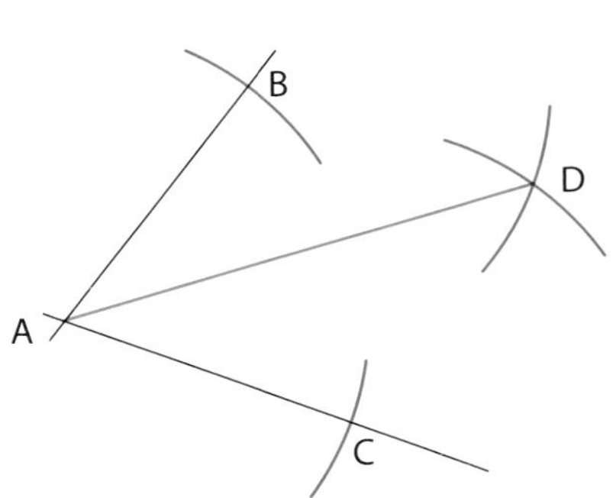

Exercise 2

1 Use a ruler and compasses to construct a right angle and then bisect it.

2 Use a ruler and compasses to construct an equilateral triangle with sides 10 cm long. Bisect one of the angles of the triangle.

Measuring angles

Exercise 3

1 Make an angle measurer from a piece of paper.

You will need plain paper, a sharp pencil, a ruler, compasses and scissors.

- On plain paper, construct an equilateral triangle with sides 12 cm long.
- Carefully cut out the triangle and fold two of its edges together.

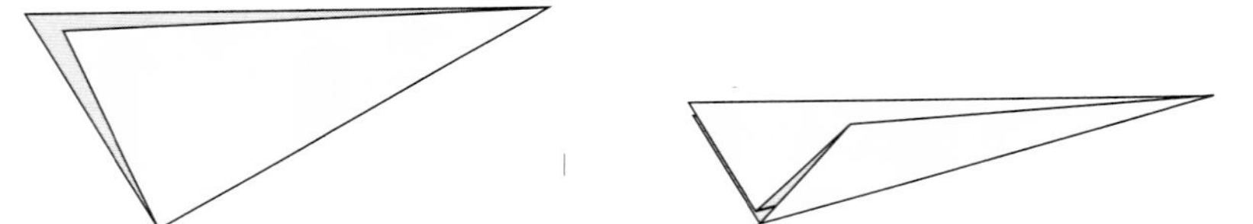

- Fold it again.

Use the smallest angle to measure with.

2 How many times will your angle measurer fit into each of these angles?

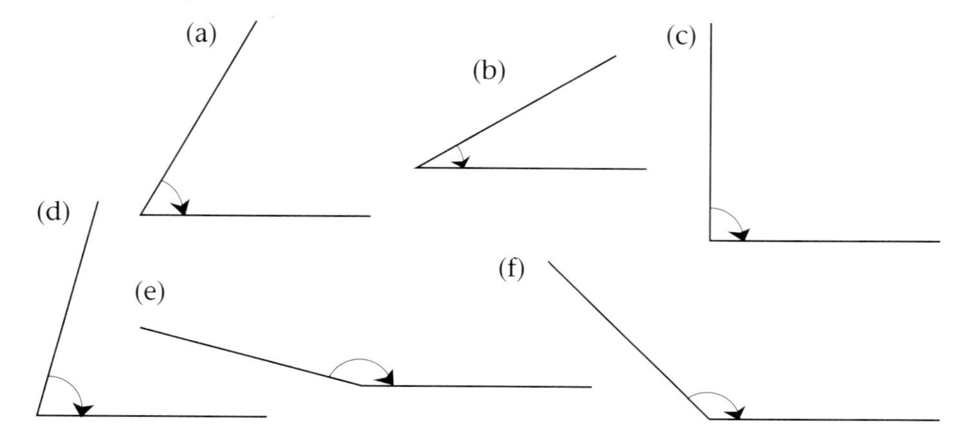

3 How many times will your angle measurer fit into each of these angles?

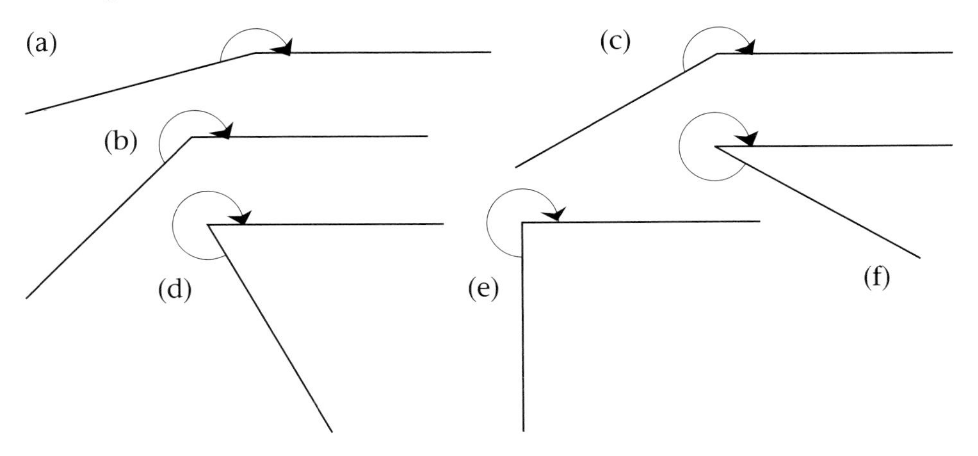

4 How many times will your angle measurer fit into a right angle?

Degrees

The ancient Babylonians divided a complete turn into 360 equal parts. Each of these parts is now called one **degree**.

Exercise 4

1 How many degrees are there in a right angle? Write down how you worked out your answer.

2 (a) Describe the two parts which make up this angle.

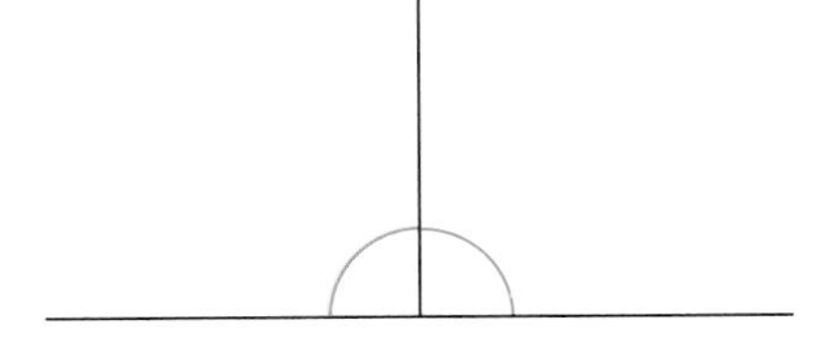

(b) How many degrees are there in the whole angle?

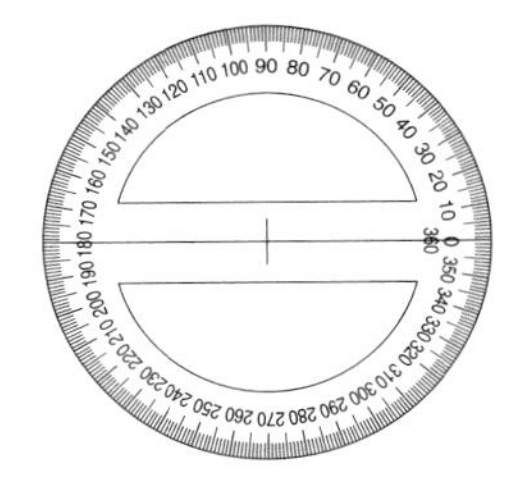

A protractor is used to measure angles in degrees.

Assignment 2 Angles as measures of turning

You will need a sharp pencil, tracing paper and a protractor.

- Trace the red line. Keep the tracing positioned over the line.
- Put your pencil point carefully on point A.
- Turn the tracing paper in the direction of the arrow until the traced line is on top of the blue line.

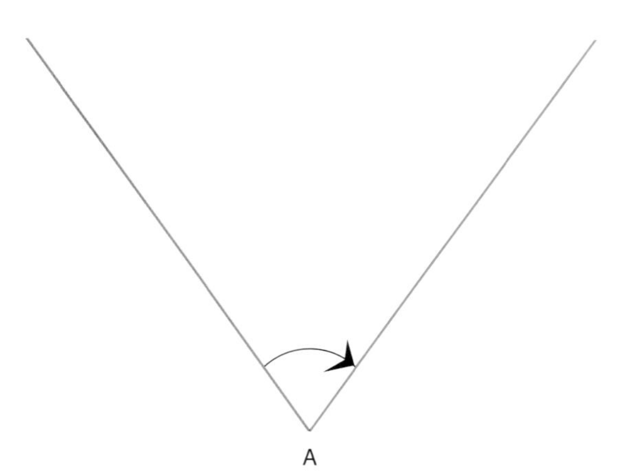

The traced line has turned through the angle marked on the diagram. At the start the line had not turned at all. The angle got bigger as you turned the tracing paper.

- Which way did you turn the paper, clockwise or anti-clockwise?
- Now look at your protractor.
- Look at the two sets of numbers. Which set gets bigger in the same direction that you turned the paper? Is it the set at the outside edge, or the inner set?

The sets of numbers around the edge of the protractor are the scales.

- Place your protractor on the diagram so that the cross at its centre is exactly on point A and the line through 0 is exactly on the red line.
- Remember which set of numbers you chose. Which number on that scale does the blue line go through?
- Describe clearly how you would measure the angle if you started with the 0 on the blue line.
- Write down the size of the angle, using the symbol for degrees.

When writing angles, we use the ° symbol for degrees.

Exercise 5

Key fact

A right angle is 90°.

90°

1 Measure the following angles carefully with your protractor. Write down your answer using the symbol for degrees.

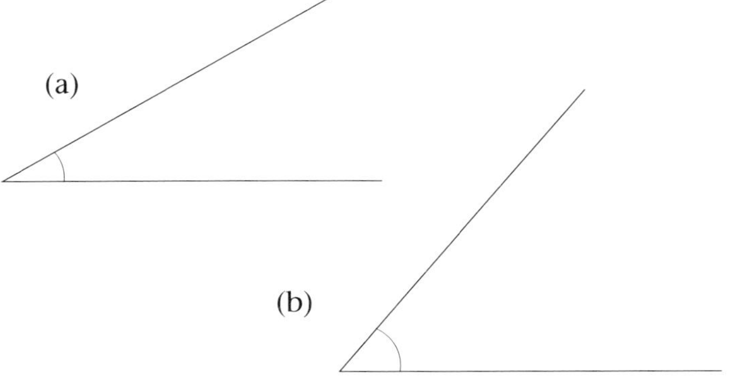

Key fact

A half turn, or straight angle, is 180°.

180°

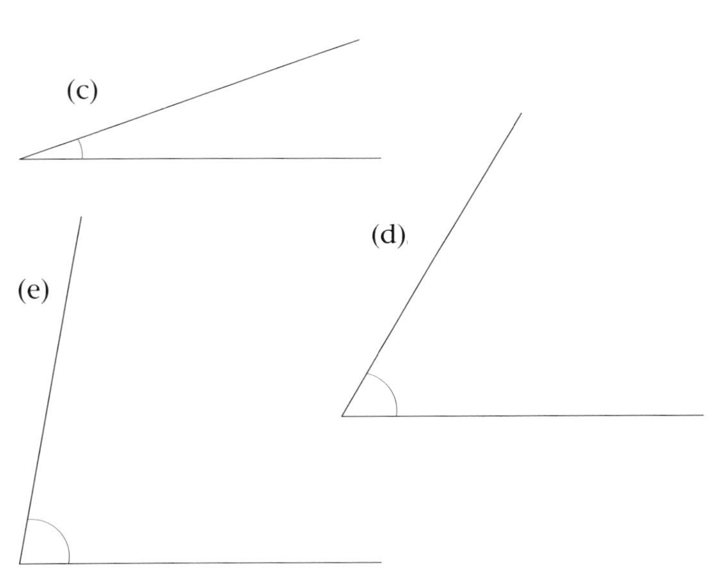

(f)

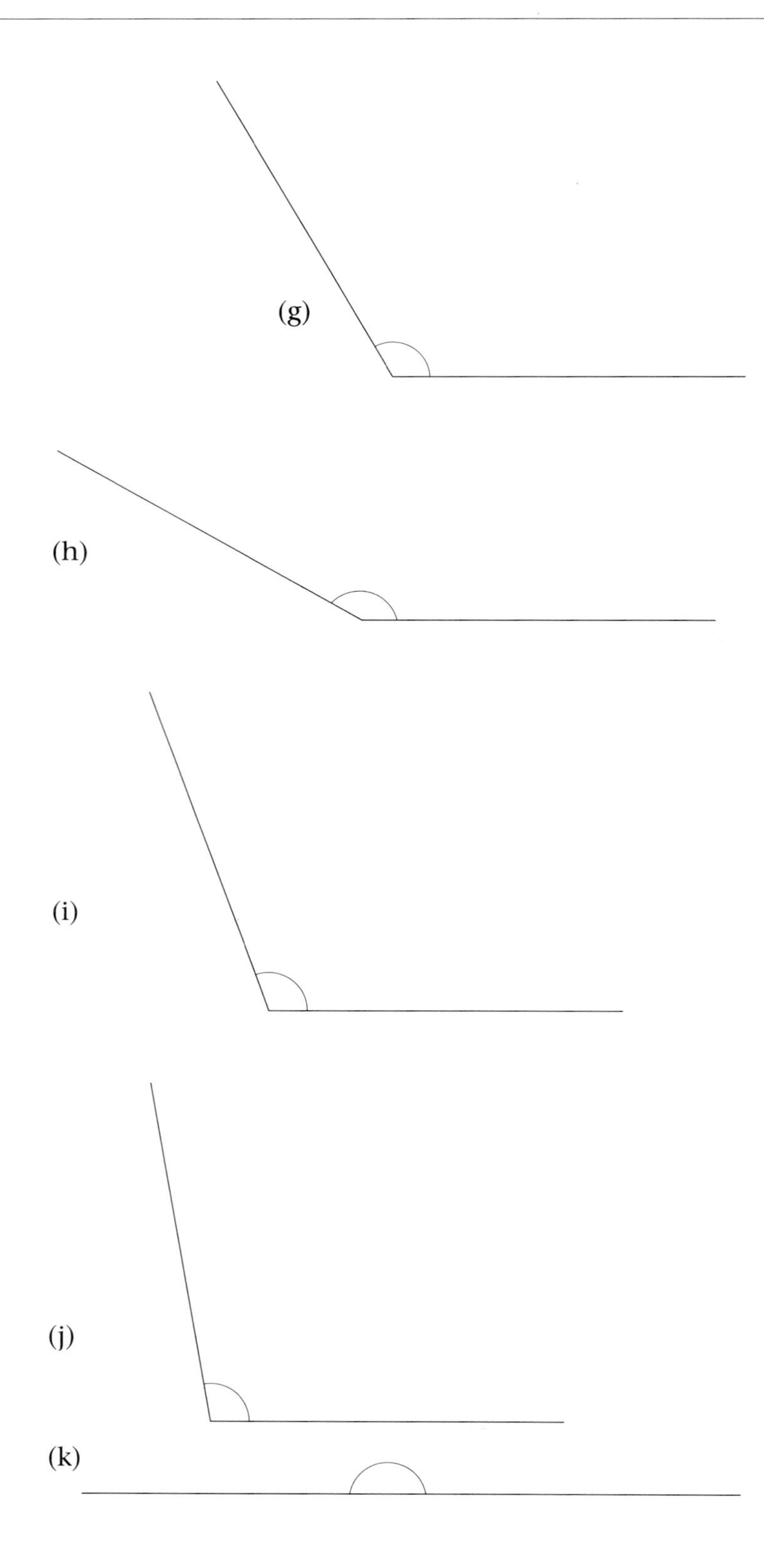
(g)
(h)
(i)
(j)
(k)

2 Measure these angles carefully with your protractor.

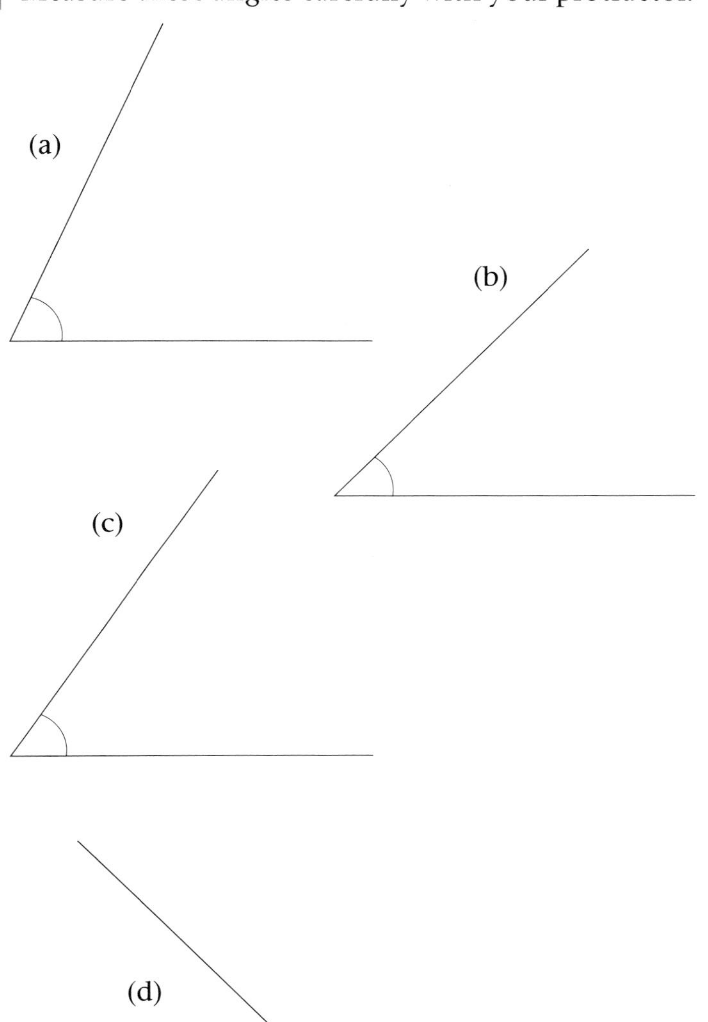

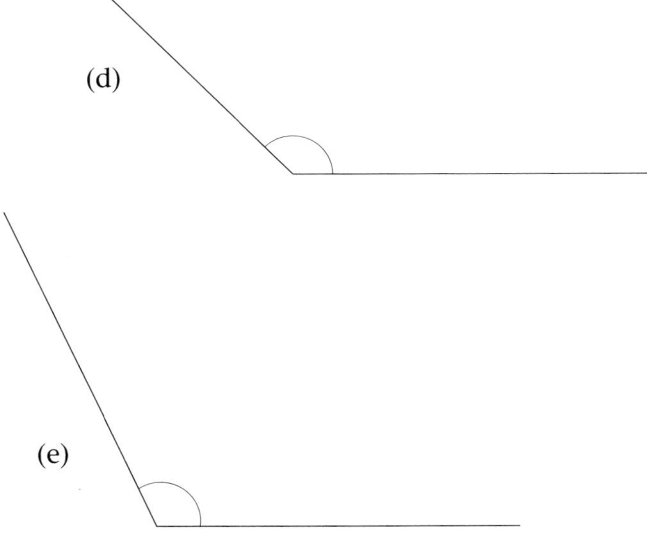

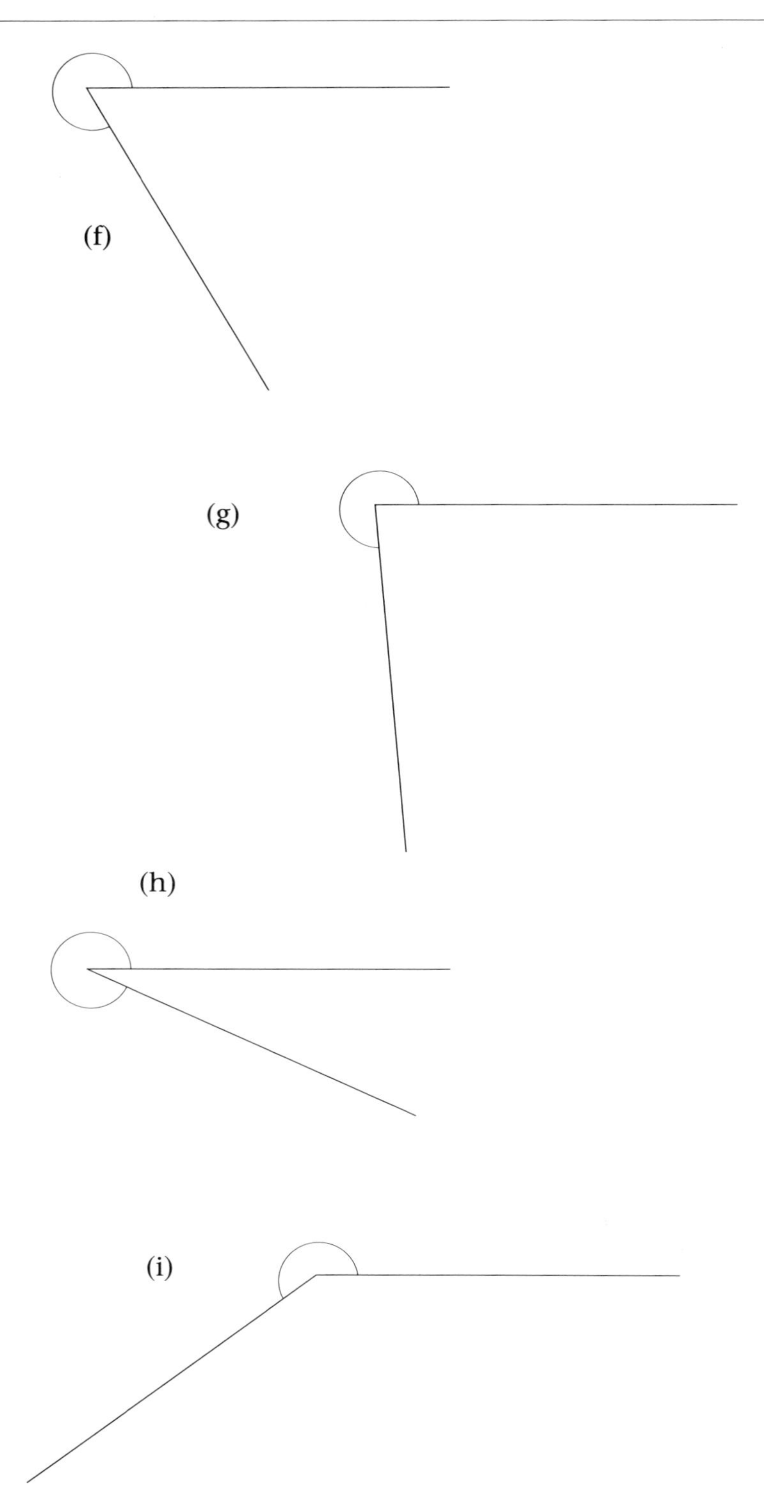
(f)
(g)
(h)
(i)

3 Copy this table.

Angle	Estimate	Accurate measurement
a		
b		
c		
d		
e		
f		
g		
h		
i		
j		

Estimate the size of each of these angles. Write your estimate in the middle column of your table.

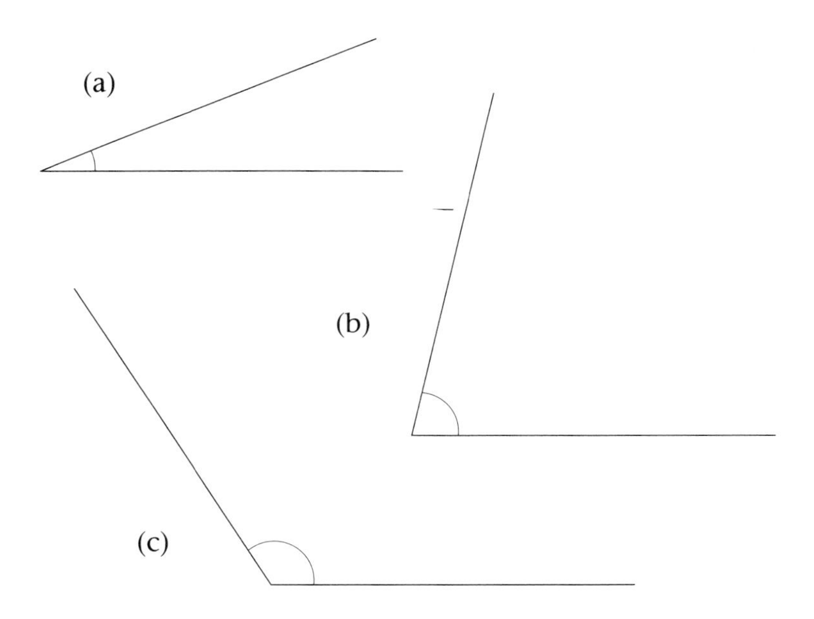

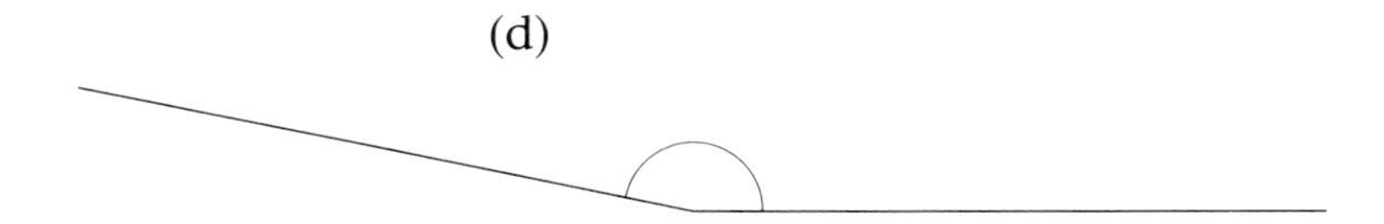

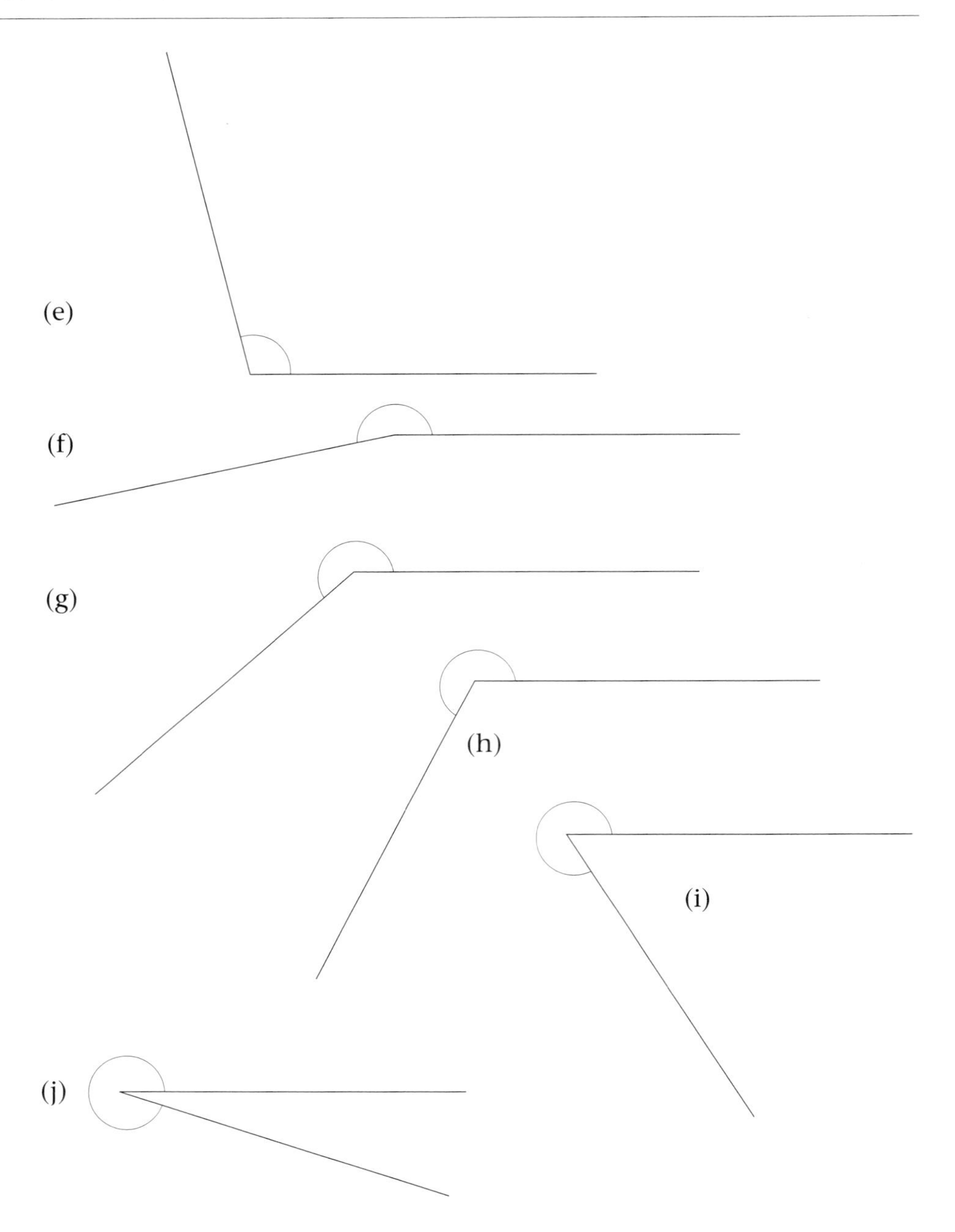

When you have written down all your estimates, measure the angles accurately with a protractor. Write the accurate size of the angle in the third column of your table.

Assignment 3 Drawing an angle

You will need plain paper, a sharp pencil and a protractor.

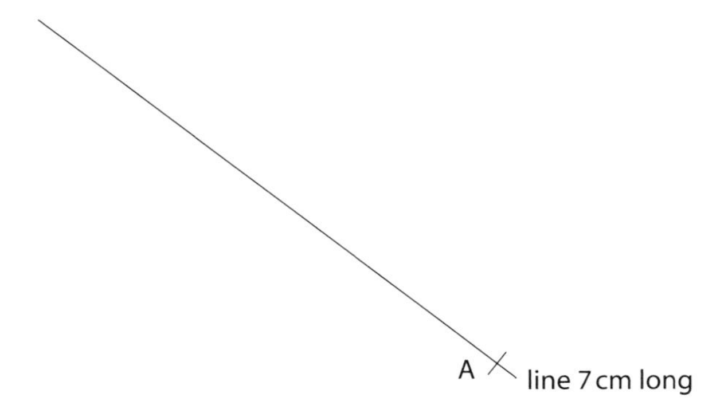

- Copy this line onto plain paper.
- Place your protractor on the line so that the centre is exactly on the point A and 0 on the protractor is on the line.
- Find 50° **clockwise** round the protractor. Mark this point.

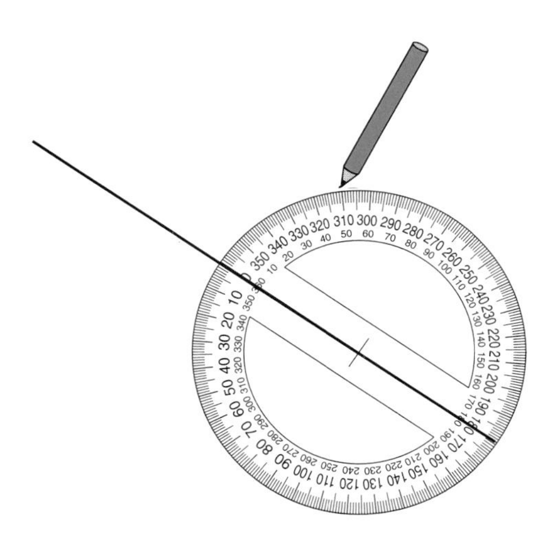

- Draw a line from A through the point.
- Draw a small arc to show the angle you have measured.

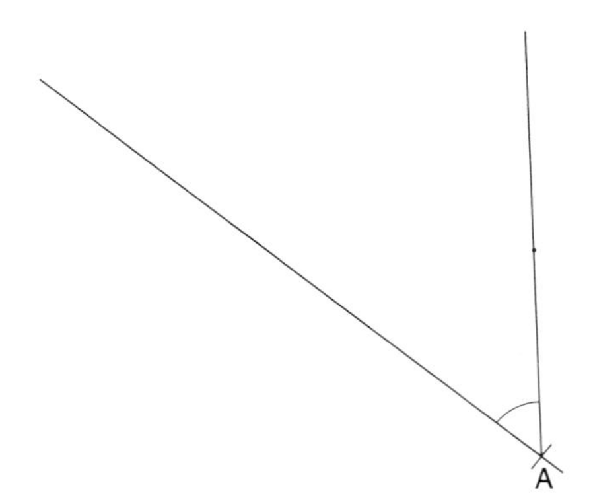

- Now draw an angle of 70° **anti-clockwise** from the original line at point A.

Exercise 6

1 Draw a horizontal line about 8 cm long. Mark a point near the right hand end of the line for the centre of the protractor. Measure and draw an angle of 40° clockwise.

2 Draw the following angles. For each one start with a line about 8 cm long and a point near the end of the line.

(a) 20° clockwise (b) 60° anticlockwise (c) 40° anticlockwise

(d) 100° clockwise (e) 130° anticlockwise (f) 45° clockwise

(g) 85° anticlockwise (h) 52° clockwise (i) 78° anticlockwise

(j) 153° anticlockwise (k) 168° clockwise (l) 136° clockwise

3 Draw a line about 8 cm long and mark a point near the end of the line. Measure and draw an angle of 200°. Mark the angle clearly.

4 Draw the following angles, mark the angle clearly.

(a) 250° (b) 300° (c) 225°

(d) 343° (e) 278° (f) 197°

What do you notice about all of these angles?

Reflex angles

Key fact

A reflex angle is an angle between 180° and 360°.

Exercise 7

1 Copy this table.

Angle	Size	Acute angle	Right angle	Obtuse angle	Reflex angle
a					
b					
c					
d					
e					
f					
g					
h					
i					
j					

Use your protractor to measure each of these angles.

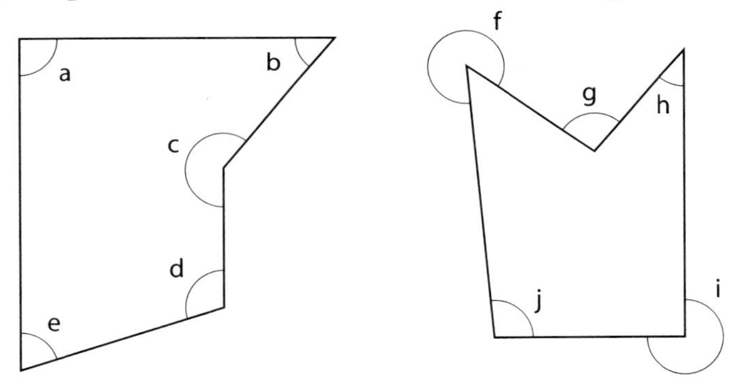

On your table, fill in the size of the angles and tick the column which you think describes the angle best.

Angles in triangles

Assignment 4 The three angles in a triangle

You will need plain paper, a sharp pencil and a ruler.

- On plain paper draw a large triangle. Mark each of the angles with a small arc.
- Cut out the triangle. Tear the triangle into three parts, so that there is an angle on each part.

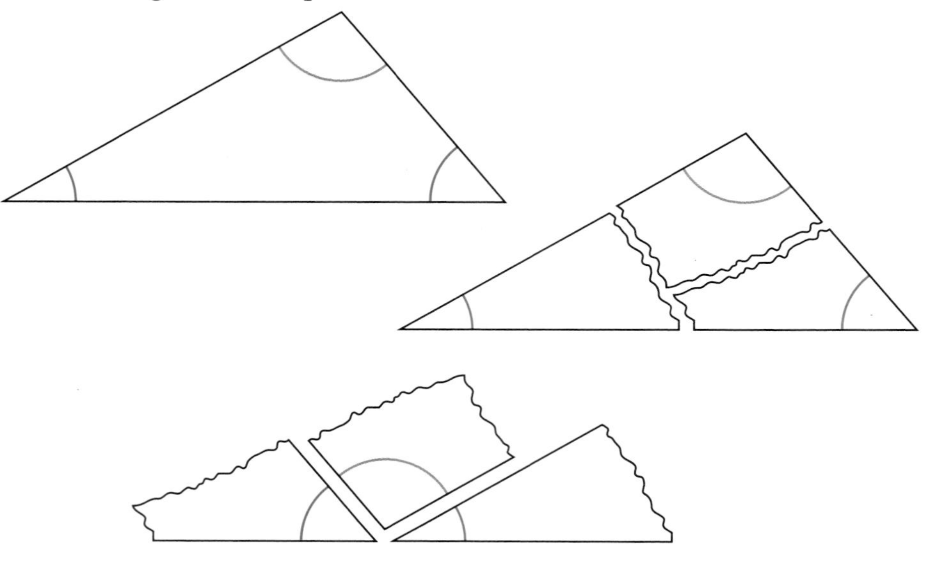

- Put the pieces together so that all three angles fit together. Write down what you notice.
- Repeat this for two other triangles.

Exercise 8

1 Copy this table.

Triangle	Angle (i)	Angle (ii)	Angle (iii)	Sum of angles
(a)				
(b)				
(c)				
(d)				
(e)				
(f)				

Carefully measure all the angles in each of these triangles.

Write your answers in your table. For each triangle, add the three angles together and write your answer in the last column.

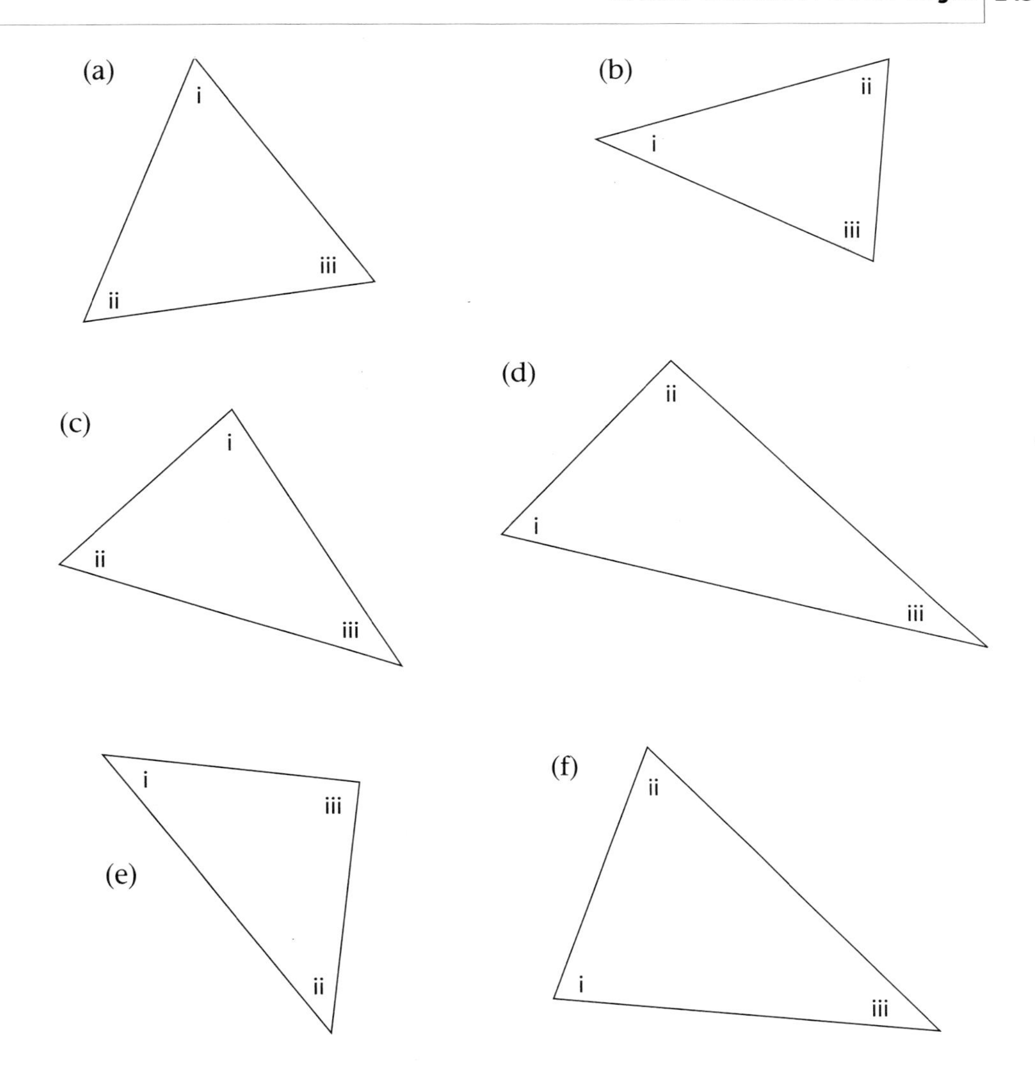

Key fact

Angles in a triangle add up to 180°.

Write down what you notice about the answers in the last column.

2 Draw a table, similar to the one you copied for question 1. Draw four triangles. Make sure each one has an obtuse angle. Using a protractor, carefully measure each of the angles in the triangles Record your answers in the table.

For each triangle add the three angles together and write your answer in the last column. Write down what you notice about the answers in the last column.

Review Exercise

1 Measure these angles with a protractor.

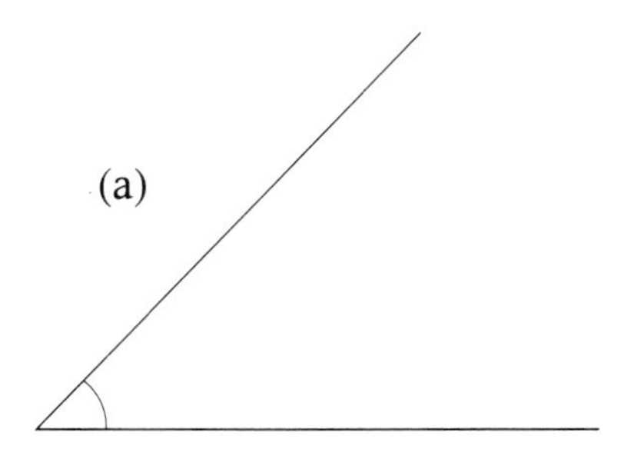

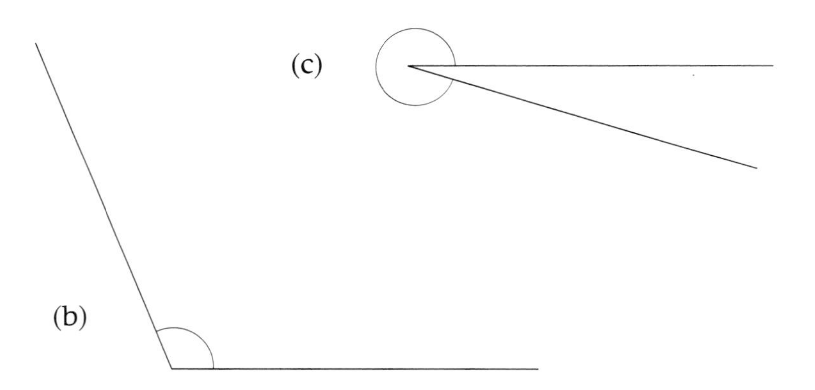

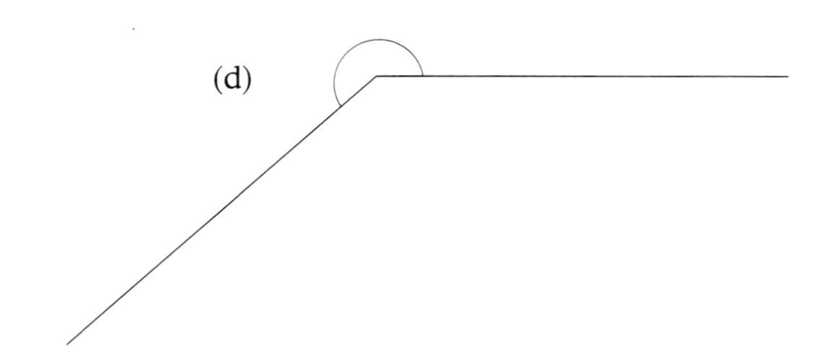

2 Draw these angles carefully, using a protractor.

(a) 55° (b) 15° (c) 100° (d) 135°

(e) 120° (f) 230° (g) 300° (h) 355°

3 Write down the words which describe each of these angles.

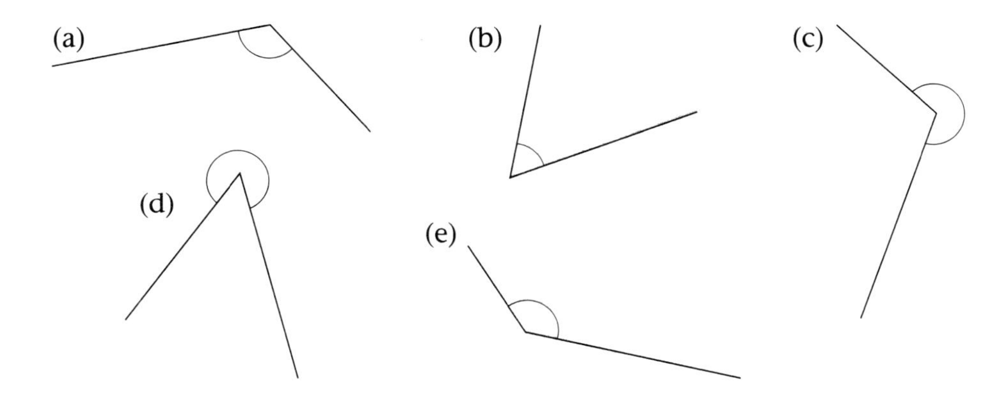

4 You will need plain paper, a sharp pencil, a ruler, compasses and a protractor.

- Draw two lines at right angles to each other.
- Bisect the right angles.
- Set your compasses to 5 cm. Place the point of your compasses carefully on the centre point of your diagram. Mark an arc on each line. Label the points A to H.

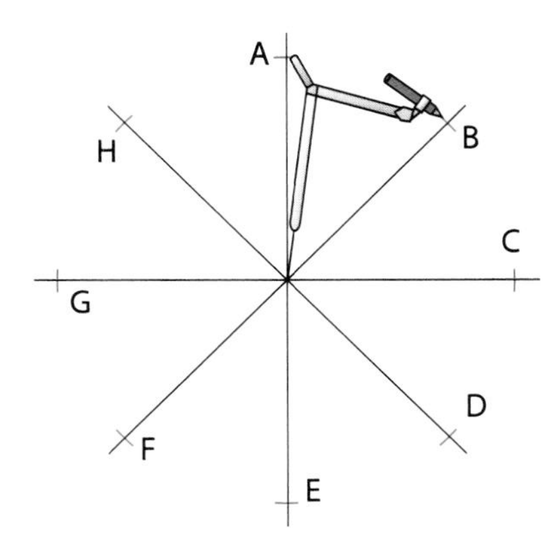

- Join A to C, B to D, C to E, D to F, E to G, F to H, G to A and H to B.
- On your diagram, mark all the 45° angles you can find.

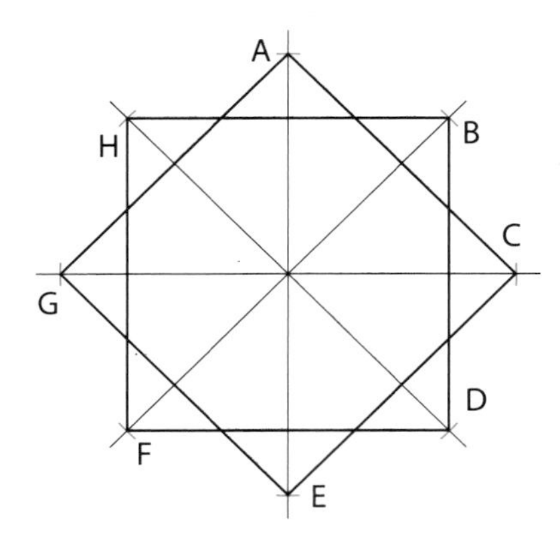

- On tracing paper, mark the points A to H and then show where you can find an eight-pointed star. Write down the size of each angle inside the star.
- Describe clearly all the other shapes you can see. Use diagrams to help you.
- What other angles can you find in the diagram? Describe clearly where they are and how you worked out their sizes. Use diagrams to help you.